PLANT GENOMICS

Plant Genomics

VOLUME 1

Themba Ndlovu
Gontse Masilo

Kruger Brentt
Publishers
2026

Kruger Brentt Publishers UK. LTD.
Company Number 9728962

Regd. Office: 68 St Margarets Road, Edgware, Middlesex HA8 9UU

ISBN: 978-1-78715-623-4 (Vol. 1)

Disclaimer:

Every possible effort has been made to ensure that the information contained in this book is accurate at the time of going to press, and the publisher and author cannot accept responsibility for any errors or omissions, however caused. No responsibility for loss or damage occasioned to any person acting, or refraining from action, as a result of the material in this publication can be accepted by the editor, the publisher or the author. The Publisher is not associated with any product or vendor mentioned in the book. The contents of this work are intended to further general scientific research, understanding and discussion only. Readers should consult with a specialist where appropriate.

Every effort has been made to trace the owners of copyright material used in this book, if any. The author and the publisher will be grateful for any omission brought to their notice for acknowledgement in the future editions of the book.

For information on all our publications visit our website at http://krugerbrentt.com/

PREFACE

Plant Genomics presents a comprehensive overview of the rapidly advancing field that integrates molecular biology, genetics, and biotechnology to explore the structure, function, and evolution of plant genomes. With the global demand for food security, climate resilience, and sustainable agriculture, plant genomics plays a crucial role in understanding genetic diversity, improving crop productivity, and developing stress-tolerant and nutritionally enriched varieties. This book is designed for students, researchers, and professionals in plant sciences, biotechnology, agriculture, and genetics, offering both foundational concepts and cutting-edge developments in the field.

The book begins with an introduction to genome organization, sequencing technologies, and bioinformatics tools essential for analyzing plant genomes. It then explores functional genomics, transcriptomics, proteomics, and metabolomics, shedding light on how genetic information translates into phenotypic traits. Special emphasis is placed on comparative genomics, genome mapping, and molecular markers that aid in crop improvement and breeding strategies.

A significant focus is given to applications of genomics in addressing agricultural challenges, including developing high-yield crops, disease- and pest-resistant varieties, drought- and salinity-tolerant plants, and biofortified crops for enhanced nutrition. The book also highlights the role of CRISPR-Cas and other genome editing technologies in revolutionizing plant breeding and precision agriculture.

Case studies and practical examples demonstrate how genomic insights are being applied to major crops worldwide, showcasing the direct link between research and real-world agricultural advancement. The book concludes with discussions on ethical considerations, data sharing, and future prospects of plant genomics in ensuring global food security and sustainability.

We hope that Plant Genomics will serve as a valuable resource, bridging theoretical knowledge with practical applications, and inspiring future research and innovation in the dynamic field of plant science.

Themba Ndlovu
Gontse Masilo

CONTENTS

VOLUME 2

1 Chapter

GENOME-WIDE ASSOCIATION STUDY AND SELECTION SIGNATURES IDENTIFY GENOMIC REGIONS LINKED TO SEED YIELD AND OIL QUALITY IN FLAX

1. INTRODUCTION: AN OVERVIEW

Flax (Linum usitatissimum L., 2n = 2x = 30) is a self-pollinating annual plant belonging to the Linaceae family. This is a dual-purpose crop cultivated for its seed oil or stem fiber, yielding two morphotypes: linseed and fiber. The linseed or flaxseed morphotype is rich in oil (40–50%) containing five main fatty acids: palmitic (PAL, C16:0, ~6%), stearic (STE, C18:0, ~2.5%), oleic (OLE, C18:1Δ9, ~19%), linoleic (LIO, C18:2Δ9, 12, ~13%), and linolenic (LIN, C18:3Δ9, 12, 15, ~55%) [1,2]. Because of its high LIN content, linseed is the richest plant source of omega-3 fatty acid which is beneficial for reducing blood cholesterol levels and mitigating heart diseases in humans [3,4]. The same attributes make it ideal as industrial oil for use in paints, linoleum flooring, inks, soaps and varnishes [4].

Linseed breeding has focused on high seed yield (YLD), high oil content (OIL), and either high or low LIN content. Low LIN (2–4%) and high LIO (65–70%) lines have been developed through mutation breeding. NuLin™ 50 with 67.8% LIN (http://www.viterra.ca) and Omégalin with 65.8% (http://www.terredelin.com) are examples of high LIN linseed cultivars currently registered. Extremely low LIN lines such as LinolaTM or SolinTM improve oxidative stability, making such cultivars suitable for the fabrication of margarine [3]. Since 1910, a total of 82 flax cultivars have been released in Canada [5]. These cultivars and elite breeding lines provide diverse genetic materials for dissecting the genetic architecture of oil biosynthesis and yield related traits in linseed.

Several methods can be used to dissect the genetic architecture of crop traits. QTL or linkage mapping uses bi-parental populations to identify loci responsible for trait variation between parents based on a recombination-based genetic linkage map [6]. Bi-parental populations, such as F2, recombinant inbred line (RIL), doubled haploid (DH) and backcross (BC) populations, are the most widely used genetic resources for mapping QTL for traits of interest in self-fertilizing crops, including flax [7–12]. While bi-parental populations are easy to develop and have power for QTL detection, only the a limited number of alleles from the parental genotypes are analyzed in a single population, resulting in a narrow genetic base and low representation of allelic diversity [13]. In addition, genetic recombination is limited in these populations [14]. To increase the QTL dissection power, attempts have been made to expand the genetic diversity through other multiple-parent population types, such as nested association mapping (NAM) populations [15–17] and multi-parent advanced generation intercross (MAGIC) populations [18–25], while retaining the advantages of association mapping and bi-parental populations. However, the development of such populations requires careful planning and time. Natural populations that possess tremendous phenotypic diversity can be advantageous in genome-wide association study (GWAS) with various molecular markers in plants and animals [26–31]. Association mapping using a diverse germplasm panel overcomes the phenotypic diversity limitation of bi-parental populations, thereby increasing the QTL mapping power [32] but is impeded by low detection power of association of rare alleles. GWAS usually uses a natural population to investigate wider phenotypic variation for complex traits by taking advantage of ancient genetic recombination events in populations [33].

GWAS may be complemented by performing genome-wide selective sweep scan (GW3S) which identifies selection signatures that are beneficial for plant adaptation. A selective sweep is the reduction or elimination of variation among the nucleotides near a new beneficial mutation. Following strong positive natural selection or artificial selection during domestication or breeding, selective sweeps affect nearby linked alleles [34]. Ancient selective sweeps are relevant to natural evolution and domestication of crop species that are subjected to natural and artificial selective pressures and forced to adapt rapidly to new environments and thus drive speciation [35]. Breeding selects favorable alleles and retains them in new cultivars. These signatures of selection can be detected by a cross-population comparison approach [34]. Recent studies demonstrated that genomic regions that exhibit selection signatures are also enriched for genes associated with biologically important traits [36–40]. Thus, detection of selection signatures is emerging as an additional approach to identify and validate novel gene-trait associations [41].

Genetic regions associated with storage oil biosynthesis in flax have been studied based on QTL mapping using bi-parental populations. Several QTL responsible for oil content and fatty acid composition have been mapped in independent studies including the three populations used herein. The first population (BM) of 243

F2:6 recombinant inbred lines (RILs) from a cross between the Canadian linseed varieties CDC Bethune and Macbeth was used for a linkage mapping study and detected three QTL each for OLE and STE, two each for LIO and IOD, and one each for PAL, LIN and OIL with several QTL co-locating at the same locus [8]. The second population (EV) was a cross between E1747 and Viking. The third population (SU) was a cross between SP2047 (a yellow-seeded SolinTM line with 2–4% LIN) and UGG5-5 (a brown-seeded flax line with 63–66% LIN) and comprised of 78 lines generated through DH method. It was used in a linkage mapping study using simple sequence repeat (SSR) markers which identified QTL for LIO, LIN and iodine value (IOD) co-locating on LG7 and LG16, and a QTL for PAL on LG9 [7]. The linkage-based studies from these populations provided numerous QTL for important traits but the QTL were generally far from the markers and poorly delimited because of the low resolution of the genetic maps [18,19,42]. The three bi-parental populations were utilized to create a consensus genetic map [43] and to conduct genomic selection [44], predominantly employing SSR markers. Given that the three populations have been concurrently phenotyped for various common agronomic and seed oil quality traits in identical environments (years/locations), we devised this study to evaluate the efficacy of the combined bi-parental population approach for GWAS and GW3S in identifying genomic regions linked to seed yield and seed oil quality traits through genotyping by sequencing (GBS).

2. RESULTS

2.1 Re-Sequencing and Genome-Wide SNPs

In the present study, a set of 260 genotypes (97 from the recombinant inbreeding line (RIL) population from a cross between CDC Bethune and Macbeth (BM), 91 from the RIL population from a cross between E1747 and Viking (EV) and 72 from the doubled haploid population from a cross between SP2047 and UGG5-5 (SU) along with the 5 of 6 parents except for the reference CDC Bethune) were re-sequenced using GBS to identify genome-wide single nucleotide polymorphism (SNP) markers on the chromosome-based flax pseudomolecules [45]. An average of ~57.7 million paired end reads were generated for each individual, corresponding to 5754 Mb sequences or 19.2× genome equivalents of the reference scaffolds (~302 Mb) [46] . Paired-end reads of each genotype were aligned to the flax scaffolds [46], resulting in a total of 536,186 SNPs. After filtering off SNPs with minor allele frequency (MAF) <0.05 and genotyping rate <60% [47,48], 17,288 SNPs were retained on the flax pseudomolecules [45] (Table S2). Out of these, 15,284 segregated in BM, 15,397 in EV and 7568 in SU. The SNPs were mostly uniformly distributed across all 15 chromosomes (chr), ranging from 601 on chr11 to 1572 on chr13 (Figure 1). Approximately 71.1% of all SNPs were located in intergenic regions, 16.2% were in introns and 12.7% were in exons. These SNPs were used for further population structure analysis, GWAS and GW3S.

2.2 Whole-Genome Pattern of LD

The LD and LD decay rates were analyzed for each population separately as well as the merged population using the filtered SNP data. The physical distances of pair-wise SNPs at which the LD r2 dropped to half were 1242, 223, 728 and 272 kb for BM, EV, SU and merged populations respectively. This indicated substantial variation in LD decay rate across populations (Figure 2). The average LD r2 of BM, EV, SU, and merged populations were 0.37, 0.26, 0.28 and 0.30, respectively, with the number of haplotype blocks for each population estimated at 599, 648, 206 and 1205, respectively.

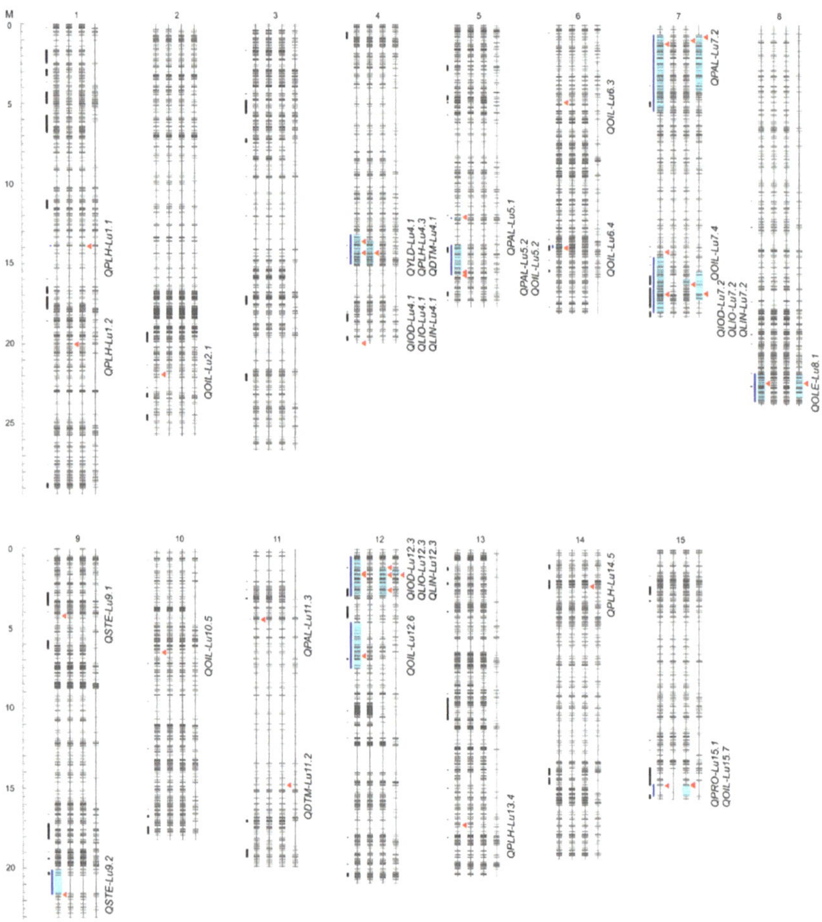

Figure 1. Distribution of 17,288 SNPs, 114 selective sweeps and 33 QTL on the 15 chromosomes of flax for each of three bi-parental populations BM, EV and SU and, for the merged population (BM + EV + SU). Four vertical bars from left to right for each chromosome represent the BM + EV + SU, BM, EV and SU populations, respectively. Short horizontal lines on bars represent SNPs. QTL regions are highlighted in cyan and by vertical blue lines. Red triangles identify QTL's peak SNP. Selective sweeps are represented by short vertical black lines.

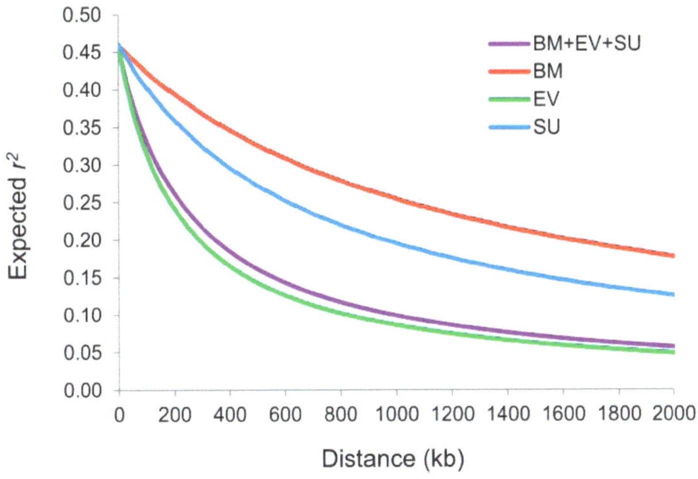

Figure 2. Intra-chromosome LD (r2) decay of SNP pairs over the entire flax genome as a function of physical distances (kb) of pair-wise SNPs for the three individual and merged populations. The curves are drawn based on a fitted non-linear model (see Section 4.2).

2.3 Genetic Diversity and Population Structure

Nucleotide diversity (π) was estimated at 41.52, 38.26 and 3.95 for the BM, EV and SU populations, respectively (Table 1), and was consistent with the number of SNPs identified from the three populations. A strong population-differentiation (Fst) was observed at 0.44 between BM and SU and 0.48 between EV and SU. However, Fst was weaker at 0.04 between the BM and EV (Table 1).

Table 1. Genetic differentiation (Fst) between three bi-parental (upper triangle elements) and nucleotide diversity (π) within these populations (diagonal elements).

Population	BM	EV	SU
BM	41.52	0.04	0.44
EV		38.26	0.48
SU			3.95

BM: CDC Bethune/Macbeth; EV: E1747/Viking; SU: SP2047/UGG5-5.

The genetic structure within the merged population was assessed based on the 17,288 SNP loci from the 260 individuals using two methods: principal component analysis (PCA) and discriminant analysis for principal components (DAPC). Bi-plots of the first three principal components of the PCA showed five distinct clusters (Figure 3a,b). The BM (BM1 and BM2) and EV (EV1 and EV2) populations each contained two sub-populations, while SU produced a single cluster. DAPC corroborated the same five clusters (Figure 3c,d). Therefore, a DAPC Q matrix based on the five clusters was generated and used as covariates to assess the population stratification in GWAS and phenotypic variation explained by the SNPs.

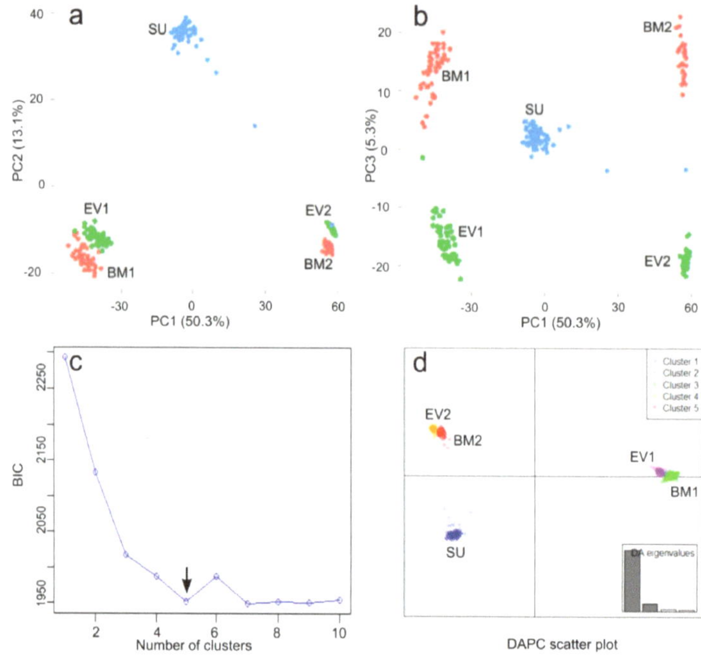

Figure 3. Principal component analysis (PCA) and discriminant analysis of principal components (DAPC) of the 260 individuals in three bi-parental populations (BM, EV and SU) based on 17,288 SNPs.

a. Bi-plot of the first and second principal components (PCs);
b. Bi-plot of the first and third PCs;
c. k-means clustering analysis based on 100 chosen PCs shows that the optimal number of clusters (k) is 5, that is where the Bayesian information criterion (BIC) is lowest (arrow);
d. DAPC scatter plot. Percentages in parentheses in the axis titles of (a) and (b) represent the variance explained by the respective PCs. Individuals from the BM and EV populations grouped into two subpopulations each, BM1 and BM2, and EV1 and EV2, respectively.

2.4 $h2_{SNP}$

Phenotypic variation of traits was largely explained by SNPs in the three individual and the merged populations (Table 2). The average $h2_{SNP}$ for all 11 traits was 0.51. The largest $h2_{SNP}$ values among the four populations ranged from 0.45 (YLD) to 0.90 (PAL). More than 80% of the phenotypic variation in one of the populations was explained by identified SNPs for days to maturity (DTM), IOD, PAL, STE, LIO and LIN. The $h2_{SNP}$ varied from one population to another depending on the genetic variation between the two parents. For SU, little or no phenotypic variation was explained by SNPs for DTM, plant height (PLH) and STE. For EV, a relatively low phenotypic variation ($h2_{SNP} < 0.1$) was explained by SNPs for STE and OLE.

Table 2. Phenotypic variation explained by all SNPs ($h2_{SNP}$) and identified QTL ($h2_{GWAS}$) for 11 traits in different populations.

Trait	Population	$h2_{SNP}$ ± s	No. QTL	$h2_{GWAS}$ ± s
YLD	BM + EV + SU BM EV SU	0.43 ± 0.12 0.22 ± 0.25 0.15 ± 0.24 0.45 ± 0.21	1	0.14 ± 0.09 §
PLH	BM + EV + SU BM EV SU	0.53 ± 0.12 0.76 ± 0.12 0.76 ± 0.14 0.06 ± 0.20	1 2 2	0.08 ± 0.11 0.21 ± 0.15 0.22 ± 0.18
DTM	BM + EV + SU BM EV SU	0.43 ± 0.13 0.81 ± 0.11 0.36 ± 0.24 0.00 ± 0.20	1 1 1	0.10 ± 0.07 0.18 ± 0.13 0.18 ± 0.22
PRO	BM + EV + SU BM EV SU	0.51 ± 0.11 0.52 ± 0.20 0.34 ± 0.23 0.58 ± 0.19	1 1	0.12 ± 0.16 0.09 ± 0.12
OIL	BM + EV + SU BM EV SU	0.66 ± 0.09 0.46 ± 0.22 0.39 ± 0.21 0.70 ± 0.15	7 1	0.62 ± 0.14 0.08 ± 0.08
IOD	BM + EV + SU BM EV SU	0.80 ± 0.06 0.49 ± 0.19 0.78 ± 0.12 0.66 ± 0.17	3 2 2	0.57 ± 0.10 0.51 ± 0.14 0.35 ± 0.18
PAL	BM + EV + SU BM EV SU	0.79 ± 0.06 0.12 ± 0.26 0.55 ± 0.20 0.90 ± 0.07	4 1 1	0.48 ± 0.11 0.09 ± 0.11 0.56 ± 0.18
STE	BM + EV + SU BM EV SU	0.21 ± 0.15 0.85 ± 0.09 0.02 ± 0.14 0.00 ± 0.22	2 1	0.41 ± 0.19
OLE	BM + EV + SU BM EV SU	0.55 ± 0.10 0.36 ± 0.22 0.09 ± 0.25 0.72 ± 0.16	1 1	0.16 ± 0.13 0.20 ± 0.19 *GWAS*
LIO	BM + EV + SU BM EV SU	0.80 ± 0.06 0.54 ± 0.20 0.75 ± 0.13 0.66 ± 0.17	3 2 2	0.66 ± 0.17 0.54 ± 0.14 0.36 ± 0.18
LIN	BM + EV + SU BM EV SU	0.80 ± 0.06 0.49 ± 0.19 0.76 ± 0.13 0.66 ± 0.17	3 2 2	0.56 ± 0.09 0.55 ± 0.14 0.36 ± 0.18

YLD: seed yield; PLH: plant height; DTM: days to maturity; PRO: protein content; OIL: oil content; IOD: iodine value; PAL: palmitic acid content; STE: stearic acid content; OLE: oleic acid content; LIO: linoleic acid content; LIN: linolenic acid content; BM: CDC Bethune/Macbeth; EV: E1747/Viking; SU: SP2047/UGG5-5. § $h2_{GWAS}$ of YLD was estimated based on the phenotypes in a single environment (Morden/2012). For all other traits, $h2_{GWAS}$ was estimated based on the BLUP estimates of phenotypes.

2.5 QTL Identified from 11 Traits

Using the best linear unbiased prediction (BLUP) values of phenotyping data collected from six to eight year/location environments with both generalized linear model (GLM) and mixed linear model (MLM), we identified a total of 33 QTL for 11 traits, one for YLD, eight for OIL, five for PLH, four for PAL, three each for IOD, LIO, and LIN, two each for DTM and STE, and one each for protein content (PRO) and OLE (Table 3, Figure 1, Figures S1 and S2). Thirty-one of the 33 QTL were detected using GLM and 13 with MLM (Tables S4 and S5). Of these latter 13, two QTL (QTL 18 for IOD and QTL 31 for LIN) were detected only by MLM, while the remaining 11 were identified by both MLM and GLM.

Out of 33 QTL identified, 12, 6, 3 and 27 were from EV, SU, BM and merged population, respectively. Only six QTL were detected exclusively from two individual populations, including four (QTL 2 and 6 for PLH, QTL 8 for DTM and QTL 17 for OIL) from EV and two (QTL 3 and 4 for PLH) from BM. Eighteen were identified exclusively from the merged population. Ten QTL were detected simultaneously from the merged population and one or more individual populations (Tables S4 and S5).

QTL for YLD (QTL 1) was identified only in two environments (2010/Morden and 2012/Saskatoon) (Figure S2) but not in other environments or using BLUP estimates over the six year/location environments. We also performed GWAS for all other traits with phenotypic data from individual environments and obtained similar results with the QTL identified using BLUP values over multiple environments (Table S6).

2.6 QTL Effect Significance

To validate the QTL, we tested statistical significance of difference of phenotypes between two contrasting haplotype pairs for each QTL in the merged and individual populations and in different year/location environments. QTL effect differences between two contrasting haplotype pairs for all 33 QTL were significant (Figure 4, Table S7). We also assessed relationship of the number of pyramiding positive-effect QTL in individuals with trait phenotypes. Significant linear relations for all eight traits which had two or more QTL identified in this study were observed, showing primarily additive or accumulative QTL effects (Figure 5).

Table 3. QTL and associated gene candidates.

Trait	QTL No.	QTL Name	Chr.	Start Position (bp)	End Position (bp)	XP-CLR Score	Known QTL or Marker	Candidate Gene IDs	Candidate Gene Location (bp)	Candidate Gene Name	Gene Annotation
YLD	1	QYLD-Lu4.1	4	13,594,936	14,958,389	12.54	QYld. BM.crc-LG4 a				
	2	QPLH-Lu1.1	1	13,887,715	13,930,292						
	3	QPLH-Lu1.2	1	20,012,490 20,012,490				Lus10020835	19,610,837	*BRI1* [49]	Leucine-rich receptor-like protein kinase family protein
								Lus10020865	19,790,777	*GA2* [49]	Terpenoid cyclases/Protein prenyltransferases superfam-ly protein
PLH	4	QPLH-Lu4.3	4	14,305,982	15,042,104	12.54		Lus10034358	14,006,288	*BIM2* [49]	BES1-interacting Myc-like protein 2
								Lus10041435	14,157,752	*MYB62* [49]	Mybdomainprotein62
								Lus10041481	14,398,338	*LMCO4* [49]	Laccase/Diphenoloxidase familyprotein
								Lus10041794	15,920,170	*ROT3* [49]	CytochromeP450superfamily protein
								Lus10041801	15,948,434	*WAT1* [49]	WallsAreThin1
	5	*QPLH-Lu13.4*	13	17,243,884	17,243,884			Lus10030567	18,680,474	*GA2OX8* [49]	Gibberellin 2-oxidase 8
	6	*QPLH-Lu13.5*	14	2,320,469	2,320,469	40.61		Lus10021395	3,647,029	*HCT* [49]	Hydroxycinnamoyl-CoA shi-kimate/quinate Hydroxycinnam-oyl transferase
DTM	7							Lus10015766	13,094,864	*FLC* [50]	K-boxregionandMADS-box transcription factor family protein
								Lus10034461	13,434,121	*CDF3* [50]	Cycling DOF factor 3
								Lus10034370	13,933,421	*AP1* [50]	K-box region and MADS-box transcription factor family protein

								Lus10041483	14,411,103	*PFT1* [50]	Phytochrome and flowering time regulatory Protein (PFT1)
								Lus10041500	14,512,085	*ATAN11* [50]	Transducin/WD40repeat-like superfamily p r o t e i n
								Lus10041540	14,716,950	*RGL1* [50]	RGA-like1
								Lus10041595	14,966,739	*AP2* [50]	Integrase-type DNA-binding superfamily protein
	8	*QDTM-Lu11.2*	11	14,768,686							
PRO	9	*QPRO-Lu15.1*	15	14,746,288	14,746,310	8.50		Lus10030671	22,732,660	*WRI* [50]	Integrase-type DNA-binding superfamily protein
OIL	10	*QOIL-Lu2.1*	2	21,913,720	21,913,720						
	11	*QOIL-Lu5.2*	5	15,704,607	15,705,039						
	12	*QOIL-Lu6.3*	6	4,879,632	4,879,632						
	13	*QOIL-Lu6.4*	6	13,799,180	13,970,951	50.58					
	14	*QOIL-Lu7.4*	7	14,209,179	14,209,179						
	15	*QOIL-Lu10.5*	10	6,517,448	6,517,448						
	16	*QOIL-Lu12.6*	12	4,591,214	7,491,405	27.77					
	17	*QOIL-Lu15.7*	15	14,665,900	15,429,055	8.89		Lus10039906	19,833,852	*KCS14-2* [51]	3-ketoacyl-CoA synthase
	18	*QIOD-Lu4.1*	4	19,909,467	19,909,467			Lus10039906	19,833,852	*KCS14-2* [51]	3-ketoacyl-CoA synthase
	20	*QIOD-Lu12.3*	12	489,561	2,981,642	106.22	*QIOD.crc-LG16* b	Lus10036184	1,035,336	*FAD3b* [52]	Fatty acid desaturase
								Lus10023359	1,729,292	*FAH1* [50]	Fatty acid hydroxylase 1
	21	*QPAL-Lu5.1*	5	12,062,376	12,182,441			Lus10029880	12,062,376	*KCS12-3* [51]	3-ketoacyl-CoA synthase

	22	*QPAL-Lu5.2*	5	13,797,851	15,658,995		*QPAL.crc-LG9* b	Lus10028925	1,085,389	*KAS IIIb-2* [51]	3-ketoacyl-acyl carrier protein synthase III
	23	*QPAL-Lu7.3*	7	624,461	5,423,691		*c79-s540_Lu2534 c*	Lus10028885	1,262,079	*SUN1* [50]	SAD1/UNC-84 domain protein 1
	24	*QPAL-Lu11.4*	11	4,417,685	4,429,424			Lus10026345	4,333,672	*KCS7-1* [51]	3-ketoacyl-CoA synthase
								Lus10006636	22,165,534	*KCS9-1* [51]	3-ketoacyl-CoA synthase
OLE	25	*QOLE-Lu8.1*	8	21,782,841	23,527,563	12.64		Lus10006637	22,174,324	*KCS1-1* [51]	3-ketoacyl-CoA synthase
								Lus1001848	23,111,453	*DES-1-LIKE* [50]	Fatty acid desaturase family
STE	27	*QSTE-Lu9.2*	9	20,080,531	21,636,823	27.55		Lus10011877	20,059,127	*DES-1-LIKE* [50]	Stearoyl acyl carrier protein desaturase
								Lus10011839	20,227,416	*SAD1* [51]	FatA acyl-ACP thioesterase
	28	*QLIO-Lu4.1*	4	19,909,467	19,909,467			Lus1003990	19,833,852	*FatA2-2* [51]	3-ketoacyl-CoA synthase
LIO	29	*QLIO-Lu7.2*	7	14,540,706	17,977,459	45.70	*QLIO.crc-LG7* b *c281-s185_Lu566 c*	Lus10038321	16,089,922	*FAD3a* [52]	Fatty acid desaturase
	30	*QLIO-Lu12.3*	12	489,561	2,981,642	106.22	*QLIO.crc-LG16* b *Llio-LG12.3 c*	Lus10036184	1,035,336	*FAD3b* [52]	Fatty acid desaturase
	31	*QLIN-Lu4.1*	4	19,909,467	19,909,467			Lus10039906	19,833,852	*KCS14-2* [51]	3-ketoacyl-CoA synthase
LIN	32	*QLIN-Lu7.2*	7	14,590,719	17 977,459	45.70	*QLIN.crc-LG7* b *c281-s185_Lu566* c	Lus10038321	16,089,922	*FAD3a* [52]	Fatty acid desaturase
	33	*QLIN-Lu12.3*	12	489,561	1,981,645	106.22	*QLIN.crc-LG16* b	Lus10036184	1,035,336	*FAD3b* [52]	Fatty acid desaturase
							Llin-LG12.3 c	Lus1002335	1,729,292	*FAH1* [50]	Fatty acid hydroxylase

a QTL identified in [8]; b QTL identified in [7]; c QTL identified in [53]. All candidate genes are labelled by references.

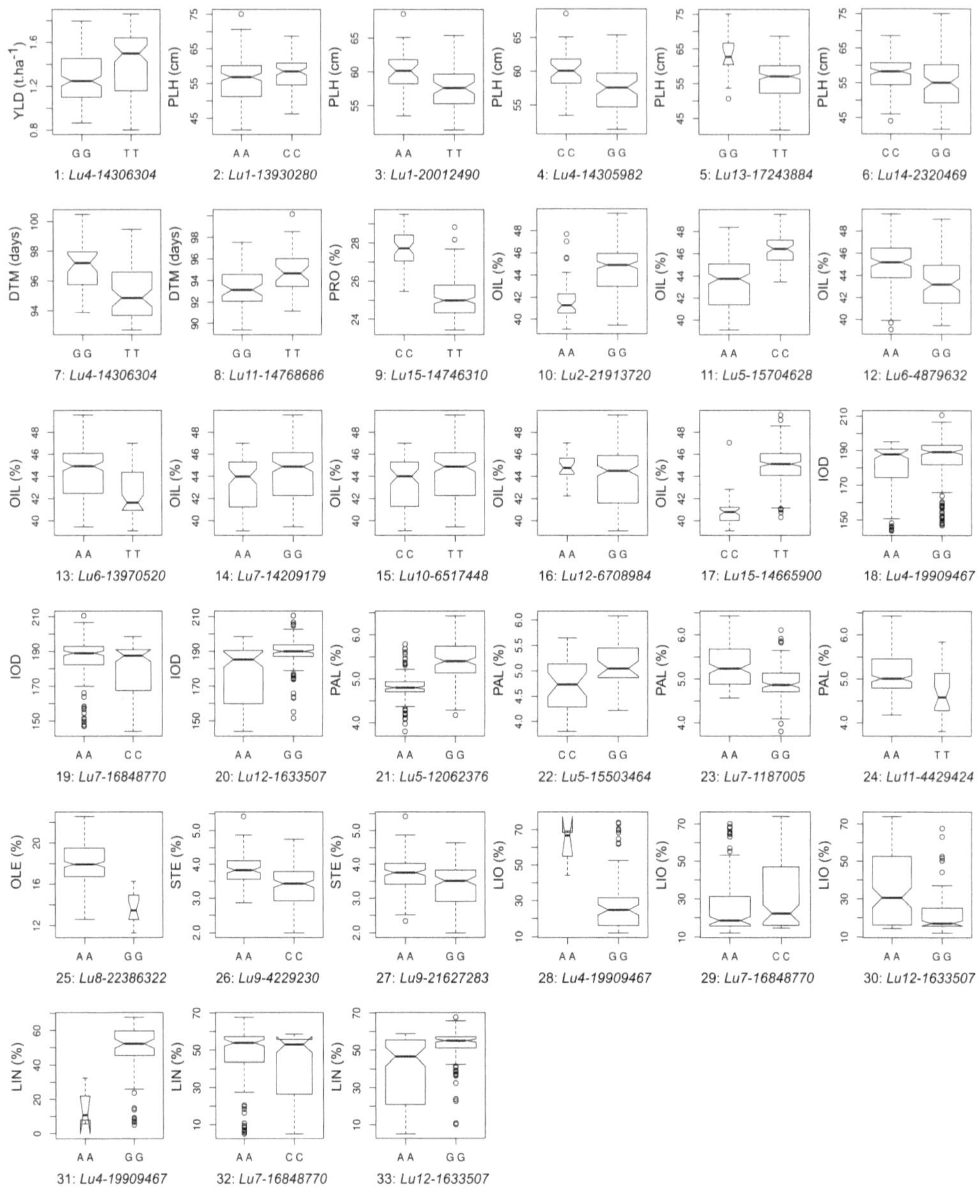

Figure 4. Trait performance of two contrasting haplotype pairs for each of 33 QTL identified from 11 traits. A QTL is represented by the peak SNP identified in the association study. The numbers of QTL correspond to QTL No in Table 3. The BLUP values of the 11 traits in the merged population were used except for PLH/QTL 3 and DTM/QTL 7 for which BM population was used, DTM/QTL 8 for which EV population was used, and PAL/QTL 22, LIO/QTL 28 and LIN/QTL 31 for which SU population was used. The box width is proportional to the size of the subpopulations. Phenotype differences between two contrasting haplotype pairs for each QTL are shown by boxes' notches. For any given QTL, boxes' notches that do not overlap indicate significant median differences at 95% confidence level.

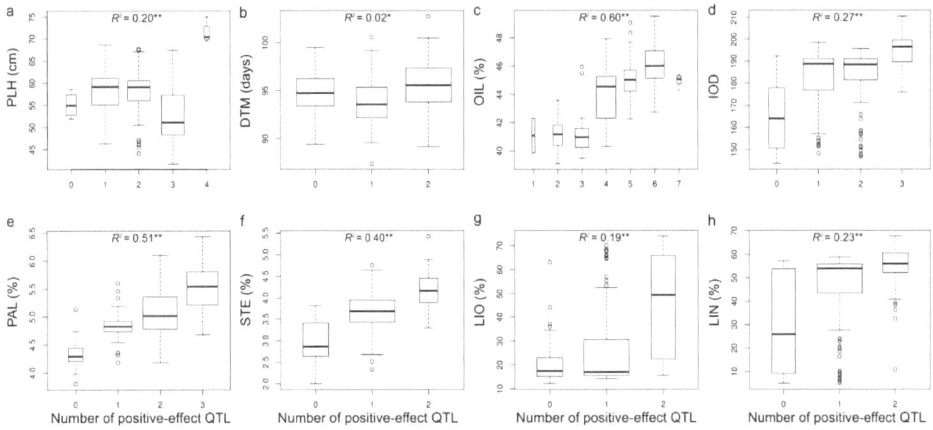

Figure 5. The relationship of phenotypes with the number of positive-effect QTL in individuals. Eight traits with two or more QTL identified were analyzed: (a) plant height (PLH), (b) days to maturity (DTM), (c) oil content (OIL), (d) iodine value (IOD), (e) palmitic acid content (PAL), (f) steric acid content (STE), (g) linoleic acid content (LIO), and (h) linolenic acid content (LIN). The BLUP values of the eight traits in the merged population were used. The correlation of phenotypes with the number of positive-effect QTL was calculated. * and ** represent statistical significance at 0.05 and 0.01 probability level.

2.7 Pleiotropy of QTL

Sixteen of the 33 QTL co-located at six genomic regions concerning nine traits (Figures 1 and 6, Table S8). QTL for PLH, DTH and YLD co-located on chr4. QTL for IOD, LIO and LIN co-located on chr4, 7 and 12. Chromosome 15 harbored QTL for OIL and PRO while chr5 had QTL for OIL and PAL.

2.8 Phenotypic Variation Explained by QTL

Phenotypic variations explained by individual QTL ($h2_{QTL}$) were estimated (Table S4). Overall, the QTL explained 4 to 66% of the total phenotypic variation, with an average of 32.5% which is more than half of the average $h2_{SNP}$ (51%). For five traits (IOD, LIO, LIN, PAL and OIL), QTL explained an average of 61% of the variation (Table 2 and Table S4). We also estimated the phenotypic variation explained by all QTL for a trait ($h2_{GWAS}$) (Table 2). In the merged population, the QTL explained 48–73% of the phenotypic variation for OIL, IOD, PAL, LIO and LIN but only 8–14% for PLH, DTM and YLD.

2.9 Candidate Genes Underlying QTL

Based on the GWAS results, we investigated the genes annotated in the flax genome [54] in an attempt to predict candidate genes from loci significantly associated with each trait. The genomic locations of SNP markers at the peaks of the QTL were scanned within a 500 Kb window in either direction to constitute a subset of genes from which we deduced a candidate gene list based on a priori knowledge of their function(s). Candidate genes were identified for every QTL except for the YLD QTL (Table 3). We discovered seven candidate genes underlying QTL for DTM on chr4. The QTL for PLH harbors five candidate genes of completely different function. The genes underlying QTL for fatty acid composition include KCS14-2, FAD3a, and FAD3b for IOD/LIN/LIO, KCS12-3 and KAS Ic-1 for PAL, KCS9-1 and KCS1-1 for OLE, and KCS18-2 and SAD1 for STE.

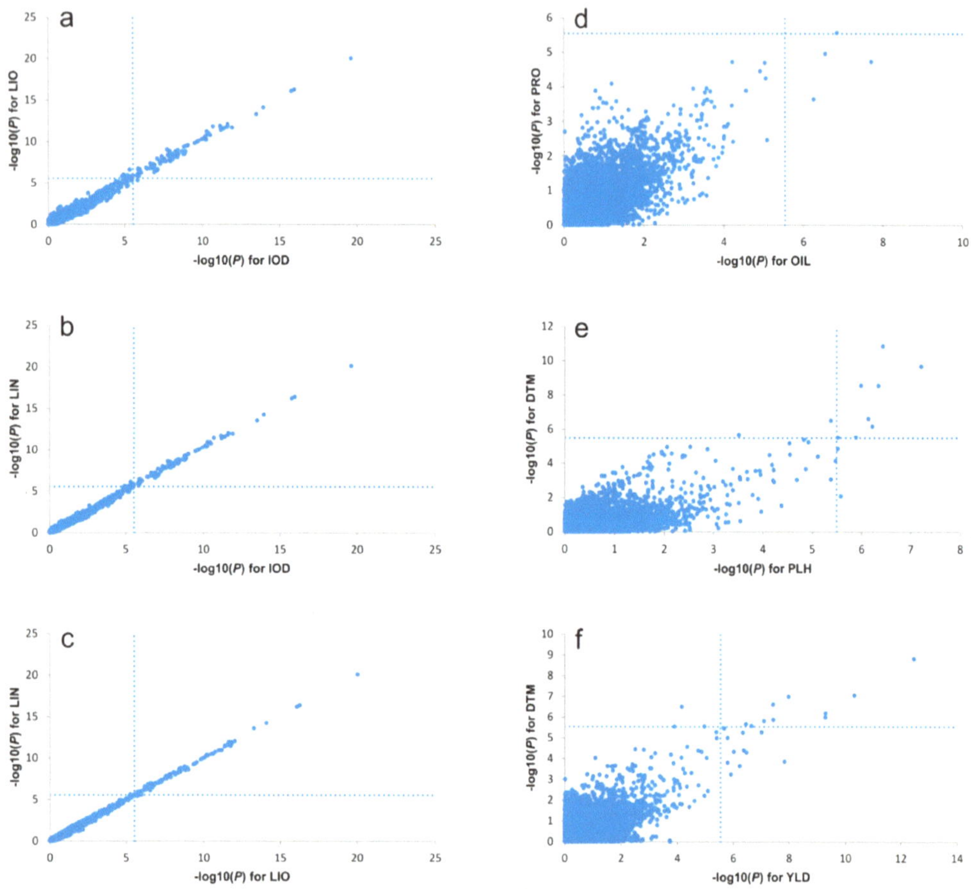

Figure 6. Relations of -log10(P) values of SNP markers between two traits showing pleiotropy or linkage relationship of SNP markers in different pairs of traits. (a) IOD vs. LIN; (b) IOD vs. LIO; LIN vs. LIO; (d) OIL vs. PRO; (e) PLH vs. DTM; (f) DTM vs. YLD. Results of the GWAS using a GLM and data from the BM + EV + SU population for IOD, LIO, and LIN (a–c), the EV population for OIL and PRO (d), the BM population for PLH and DTM (e) and the BM + EV + SU population for DTM and YLD (f) are shown. The vertical and horizontal dashed lines show the cut-off value of significant SNP markers associated with a trait. YLD: seed yield (t ha-1); DTM: days to maturity; OIL: oil content (%); PRO: protein content (%); IOD: iodine value; LIO: linoleic acid content (%); LIN: linolenic acid content (%).

2.10 Selection Signatures in Bi-Parental Populations

A GW3S was performed to identify potential selection signatures during breeding improvement using XP-CLR [34]. Due to the high genetic diversity in BM and EV (Table 1) and large phenotypic differences between them (Table S9), GW3S between BM and EV was conducted. A total of 114 selection signatures with an average size of 226.3 kb were identified (Figures 1 and 7, Table S10), accounting for 7.82% of the flax pseudomolecules (~316 Mb). These putative selection signatures overlapped with 11 GWAS-detected genomic regions associated with 18 QTL (Figures 1 and 7).

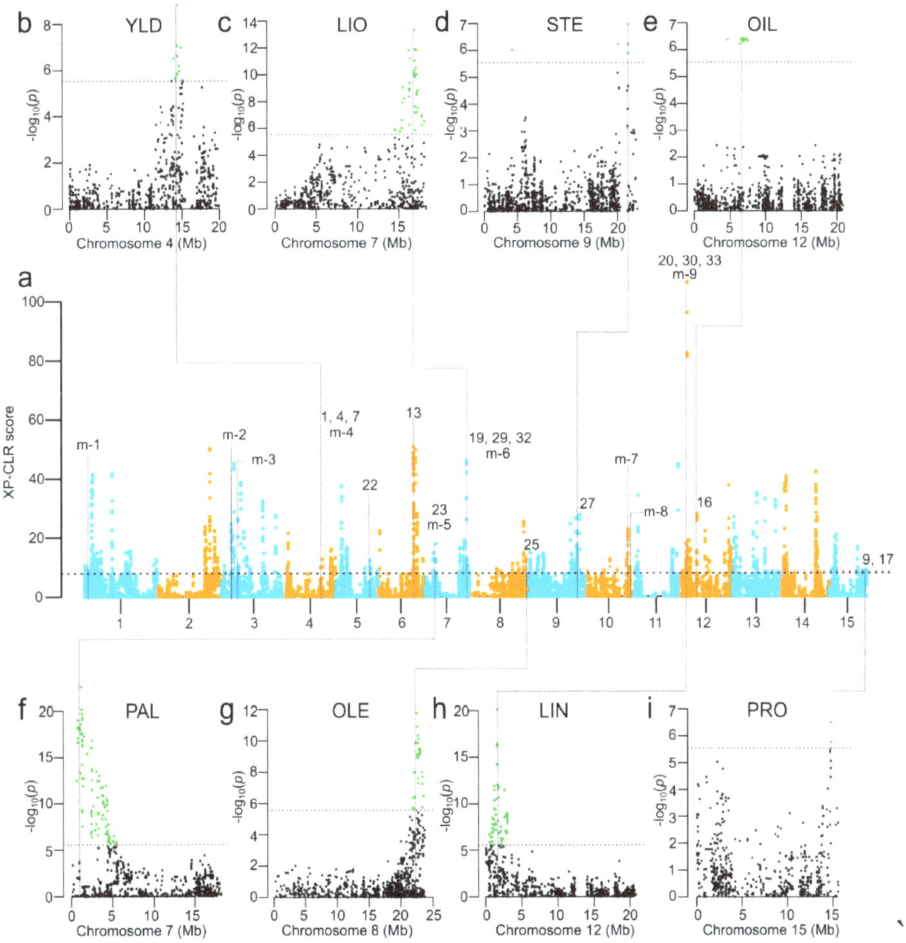

Figure 7. Genome-wide selective sweep scan using XP-CLR between BM and EV (a), and Manhattan plots of QTL overlapping with selective sweeps for (b) seed yield (YLD), (c) linoleic acid content (LIO), (d) steric acid content (STE), (e) oil content (OIL), (f) palmitic acid content (PAL), (g) oleic acid content (OLE), (h) linolenic acid content (LIN), and (i) protein content (PRO). QTL associated with selective sweeps are also labeled on peaks of selective sweeps. The numbers represent the QTL numbers listed in Table 3. Multiple numbers on the same peak represent genomic regions co-located with more than one trait. The labels 'm-#' represent

the genomic regions associated with QTL previously identified and listed in Table S11. The green dots on Manhattan plots represent significant SNPs.

Some selection signatures were also associated with previously identified QTL (Table S11). For example, the selection signatures were associated with 10 previously reported QTL (Figure 7). The signatures at position 2.45–2.46 Mb on chr1 overlapped with SNP marker Lu1_2670961 linked to QTL QSte.BM.crc-LG1 for STE; the ones at 4.74–4.77 Mb on chr3 overlapped with Lu3_5950394, a SNP linked to QTL QOle.BM.crc-LG3-1/QLio.BM.crc-LG3 for OLE and LIO; signatures at 7.24–7.25 Mb on chr3 overlapped with SNP Lu3_8415336 linked to QTL QSte.BM.crc-LG3 for STE [8]; position 16.80–16.81 Mb on chr10 harbors signatures that overlap with SSR Lu2262 linked to an unnamed QTL for OIL; finally, position 17.52–17.53 Mb on chr10 has selection signatures that coincide with SSR Lu2746 linked to an unnamed QTL for LIN/IOD [53].

3. DISCUSSION

3.1 QTL Associated with Seed Yield and Seed Oil Quality Traits

Thirty-three QTL were identified in the current study. Of which, nine QTL were identified in previous studies [7,8] for the same traits, including seed yield and seed oil quality traits. Cloutier et al. [7] detected six major QTL for LIO, LIN and IOD in SU population. These six QTL correspond to the two underlying genes, FAD3a and FAD3b. Some of these QTL were in close proximity on the same chromosome. We identified the same QTL by association mapping that were previously detected by linkage mapping [7] using the same phenotype and SNP genotype data in the SU population (Table 3). The refinement of flax pseudomolecule [45] between the linkage study and our current association study allowed reassignment of chr12 for LIO, LIN and IOD QTL which were previously assigned to LG16 [8]. In addition, the same QTL were also detected in the EV population as well as the merged population. Our association study also validated three QTL for YLD, DTM and PAL which were previously identified using linkage mapping using SSRs and SNPs [8,9] and from the association mapping using a flax core collection population with SSR markers [53] (Table 3). These verified QTL for fatty acid composition, seed yield and maturity demonstrate the feasibility of the association mapping method to detect QTL in a bi-parental population as well as a multi-parent population.

An additional 24 novel QTL were detected in our current study which were not discovered in previous studies using individual BM or SU populations. These new QTL were detected using the merged population which greatly increased the population size, thereby enhancing the association power and resolution for QTL detection. We noted that only two QTL were discovered from the BM population alone. This is likely the result of significantly reduced representation of lines re-sequenced from BM population [8]. The discovery of new QTL demonstrates that GWAS using multiple bi-parental populations is equally or more efficient for QTL detection than QTL mapping using single bi-parental populations alone.

We tested the statistical significance of QTL effects for all 33 QTL identified for the 11 traits and found that all effect differences were significant. We also observed significant positive correlation between the number of positive-effect QTL and corresponding trait phenotypes in individuals for eight traits from which had two or more QTL were identified (Figures 4 and 5, Table S7). These results not only corroborate the significance of the QTL but also demonstrate that effects of QTL in an individual performed additively, suggesting that marker-assisted selection (MAS) for these QTL would be effective in breeding. Thus, we listed the flanking sequences of these QTL in Table S12 for MAS purpose.

3.2 Pleotropic QTL Associated with Seed Yield and Quality Traits

Six genomic regions associated with more than one trait were identified. QTL for IOD, LIO, and LIN were concurrent on chromosomes 4, 7 and 12; QTL for YLD, PLH, and DTM co-located on chr4; QTL for PRO and OIL were on chr15 and QTL for PAL and OIL were on chr5 (Figures 1 and 6, Table S8).

IOD is a measure of the degree of unsaturation of the oil that is calculated from the GC-derived fatty acid composition. Thus, breeding lines with high LIN normally show high IOD [7] due to the high correlation between IOD, LIO, and LIN [44] (Table S13). QTL co-located at the same genomic regions indicate that the traits may be controlled by the same gene or tightly linked genes. The two genomic regions on chromosomes 7 and 12 harbor the two fatty acid desaturase genes, FAD3a and FAD3b. These genes are responsible for linoleic and linolenic acid composition [52,55].

PLH and DTM are complex traits that considerably impact the adaptability, biomass, and economic yield of agricultural crops [56,57]. In soybean, one QTL that strongly associated with both PLH and DTM traits was identified with an SNP at 45.0 Mb position on chromosome 19 and it harbors the candidate gene DT1, which is homolog to Arabidopsis terminal flower 1 (TFL-1, AT5G03840) [56]. Based on in silico gene annotation, the DT1 homolog are located on chromosomes 6 and 8 in flax but no QTL for either PLH or DTM were identified on these two chromosomes. This could be due to the lack of functional polymorphism(s) at those loci among the parents of our three populations. However, a different genomic region on chr4 harbours five candidate genes for PLH and seven for DTM, raising the possibility that PLH and DTM are controlled by tightly linked genes in flax. The same genomic region was also associated with YLD. Because plant height and maturity affect seed yield, it is not surprising that QTL for PLH, DTM and YLD were mapped to the same locus. This pleiotropic relationship between YLD and DTM was previously validated [8] (Table 3).

Inheritance of seed oil content is complicated due to its quantitative nature. The seed oil content was directly affected by fatty acid composition traits, such as PAL, STE, OLE, LIO, and LIN, or indirectly by several major agronomic traits, such as seed yield and protein content [58]. Significant correlations of OIL were observed with PAL (-0.57; p = 0) and PRO (-0.70; p = 0) (Table S13). OIL is also usually negatively correlated with PRO in oilseed crops [59]. Of the eight QTL associated with oil content, two co-located with QTL for PAL on chr5 and for PRO on chr15, respectively.

3.3 Phenotypic Variation Explained by SNPs and QTL

SNP heritability ($h2_{SNP}$) for a trait is the total proportion of phenotypic variance explained by additive contributions from genome-wide SNPs. A method for estimating $h2_{SNP}$ for a complex trait was initially proposed in 2011 [60,61] and implemented in GCTA (Genome-wide Complex Trait Analysis) software [61]. Since then, the method has been applied to many quantitative traits largely in human and animal genetic studies [62,63]. The method was also used to estimate phenotypic variance explained by a subset of SNPs selected by p-values from GWAS in an independent sample [64]. However the estimate of variance explained by the SNP subsets ascertained by the p-values from GWAS in the same sample may be inflated due to positive correlation between true SNP effects and estimation errors (personal communication to the GCTA author, Jian Yang). However, as the GCTA-based heritability estimation method includes the population structure effect in the linear model and also considers heritability estimates to be irrelevant to the number of SNPs used [60,61], the accuracy of estimates should be higher than those obtained simply using the simple multivariate regression adopted in most GWAS of plant traits. In the current study, for the first time we applied this method to estimate h2 for 11 agronomic and seed quality traits in three bi-parental populations and a merged population. As the number of SNPs identified from a population depends on its genetic variation for the traits, the trait-associated $h2_{SNP}$ estimates vary across populations and traits. Overall, seed yield had a lower $h2_{SNP}$ than seed quality traits as expected considering the extent of genetic complexity of the former(Table 2). We also used the same method to estimate phenotypic variation explained by individual QTL ($h2_{QTL}$) and by all QTL for a specific trait ($h2_{GWAS}$ measures the extent of the phenotypic variation explained by QTL compared to that of all SNPs. This comparison led to the conclusion that many QTL for PLH, DTM and YLD were not detected in our study but the QTL for seed quality traits identified herein likely represent major genetic regions or genes controlling these traits.

3.4 Selection Signatures Associated with Seed Yield and Seed Quality Traits

GW3S has been used for screening putative genomic regions under selection pressure caused by domestication or artificial selection [36,38]. Usually, contrasting genetic populations are compared (such as wild accessions vs. cultivated accessions, landraces vs. breeding lines) to identify the allele frequency differentiation between different populations. In this study, we alternatively used two contrasting bi-parental mapping populations and identified 114 selection signatures with an average size of 226.3 kb. Some of these selection signatures support nearly 50% of the 23 GWAS-detected genomic regions associated with 33 QTL. Some of the QTL identified by GWAS have no overlapping selection signatures, partially because the regions of QTL had XP-CLR (Cross Population Composite Likelihood Ratio) scores less than the predetermined cut-off values. On the other hand, many selection signatures have high XP-CLR scores but no associated QTL (Figure 7). These significant selection signatures may be associated with QTL for traits not included in this study. This is suggested by the fact that five previously identified genomic regions related to seven QTL overlapped with the selection signatures identified in our current study comparing BM and EV (Table S10). These putative selection signatures provide useful candidates for further QTL-trait association study. Our results combined with previous studies demonstrate that GW3S combined with GWAS is a powerful approach for dissecting genetic structure of breeding populations and for the identification of underlying genomic regions for breeding improvement. The application of GWAS with bi-parental populations and selection signatures facilitated the cross-validation of QTL previously revealed using alternative mapping techniques and laid the groundwork for genomic-assisted breeding in flax.

4. MATERIALS AND METHODS

4.1 Plant Materials

Three bi-parental mapping populations of different genetic backgrounds served as genotype panel for the association study. The first population (BM) consisted of 243 F6-derived RILs generated by single seed descent from a cross between CDC Bethune and Macbeth. Its two parents are Canadian high-yielding conventional linseed cultivars with 55–57% LIN [65,66]. The second population (EV) contained 90 F6-derived RILs from a cross between E1747, an ethyl methanesulfonate (EMS)-induced low LIN breeding line [67], and Viking, a French fiber flax cultivar with ~55% LIN that was grown extensively in the 2000's but deregistered in 2012. The third population (SU) is an F1-derived DH population of 78 lines obtained from a cross between the breeding line SP2047, from which a yellow-seeded SolinTM 2047 with only 2–3% LIN has been derived, and breeding line UGG5-5, which is a high LIN line with 63–66% LIN [7,55]. BM was designed to study yield-related traits while EV and SU were intended for QTL identification for fatty acid composition and fiber traits.

4.2 Whole Genome Resequencing, SNP Calling, SNP Imputation and LD Analysis

Three populations consisting of 97 randomly selected lines from BM, 91 from EV, 72 from SU including five parents (one parent is the reference genome) were grown in growth cabinets with a 16-h light and 8-h dark cycle at 20/18 °C. DNA was extracted from young leaf tissue using the DNeasy 96 Plant kit (Qiagen, Mississauga, ON, Canada) according to the manufacturer's instructions. The DNA was subsequently restricted, size-selected and quantified prior to the construction of the reduced representation libraries used for Illumina sequencing as previously described [47]. Reduced representation libraries from a total of 260 individuals of the three populations, i.e., 96 randomly selected from BM, 89 from EV, 70 from SU, and five parents (One parent CDC Bethune of BM is used as a reference genome) were re-sequenced by the Michael Smith Genome Sciences Centre of the BC Cancer Agency, Genome British Columbia (Vancouver, BC, Canada) using 100-bp paired-end reads on an Illumina HiSeq 2000 platform (Illumina Inc., San Diego, CA, USA).

SNP calling was performed using the revised AGSNP pipeline [47,48,68]. The flax scaffold sequences of cultivar CDC Bethune [46] were used as reference for read mapping. Then SNPs were called using SAMtools [69] and further filtered using a set of criteria such as mapped read depth, consensus base ratio, mapping quality score and homopolymers with a validation rate of 96.8% for the called SNPs as described in detail [47]. Finally SNPs with a MAF < 0.05 and a genotyping rate <60% were removed for further analysis. The coordinates of all SNPs were extracted from the chromosome-based flax pseudomolecules v2.0 [45]. Missing data for these filtered SNPs were imputed using Beagle v.4.2 [70].

Intra-chromosome LD (r2) was calculated using plink ver. 1.9 [71] with the parameters "-r2 -ld-window-kb 2000 -ld-window-r2 0". Before LD calculation, SNP data were pruned using the parameter "–indep-pairwise 2000 50 0.9" to remove SNPs with high r2 (>0.9) in a 2000 kb window with step size of 50 SNPs. Pair-wise r2 values were plotted against the base pair distance, and curves of LD decay with distances of paired SNPs were fitted using a non-linear regression model [72] as follows:

$$r^2 = \frac{10+cd}{(2+cd)(11+cd)} \times \left\{1 + \frac{(3+cd)\left(12+12cd+(cd)^2\right)}{n(2+cd)(11+cd)}\right\}, \quad (1)$$

where c is the coefficient to be estimated, d is the distance between pair-wise SNPs, and n is the number of total gametes, corresponding to twice the number of individuals in a population. The R function nls was used to fit the model. The rate of LD decay for each population was determined from the fitted model at the half of the maximum LD (r2). Haplotype blocks were calculated using plink with the parameters "–blocks no-pheno-req –blocks-max-kb 2000".

4.3 Differentiation and Stratification

Nucleotide diversity (π) of three bi-parental populations and genetic differentiation (Fst) between the populations were estimated using the R package "PopGenome" [73]. The genetic structures of the three separate inbreeding populations and the combined population were assessed using both PCA and DAPC [74]. Analyses with DAPC included several steps: (1) PCA was conducted using the imputed SNPs. According to the curve of accumulative variances against the number of principle components (PCs), the optimum number of PCs was chosen at which the cumulative variance had no obvious increase; (2) k-means clustering analysis was performed based on the chosen PCs. To identify the optimal number of clusters, k-means was run sequentially with increasing k values and the Bayesian information criterion (BIC) was calculated for each k. The optimum k corresponded to the lowest BIC; (3) Discriminant analysis was conducted based on the chosen clusters and individuals were reassigned to the different clusters. The posterior probability of cluster membership was calculated based on the retained discrimination functions and used as the Q matrix for GWAS and heritability estimation. PCA was performed using the function implemented in TASSEL while DAPC was conducted using the R package "adegenet" 2.0 [75].

4.4 Phenotyping of Bi-Parental Populations

Individuals from the three populations were evaluated in field trials over four years (2009–2012) at two sites, Morden Research and Development Centre, Manitoba (MD) and Kernen Crop Research Farm near Saskatoon, Saskatchewan (SAS) in Canada. A type-2 modified augmented design (MAD) [76] was used for the field experiments from which phenotypic data were collected. The detailed experimental design was previously described [44,77]. All 243 individuals of the BM population were phenotyped in four years (2009–2012) and two sites (MD and SAS), while 86 individuals of the EV population and 72 individuals of the SU population were evaluated in three years (2010–2012) and two sites (MD and SAS).

Eleven common traits were evaluated in the three populations, including YLD, PLH, DTM, PRO, OIL, IOD and five fatty acid composition traits (OLE, PAL, STE, LIO, and LIN). PLH was measured from ground to the uppermost part of the plant at maturity. DTM was recorded from sowing to 95% of capsule maturity (seeds rattling in the capsules or bolls). Seed yield data were measured by harvesting two 0.5-m sections from rows located in the central part of each subplot (0.2 m2). A total of 1 g of seed from each line at each environment was sampled for OIL measurement and fatty acid composition. Methyl esters of fatty acids were prepared according to the American Oil Chemists' Society (AOCS) Official Method Ce 2-66 [78] and fatty acid composition was measured by gas chromatography (GC) following AOCS Official Method Ce 1e-91. OIL was determined by nuclear magnetic resonance calibrated against the FOSFA extraction reference method. PRO was measured using near-infrared spectroscopy calibrated against the combustion analysis reference method and expressed on an N × 6.25 dry basis. Phenotyping of these seed quality traits has been previously described [53]. All phenotypic data from the field experiments and laboratory measurements were adjusted for soil heterogeneity as previously described based on the MAD pipeline [77]. The BLUP values over multiple environmental phenotypes estimated using R package "lme4" [79] were used for further association study analyses. The Shapiro-Wilk normality test was performed for all traits using the R function "shapiro.test". All 11 traits followed approximately a normal or mixed normal distribution (Figure S3). Simple correlations among 11 traits were calculated using the function "rcorr" of the R package "Hmisc".

4.5. Phenotypic Variation Explained by All SNPs

The phenotypic variation explained by all SNPs, denoted as h2 *SNP* , was estimated for all traits based on the following mixed linear model [60] implemented in the GCTA software [61]:

$$y = X\beta + g + \varepsilon \text{ with its variance } V = A\sigma_g^2 + I\sigma_\varepsilon^2 \quad (2)$$

where y is an n × 1 vector of phenotypes with n individuals in a population, X is the n × n. structure matrix, β is a vector of fixed effects of population structure, including posterior probabilities of an individual assigning to a cluster in DAPC, g is an n × 1 vector of the total genetic effects of the individuals with g ~N (0, Aσg2), and ε is a vector of residual effects with ε ~N (0, Iσε2), A is interpreted as the genetic relationship matrix (GRM) between individuals and estimated from SNPs.

σg2 is estimated using the restricted maximum likelihood (REML) method based on the GRM estimated from all SNPs. Thus, SNP heritability h2*SNP SNP* was estimated.

4.6 Genome-Wide Association Study

GWAS was performed with the GLM and compressed MLM [80,81] implemented in TASSEL (v5.2) [82], which employs the EMMA and P3D algorithms to reduce computing time. For MLM, the kinship matrices for the merged population and the three single populations were calculated using TASSEL (v5.2) [82]. Manhattan plots and quantile-quantile (Q-Q) plots of GWAS were obtained using the R package "qqman" [83].

SNP markers for candidate QTL were determined based on the p-value for each marker estimated in the GLM or MLM analysis. The p-values were adjusted by the Bonferroni correction, being α (0.05)/No. of SNPs used in the analyses. Allele effects of significant markers were calculated as the difference between the average phenotypic values of homozygous alleles which were obtained directly from the TASSEL outputs. Candidate QTL were defined based on peaks of SNPs exceeded the significance threshold for the trait. The genomic region for a QTL was defined as a genome block spanning all significant SNPs.

The amount of phenotypic variation explained by significant QTL was estimated for all SNP markers within the QTL regions using the same method as described above [61], denoted as $h2_{QTL}$. We similarly estimated phenotypic variation explained by all significant QTL for a single trait and denoted it $h2_{GWAS}$.

4.7 Candidate Gene Mining

A comprehensive genome-wide scan across chromosomes for significant quantitative trait loci (QTL) was conducted to delineate the relevant genomic areas and identify potential candidate genes. Initially, all orthologous genes of the model organism Arabidopsis thaliana were aligned to the flax genome utilizing BLASTP with flax protein sequences.A. thaliana protein sequences at an E-value of 1.0 × 10–10. A total of 15,323 unique A. thaliana genes were mapped. A list of known flax or A. thaliana genes associated with our studied traits and their associations was drawn based on literature and database searches [49,51,84]. We investigated candidate genes within QTL regions or within a 500 kb window upstream and downstream of the peaks depending on the LD decay estimates. In addition, previously identified QTL (SSR markers) in flax [7,8,53] were mapped to the flax pseudomolecules to validate the QTL results from this study.

4.8 QTL Validation

Three approaches were applied to validate QTL identified by GWAS. The first approach was to compare our QTL with previously identified QTL as described above. The same QTL was inferred if two QTL were mapped to the same recombination block or haplotype block. The second approach tested the significance of difference of phenotypes between two contrasting haplotype pairs of a QTL in the populations. Statistically significant differences served to validate significant QTL. Both t and Wilcox non-parametric tests were performed using the R functions "t.test" and "wilcox.test" for each QTL in the merged and individual populations and in different year/location environments. To test the positive correlations of a trait upon pyramiding of QTL, a simple regression of the number of positive-effect

QTL on phenotypic values of a trait was calculated. A positive-effect QTL in an individual meant that this individual possessed a positive effect allele for the QTL. The last approach was to perform genome-wide selective sweep scans to confirm QTL associated genomic regions as described below.

4.9 Genome-Wide Selective Sweep Scan

A WG3S was performed along chromosomes across two populations using the program XP-CLR [34]. Comparisons between BM and EV using XP-CLR were conducted. The genetic distances (cM) between SNPs were estimated using the integrated flax consensus genetic map [43], assuming uniform recombination between SNPs. For each chromosome, XP-CLR was executed with the parameters "-w1 0.005 100 100 1 -p1 0.7" to estimate XP-CLR scores for 100-bp windows. Each chromosome was then divided into 10-kb segments and the highest XP-CLR score from windows with at least one SNP were assigned to each 10-kb segment (xmax,i). If the XP-CLR scores (xmax, i and xmax, i+1) of two adjacent 10-kb segments were greater than the 80th percentile (xmax,80th) of the genome-wide scores of all 10-kb fragments, then they were grouped as a single putative selective sweep. In addition, putative selective sweeps were also merged if they were separated by no more than one low score (< xmax,80th) segment. Merged selective sweeps were assigned the highest score from their merged 10-kb segments. These merged segments were further combined into a larger region if these segments belonged to the same peak in the genome-wide selective sweep plot (Figure 5a). Finally, the combined regions falling in the highest 10th percentile of all putative selective sweeps were considered differentially selected regions or selection signatures.

The selection signatures were examined in respect to our found QTL and previously known QTL at the genetic loci to establish correlations. The sites of the QTL-related markers were extended by 100 kb on either side and thereafter compared with the locations of the selection signatures. The QTL and selection signatures were considered associated due to their overlap.

REFERENCES

1. Westcott, N.D.; Muir, A.D. Chemical studies on the constituents of Linum spp. In Flax, the Genus Linum; Muir, A.D., Westcott, N.D., Eds.; Taylor and Francis: New York, NY, USA, 2003; pp. 55–73.

2. Diederichsen, A.; Kusters, P.M.; Kessler, D.; Bainas, Z.; Gugel, R.K. Assembling a core collection from the flax world collection maintained by Plant Gene resources of Canada. Genet. Resour. Crop Evol. 2013, 60, 1479–1485. [CrossRef]

3. Green, A.G.; Chen, Y.; Singh, S.P.; Dribnenki, J.C.P. Flax. In Compendium Transgenic Crop Plants: Transgenic Oilseed Crops; Kole, C., Hall, T.C., Eds.; Blackwell Publishing Ltd.: Oxford, UK, 2008; pp. 199–226.

4. Tolkachev, O.N.; Zhuchenko, A.A. Biologically active substances of flax: Medicinal and nutritional properties. Pharm. Chem. J. 2000, 34, 360–367. [CrossRef]

5. You, F.M.; Duguid, S.D.; Lam, I.; Cloutier, S.; Rashid, K.Y.; Booker, H. Pedigrees and genetic base of the flax varieties registered in Canada. Can. J. Plant Sci. 2016, 96, 837–852. [CrossRef]

6. Price, A.H. Believe it or not, QTLs are accurate! Trends Plant Sci. 2006, 11, 213–216. [CrossRef] [PubMed]

7. Cloutier, S.; Ragupathy, R.; Niu, Z.; Duguid, S.D. SSR-based linkage map of flax (Linum usitatissimum L.) and mapping of QTLs underlying fatty acid composition traits. Mol. Breed. 2011, 28, 437–451. [CrossRef]

8. Kumar, S.; You, F.M.; Duguid, S.; Booker, H.; Rowland, G.; Cloutier, S. QTL for fatty acid composition and yield in linseed (Linum usitatissimum L.). Theor. Appl. Genet. 2015, 128, 965–984. [CrossRef] [PubMed]

9. Asgarinia, P.; Cloutier, S.; Duguid, S.; Rashid, K.; Mirlohi, A.; Banik, M.; Saeidi, G. Mapping quantitative trait loci for powdery mildew resistance in flax (Linum usitatissimum L.). Crop Sci. 2013, 53, 2462–2472. [CrossRef]

10. Fu, Y.-B. Genetic evidence for early flax domestication with capsular dehiscence. Genet. Resour. Crop Evol. 2011, 58, 1119–1128. [CrossRef]

11. Soto-Cerda, B.J.; Maureira-Butler, I.; Muñoz, G.; Rupayan, A.; Cloutier, S. SSR-based population structure, molecular diversity and linkage disequilibrium analysis of a collection of flax (Linum usitatissimum L.) varying for mucilage seed-coat content. Mol. Breed. 2012, 30, 875–888. [CrossRef]

12. Wiesnerova, D.; Wiesner, I. ISSR-based clustering of cultivated flax germplasm is statistically correlated to thousand seed mass. Mol. Biotechnol. 2004, 26, 207–213. [CrossRef]

13. McMullen, M.D.; Kresovich, S.; Villeda, H.S.; Bradbury, P.; Li, H.; Sun, Q.; Flint-Garcia, S.; Thornsberry, J.; Acharya, C.; Bottoms, C. Genetic properties of the maize nested association mapping population. Science 2009, 325, 737–740. [CrossRef] [PubMed]

14. Bandillo, N.; Raghavan, C.; Muyco, P.A.; Sevilla, M.A.L.; Lobina, I.T.; Dilla-Ermita, C.J.; Tung, C.-W.; McCouch, S.; Thomson, M.; Mauleon, R.; et al. Multi-parent advanced generation inter-cross (MAGIC) populations in rice: Progress and potential for genetics research and breeding. Rice 2013, 6, 11. [CrossRef] [PubMed]

15. Yu, J.; Holland, J.B.; McMullen, M.D.; Buckler, E.S. Genetic design and statistical power of nested association mapping in maize. Genetics 2008, 178, 539–551. [CrossRef] [PubMed]

16. Monir, M.M.; Zhu, J. Dominance and epistasis interactions revealed as important variants for leaf traits of maize NAM population. Front. Plant Sci. 2018, 9, 627. [CrossRef] [PubMed]

17. Ren, D.; Fang, X.; Jiang, P.; Zhang, G.; Hu, J.; Wang, X.; Meng, Q.; Cui, W.; Lan, S.; Ma, X.; et al. Genetic architecture of nitrogen-deficiency tolerance in wheat seedlings based on a nested association mapping (NAM) population. Front. Plant Sci. 2018, 9, 845. [CrossRef] [PubMed]

18. Mackay, I.; Powell, W. Methods for linkage disequilibrium mapping in crops. Trends Plant Sci. 2007, 12, 57–63. [CrossRef] [PubMed]

19. Cavanagh, C.; Morell, M.; Mackay, I.; Powell, W. From mutations to MAGIC: Resources for gene discovery, validation and delivery in crop plants. Curr. Opin. Plant Biol. 2008, 11, 215–221. [CrossRef] [PubMed]

20. Mathew, B.; Leon, J.; Sannemann, W.; Sillanpaa, M.J. Detection of epistasis for flowering time using Bayesian multilocus estimation in a barley MAGIC population. Genetics 2018, 208, 525–536. [CrossRef] [PubMed]

21. Camargo, A.V.; Mackay, I.; Mott, R.; Han, J.; Doonan, J.H.; Askew, K.; Corke, F.; Williams, K.; Bentley, A.R. Functional mapping of quantitative trait loci (QTLs) associated with plant performance in a wheat magic mapping population. Front. Plant Sci. 2018, 9, 887. [CrossRef] [PubMed]

22. Ongom, P.O.; Ejeta, G. Mating design and genetic structure of a multi-parent advanced generation intercross (magic) population of sorghum (Sorghum bicolor (L.) Moench). G3 (Bethesda) 2018, 8, 331–341. [CrossRef] [PubMed]

23. Huynh, B.L.; Ehlers, J.D.; Huang, B.E.; Munoz-Amatriain, M.; Lonardi, S.; Santos, J.R.P.; Ndeve, A.; Batieno, B.J.; Boukar, O.; Cisse, N.; et al. A multi-parent advanced generation inter-cross (MAGIC) population for genetic analysis and improvement of cowpea (Vigna unguiculata L. Walp.). Plant J. 2018, 93, 1129–1142. [CrossRef] [PubMed]

24. Ponce, K.S.; Ye, G.; Zhao, X. Qtl identification for cooking and eating quality in indica rice using multi-parent advanced generation intercross (MAGIC) population. Front. Plant Sci. 2018, 9, 868. [CrossRef] [PubMed]

25. Huang, C.; Shen, C.; Wen, T.; Gao, B.; Zhu, D.; Li, X.; Ahmed, M.M.; Li, D.; Lin, Z. SSR-based association mapping of fiber quality in upland cotton using an eight-way MAGIC population. Mol. Genet. Genom. 2018, 293, 793–805. [CrossRef] [PubMed]

26. Garrido-Cardenas, J.A.; Mesa-Valle, C.; Manzano-Agugliaro, F. Trends in plant research using molecular markers. Planta 2018, 247, 543–557. [CrossRef] [PubMed]

27. Pena, R.N.; Ros-Freixedes, R.; Tor, M.; Estany, J. Genetic marker discovery in complex traits: A field example on fat content and composition in pigs. Int. J. Mol. Sci. 2016, 17. [CrossRef] [PubMed]
28. Zhu, X.M.; Shao, X.Y.; Pei, Y.H.; Guo, X.M.; Li, J.; Song, X.Y.; Zhao, M.A. Genetic diversity and genome-wide association study of major ear quantitative traits using high-density SNPs in maize. Front. Plant Sci. 2018, 9, 966. [CrossRef] [PubMed]
29. Chen, L.; Wan, H.; Qian, J.; Guo, J.; Sun, C.; Wen, J.; Yi, B.; Ma, C.; Tu, J.; Song, L.; et al. Genome-wide association study of cadmium accumulation at the seedling stage in rapeseed (Brassica napus L.). Front. Plant Sci. 2018, 9, 375. [CrossRef] [PubMed]
30. MacGregor, S.; Ong, J.S.; An, J.; Han, X.; Zhou, T.; Siggs, O.M.; Law, M.H.; Souzeau, E.; Sharma, S.; Lynn, D.J.; et al. Genome-wide association study of intraocular pressure uncovers new pathways to glaucoma. Nat. Genet. 2018, 50, 1067–1071. [CrossRef] [PubMed]
31. Huang, X.; Wei, X.; Sang, T.; Zhao, Q.; Feng, Q.; Zhao, Y.; Li, C.; Zhu, C.; Lu, T.; Zhang, Z.; et al. Genome-wide association studies of 14 agronomic traits in rice landraces. Nat. Genet. 2010, 42, 961–967. [CrossRef] [PubMed]
32. Meng, L.; Zhao, X.; Ponce, K.; Ye, G.; Leung, H. QTL mapping for agronomic traits using multi-parent advanced generation inter-cross (MAGIC) populations derived from diverse elite indica rice lines. Field Crops Res. 2016, 189, 19–42. [CrossRef]
33. Huang, X.; Han, B. Natural variations and genome-wide association studies in crop plants. Annu. Rev. Plant Biol. 2014, 65, 531–551. [CrossRef] [PubMed]
34. Chen, H.; Patterson, N.; Reich, D. Population differentiation as a test for selective sweeps. Genome Res. 2010, 20, 393–402. [CrossRef] [PubMed]
35. Gore, M.A.; Chia, J.M.; Elshire, R.J.; Sun, Q.; Ersoz, E.S.; Hurwitz, B.L.; Peiffer, J.A.; McMullen, M.D.; Grills, G.S.; Ross-Ibarra, J.; et al. A first-generation haplotype map of maize. Science 2009, 326, 1115–1117. [CrossRef] [PubMed]
36. Xie, W.; Wang, G.; Yuan, M.; Yao, W.; Lyu, K.; Zhao, H.; Yang, M.; Li, P.; Zhang, X.; Yuan, J.; et al. Breeding signatures of rice improvement revealed by a genomic variation map from a large germplasm collection. Proc. Natl. Acad. Sci. USA 2015, 112, E5411–5419. [CrossRef] [PubMed]
37. Wen, Z.; Boyse, J.F.; Song, Q.; Cregan, P.B.; Wang, D. Genomic consequences of selection and genome-wide association mapping in soybean. BMC Genom. 2015, 16, 671. [CrossRef] [PubMed]

38. Zhou, Z.; Jiang, Y.; Wang, Z.; Gou, Z.; Lyu, J.; Li, W.; Yu, Y.; Shu, L.; Zhao, Y.; Ma, Y.; et al. Resequencing 302 wild and cultivated accessions identifies genes related to domestication and improvement in soybean. Nat. Biotechnol. 2015, 33, 408–414. [CrossRef] [PubMed]

39. He, C.; Fu, J.; Zhang, J.; Li, Y.; Zheng, J.; Zhang, H.; Yang, X.; Wang, J.; Wang, G. A gene-oriented haplotype comparison reveals recently selected genomic regions in temperate and tropical maize germplasm. PLoS ONE 2016, 12, e0169806. [CrossRef] [PubMed]

40. Jordan, K.W.; Wang, S.; Lun, Y.; Gardiner, L.J.; MacLachlan, R.; Hucl, P.; Wiebe, K.; Wong, D.; Forrest, K.L.; Sharpe, A.G.; et al. A haplotype map of allohexaploid wheat reveals distinct patterns of selection on homoeologous genomes. Genome Biol. 2015, 16, 48. [CrossRef] [PubMed]

41. Cadzow, M.; Boocock, J.; Nguyen, H.T.; Wilcox, P.; Merriman, T.R.; Black, M.A. A bioinformatics workflow for detecting signatures of selection in genomic data. Front. Genet. 2014, 5, 293. [CrossRef] [PubMed]

42. Korte, A.; Farlow, A. The advantages and limitations of trait analysis with GWAS: A review. Plant Methods 2013, 9, 29. [CrossRef] [PubMed]

43. Cloutier, S.; Ragupathy, R.; Miranda, E.; Radovanovic, N.; Reimer, E.; Walichnowski, A.; Ward, K.; Rowland, G.; Duguid, S.; Banik, M. Integrated consensus genetic and physical maps of flax (Linum usitatissimum L.). Theor. Appl. Genet. 2012, 125, 1783–1795. [CrossRef] [PubMed]

44. You, F.M.; Booker, M.H.; Duguid, D.S.; Jia, G.; Cloutier, S. Accuracy of genomic selection in biparental populations of flax (Linum usitatissimum L.). Crop J. 2016, 4, 290–303. [CrossRef]

45. You, F.M.; Xiao, J.; Li, P.; Yao, Z.; Jia, G.; He, L.; Zhu, T.; Luo, M.C.; Wang, X.; Deyholos, M.K.; et al. Chromosome-scale pseudomolecules refined by optical, physical and genetic maps in flax. Plant J. 2018, 95, 371–384. [CrossRef] [PubMed]

46. Wang, Z.; Hobson, N.; Galindo, L.; Zhu, S.; Shi, D.; McDill, J.; Yang, L.; Hawkins, S.; Neutelings, G.; Datla, R.; et al. The genome of flax (Linum usitatissimum) assembled de novo from short shotgun sequence reads. Plant J. 2012, 72, 461–473. [CrossRef] [PubMed]

47. Kumar, S.; You, F.M.; Cloutier, S. Genome wide SNP discovery in flax through next generation sequencing of reduced representation libraries. BMC Genom. 2012, 13, 684. [CrossRef] [PubMed]

48. You, F.M.; Deal, K.R.; Wang, J.; Britton, M.T.; Fass, J.N.; Lin, D.; Dandekar, A.M.; Leslie, C.A.; Aradhya, M.; Luo, M.C.; et al. Genome-wide SNP discovery

in walnut with an AGSNP pipeline updated for SNP discovery in allogamous organisms. BMC Genom. 2012, 13, 354. [CrossRef] [PubMed]

49. Sun, C.; Wang, B.; Yan, L.; Hu, K.; Liu, S.; Zhou, Y.; Guan, C.; Zhang, Z.; Li, J.; Zhang, J.; et al. Genome-wide association study provides insight into the genetic control of plant height in rapeseed (Brassica napus L.). Front. Plant Sci. 2016, 7, 1102. [CrossRef] [PubMed]

50. Lamesch, P.; Berardini, T.Z.; Li, D.; Swarbreck, D.; Wilks, C.; Sasidharan, R.; Muller, R.; Dreher, K.; Alexander, D.L.; Garcia-Hernandez, M.; et al. The Arabidopsis Information Resource (TAIR): Improved gene annotation and new tools. Nucleic Acids Res. 2012, 40, D1202–D1210. [CrossRef] [PubMed]

51. You, F.M.; Li, P.; Kumar, S.; Ragupathy, R.; Li, Z.; Fu, Y.-B.; Cloutier, S. Genome-wide identification and characterization of the gene families controlling fatty acid biosynthesis in flax (Linum usitatissimum L.). J. Proteom. Bioinf. 2014, 7, 310–326.

52. Vrinten, P.; Hu, Z.; Munchinsky, M.A.; Rowland, G.; Qiu, X. Two FAD3 desaturase genes control the level of linolenic acid in flax seed. Plant Physiol. 2005, 139, 79–87. [CrossRef] [PubMed]

53. Soto-Cerda, B.J.; Duguid, S.; Booker, H.; Rowland, G.; Diederichsen, A.; Cloutier, S. Association mapping of seed quality traits using the Canadian flax (Linum usitatissimum L.) core collection. Theor. Appl. Genet. 2014, 127, 881–896. [CrossRef] [PubMed]

54. You, F.M.; Li, P.; Ragupathy, R.; Kumar, S.; Zhu, T.; Luo, M.-C.; Duguid, S.D.; Rashid, K.Y.; Booker, H.M.; Deyholos, M.K.; et al. The Draft Flax Genome Pseudomolecules. In Proceedings of the 66th Flax Institute of the United States, Fargo, ND, USA, 31 March–1 April 2016; pp. 17–24.

55. Banik, M.; Duguid, S.; Cloutier, S. Transcript profiling and gene characterization of three fatty acid desaturase genes in high, moderate, and low linolenic acid genotypes of flax (Linum usitatissimum L.) and their role in linolenic acid accumulation. Genome 2011, 54, 471–483. [CrossRef] [PubMed]

56. Zhang, J.; Song, Q.; Cregan, P.B.; Nelson, R.L.; Wang, X.; Wu, J.; Jiang, G.L. Genome-wide association study for flowering time, maturity dates and plant height in early maturing soybean (Glycine max) germplasm. BMC Genom. 2015, 16, 217. [CrossRef] [PubMed]

57. Zhang, W.K.; Wang, Y.J.; Luo, G.Z.; Zhang, J.S.; He, C.Y.; Wu, X.L.; Gai, J.Y.; Chen, S.Y. QTL mapping of ten agronomic traits on the soybean (Glycine max L. Merr.) genetic map and their association with EST markers. Theor. Appl. Genet. 2004, 108, 1131–1139. [CrossRef] [PubMed]

58. Eskandari, M.; Cober, E.R.; Rajcan, I. Genetic control of soybean seed oil: Ii. QTL and genes that increase oil concentration without decreasing protein or with increased seed yield. Theor. Appl. Genet. 2013, 126, 1677–1687. [CrossRef] [PubMed]
59. Hwang, E.Y.; Song, Q.; Jia, G.; Specht, J.E.; Hyten, D.L.; Costa, J.; Cregan, P.B. A genome-wide association study of seed protein and oil content in soybean. BMC Genom. 2014, 15, 1. [CrossRef] [PubMed]
60. Yang, J.; Benyamin, B.; McEvoy, B.P.; Gordon, S.; Henders, A.K.; Nyholt, D.R.; Madden, P.A.; Heath, A.C.; Martin, N.G.; Montgomery, G.W.; et al. Common SNPs explain a large proportion of the heritability for human height. Nat. Genet. 2010, 42, 565–569. [CrossRef] [PubMed]
61. Yang, J.; Lee, S.H.; Goddard, M.E.; Visscher, P.M. GCTA: A tool for genome-wide complex trait analysis. Am. J. Hum. Genet. 2011, 88, 76–82. [CrossRef] [PubMed]
62. Yang, J.; Bakshi, A.; Zhu, Z.; Hemani, G.; Vinkhuyzen, A.A.E.; Lee, S.H.; Robinson, M.R.; Perry, J.R.B.; Nolte, I.M.; van Vliet-Ostaptchouk, J.V.; et al. Genetic variance estimation with imputed variants finds negligible missing heritability for human height and body mass index. Nat. Genet. 2015, 47, 1114–1120. [CrossRef] [PubMed]
63. Yang, J.; Manolio, T.A.; Pasquale, L.R.; Boerwinkle, E.; Caporaso, N.; Cunningham, J.M.; de Andrade, M.; Feenstra, B.; Feingold, E.; Hayes, M.G.; et al. Genome partitioning of genetic variation for complex traits using common SNPs. Nat. Genet. 2011, 43, 519–525. [CrossRef] [PubMed]
64. Wood, A.R.; Esko, T.; Yang, J.; Vedantam, S.; Pers, T.H.; Gustafsson, S.; Chu, A.Y.; Estrada, K.; Luan, J.; Kutalik, Z.; et al. Defining the role of common variation in the genomic and biological architecture of adult human height. Nat. Genet. 2014, 46, 1173–1186. [CrossRef] [PubMed]
65. Duguid, S.D.; Kenaschuk, E.O.; Rashid, K.Y. Macbeth flax. Can. J. Plant Sci. 2003, 83, 803–805. [CrossRef]
66. Rowland, G.G.; Hormis, Y.A.; Rashid, K.Y. CDC bethune flax. Can. J. Plant Sci. 2002, 82, 101–102. [CrossRef]
67. Rowland, G.G.; Bhatty, R.S. Ethyl meththane-sulphonate induced fatty acid mutations in flax. J. Am. Oil Chem. Soc. 1990, 67, 213–214. [CrossRef]
68. You, F.M.; Huo, N.; Deal, K.R.; Gu, Y.Q.; Luo, M.C.; McGuire, P.E.; Dvorak, J.; Anderson, O.D. Annotation-based genome-wide SNP discovery in the large and complex Aegilops tauschii genome using next-generation sequencing without a reference genome sequence. BMC Genom. 2011, 12, 59. [CrossRef] [PubMed]

69. Li, H.; Handsaker, B.; Wysoker, A.; Fennell, T.; Ruan, J.; Homer, N.; Marth, G.; Abecasis, G.; Durbin, R. The sequence alignment/map format and SAMtools. Bioinformatics 2009, 25, 2078–2079. [CrossRef] [PubMed]

70. Browning, S.R.; Browning, B.L. Rapid and accurate haplotype phasing and missing-data inference for whole-genome association studies by use of localized haplotype clustering. Am. J. Hum. Genet. 2007, 81, 1084–1097. [CrossRef] [PubMed]

71. Purcell, S.; Neale, B.; Todd-Brown, K.; Thomas, L.; Ferreira, M.A.; Bender, D.; Maller, J.; Sklar, P.; de Bakker, P.I.; Daly, M.J.; et al. Plink: A tool set for whole-genome association and population-based linkage analyses. Am. J. Hum. Genet. 2007, 81, 559–575. [CrossRef] [PubMed]

72. Hill, W.G.; Weir, B.S. Variances and covariances of squared linkage disequilibria in finite populations. Theor. Popul. Biol. 1988, 33, 54–78. [CrossRef]

73. Pfeifer, B.; Wittelsburger, U.; Ramos-Onsins, S.E.; Lercher, M.J. Popgenome: An efficient swiss army knife for population genomic analyses in R. Mol. Biol. Evol. 2014, 31, 1929–1936. [CrossRef] [PubMed]

74. Jombart, T.; Devillard, S.; Balloux, F. Discriminant analysis of principal components: A new method for the analysis of genetically structured populations. BMC Genet. 2010, 11, 94. [CrossRef] [PubMed]

75. Jombart, T. Adegenet: A R package for the multivariate analysis of genetic markers. Bioinformatics 2008, 24, 1403–1405. [CrossRef] [PubMed]

76. Lin, C.S.; Poushinsky, G. A modified augmented design (type 2) for rectangular plots. Can. J. Plant Sci. 1985, 65, 743–749. [CrossRef]

77. You, F.M.; Duguid, S.D.; Thambugala, D.; Cloutier, S. Statistical analysis and field evaluation of the type 2 modified augmented design (MAD) in phenotyping of flax (Linum usitatissimum) germplasms in multiple environments. Aust. J. Crop Sci. 2013, 7, 1789–1800.

78. Association of Official Analytical Chemists. Fat (total, saturated and unsaturated) in foods: Hydrolytic extraction gas chromatographic method. In Official Methods of Analysis of AOAC International, 18th ed.; Horwitz, W., Ed.; AOAC International: Gaithersburg, MD, USA, 2001.

79. Bates, D.; Maechler, M.; Bolker, B.; Walker, S. Fitting linear mixed-effects models using lme4. J. Stat. Softw. 2015, 67, 1–48. [CrossRef]

80. Zhang, Z.; Ersoz, E.; Lai, C.Q.; Todhunter, R.J.; Tiwari, H.K.; Gore, M.A.; Bradbury, P.J.; Yu, J.; Arnett, D.K.; Ordovas, J.M.; et al. Mixed linear model approach adapted for genome-wide association studies. Nat. Genet. 2010, 42, 355–360. [CrossRef] [PubMed]

81. Yu, J.; Pressoir, G.; Briggs, W.H.; Vroh Bi, I.; Yamasaki, M.; Doebley, J.F.; McMullen, M.D.; Gaut, B.S.; Nielsen, D.M.; Holland, J.B.; et al. A unified mixed-model method for association mapping that accounts for multiple levels of relatedness. Nat. Genet. 2006, 38, 203–208. [CrossRef] [PubMed]
82. Bradbury, P.J.; Zhang, Z.; Kroon, D.E.; Casstevens, T.M.; Ramdoss, Y.; Buckler, E.S. Tassel: Software for association mapping of complex traits in diverse samples. Bioinformatics 2007, 23, 2633–2635. [CrossRef] [PubMed]
83. Turner, S.D. Qqman: An R package for visualizing GWAS results using Q-Q and manhattan plots. biorXiv 2014. [CrossRef]
84. Thambugala, D.; Duguid, S.; Loewen, E.; Rowland, G.; Booker, H.; You, F.M.; Cloutier, S. Genetic variation of six desaturase genes in flax and their impact on fatty acid composition. Theor. Appl. Genet. 2013, 126, 2627–2641. [CrossRef] [PubMed]

2 Chapter

ASSESSMENT OF GENOMIC PREDICTION FOR PASMO RESISTANCE IN FLAX

1. INTRODUCTION: AN OVERVIEW

Flax (Linum usitatissimum L.) is a significant food and fiber crop planted in cooler locations throughout, including Canada [1]. Pasmo, caused by the fungus Septoria linicola, is a prevalent disease of flax that leads to decreased seed and oil yield, along with diminished fiber quality and durability[2]. Creating resistant cultivars is the most feasible and effective strategy to manage this disease, which has proliferated over all flax production regions in North America and other global places. Resistance to pasmo exhibits poor heredity [3] and is inherited quantitatively [4]. Significant disparities in pasmo disease severity were noted within the flax core collection, which can be leveraged to create resistant cultivars [3]. Phenotypic recurrent selection is commonly employed to create cultivars with enhanced resistance, with selection typically conducted based on phenotypic evaluations of resistance in field conditions [5]. Field measurement of pasmo severity (PS) in germplasm and breeding lines is expensive and significantly affected by environmental factors due to pronounced genotype × environment (G × E) interactions [3,4].

With the advancements in molecular marker development over the last decade, efforts to use marker-assisted breeding strategies have been pursued. One such strategy involves identifying quantitative trait loci (QTL) in biparental mapping populations and using markers to efficiently backcross QTL into elite breeding materials [6]. This so-called marker-assisted recurrent selection (MARS) or simply marker-assisted selection (MAS) characterizes many breeding programs that employ molecular markers to select non-phenotyped individuals for crossing and downstream selection of segregating populations [7]. This method is suitable for the selection of monogenic or oligo-genic architectures but has limited use for quantitative traits controlled by many genes of smaller effects [8]. Genomic selection (GS) or prediction (GP) is an alternative marker-assisted breeding strategy better suited to polygenic quantitative traits, especially those with low heritability, because it makes use of all marker effects across the entire genome to calculate genomic estimated breeding values (GEBVs) [9] for individual plant selection [9,10].

In GP, a training population (TP) is genotyped with genome-wide markers and phenotyped for the trait(s) under selection; statistical models that best predict the breeding values from the marker data are then applied to select non-phenotyped germplasm. GP has been used to select for disease resistance in several crops such as Fusarium head blight (FHB) in wheat, a typically quantitatively inherited trait with predominantly additive genetic variation, where GP had a significantly higher accuracy than pedigree-based information alone [11]. GP feasibility has also been studied for selection of wheat rust resistance and was found particularly effective when validation lines had at least one which is close to the reference lines [12]. The implementation of GP on northern leaf blight, a complex genetic architecture trait in maize, resulted in superior gains and reduced breeding cycle time to ≤80% of the phenotypic cycle [13]. Despite the many successful examples, the use of GP to improve disease resistance in crops has been challenging for two reasons: (i) selection for major resistance genes can be ephemeral due to changes in pathogen races; and (ii) breeding for minor resistance genes with small effects may face the remarkable complexities encountered in GP [14].

The fast-evolving genotyping platforms have been a game-changer in the implementation of GP, allowing the production of large numbers of genome-wide markers, whereas progresses in phenotyping were not associated with similar cost reduction or quantum leaps in throughput. Given the number of markers (p) and sample size (n) in a given population, there are many more p effects to be estimated than the n, leading to an infinite number of possible marker effect estimates [15], that is, the so-called "large p, small n problem" ($p >> n$) when applying markers to predict phenotypes [11]. Several GP statistical models have been proposed to address this issue [16]. For example, the ridge-regression best linear unbiased prediction (RR-BLUP) is a mixed linear model that considers markers as random effects. Covariance between markers is considered to be zero and the marker variance is assumed to be the total genetic variance divided by the number of markers. The variance is assumed to be equal for all markers, allowing many more marker effects to be estimated than there are phenotypic records [17]. Unlike RR-BLUP, the Bayesian LASSO (BL) assumes markers to have unequal variances and, performs continuous shrinkage and variable selection simultaneously, with small-effect markers shrinking more severely than larger-effect loci. In the $p >> n$ setting, LASSO will select at most $n - 1$ variables and set the effects of the remaining predictors at zero [18]. Although the problem is solved statistically in these models, improving the accuracy and efficiency of GP by reducing the number of genome-wide markers would be advantageous because any increment in the TP size comes at a cost [19–22]. A genome-wide association study (GWAS) is a method used to find genome-wide markers associated with quantitative trait loci (QTL), leading to a restricted set of advantageous genetic loci responsible for specific traits of interest [23]. The genetic potential for crown rust resistance in Lolium perenne exemplified GWAS's capacity to identify and prioritize markers, facilitating the discovery of a limited number of single nucleotide polymorphisms (SNPs) that could approximate the predictive accuracy achieved with the entire marker set [24]. The application of GWAS eliminates a significant number of irrelevant markers in the development of predictive models. The sole empirical investigation on genomic prediction in flax, utilizing bi-parental populations for yield, oil content, and fatty acid composition parameters, demonstrated that genomic prediction might enhance genetic gain per unit time in linseed breeding. The GP results markedly surpassed those from direct phenotypic selection, particularly for variables with poor broad-sense heritability [25]. Flax pasmo resistance is polygenic. Our prior research identified 500 non-redundant quantitative trait loci (QTL) for pathogen resistance (PR) from 370 distinct flax accessions within a core collection, based on five years of pasmo field evaluations; among them, 134 QTL exhibited statistical stability throughout all five years, and 67 had relatively stable and substantial effects [4]. This study aimed to assess the efficacy of QTL markers in GP and to compare the GP efficiency influenced by various markers, including genome-wide SNPs and QTL markers, to establish a realistic and precise model for germplasm evaluation and parent selection in pasmo resistance breeding.

2. RESULTS

2.1 Evaluation of Pasmo Resistance

PS ratings at green boll stage or maturity across five consecutive years were similar but on average PS ratings in 2014 and 2016 were higher than those in other years (Table 1).

Table 1. Pasmo severity of 370 flax accessions across five years in the field condition.

Data Set	x ± s	Range	CV (%)
PS-2012	5.57 ± 1.86	1.00–9.00	32.76
PS-2013	5.69 ± 1.91	2.00–9.00	33.20
PS-2014	6.86 ± 2.07	1.00–9.00	29.41
PS-2015	6.11 ± 1.55	1.00–9.00	25.44
PS-2016	6.72 ± 1.37	2.00–9.00	20.39
PS-mean	6.22 ± 1.32	1.80–9.00	21.27

x: average pasmo severity across five years; s: standard deviation; CV: coefficient of variation.

They had single peak distributions but skewed towards high PS ratings except for those in 2014 (Figure 1). Scatter plots of PS ratings between years indicated strong genotype × year interaction even though statistically significant correlations of PS ratings between years were observed (Figure 1), as shown in the variance analysis results in the previous study [4]. However, the Pearson correlations of 5-year averages of PS ratings (PS-mean) with those in individual years (r=0.72–0.83) were much higher than the Pearson correlations between individual years (r = 0.31–0.62) (Figure 1), implying that the mean PS ratings over multiple years or environments were a more suitable data set than individual year's data sets for model construction of genomic prediction.

2.2 Evaluation of Marker Sets Used in Genomic Prediction

Four marker sets were used for GP of pasmo resistance. The first marker set contained 52,347 genome-wide SNPs (SNP-52347) that were correlated to the five-year average PS and the PS of the five individual years at a 10-5 probability level [4]. The other three marker sets were the 500 unique QTL (SNP-500QTL), the 134 QTL statistically stable over five consecutive years (SNP-134QTL) and the 67 stable and relatively large-effect QTL (SNP-67QTL) sets previously identified [4]. The SNP-500QTL dataset comprises markers for all small- or large-effects, including QTL stable across environments and environment-specific QTL identified using three single-locus and seven multi-locus statistical models and all six phenotypic datasets (Figure 2). The SNP-134QTL dataset is a subset of the SNP-500QTL dataset whereas SNP-67QTL is a subset of the former; all SNP-500QTL markers were included in SNP-52347. These four marker sets explained 54%, 72%, 27% and 29% of the phenotypic variation of the five-year PS average (PS-mean), respectively; these values exceeded those of the individual year PS data (Table 2). Although SNP-500QTL was a subset of SNP-52347, this marker set explained a greater percentage of the phenotypic variation for PS than SNP-52347 for all datasets.

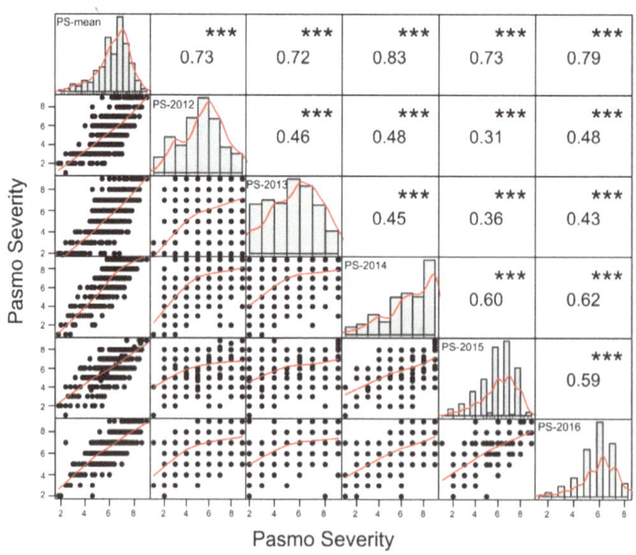

Figure 1. Dot plots (lower triangle), histograms (diagonal) and Pearson correlations (upper triangle) between six pasmo severity datasets. Best curves are fitted in dot plots and histograms. * represents significance at the <0.001 probability level.**

Table 2. Phenotypic variation of pasmo severity (PS) (h2 ± s) explained by the four marker sets.

PS Dataset	Marker Set			
	SNP-500QTL	SNP-134QTL	SNP-67QTL	SNP-52347
PS-mean	0.72± 0.04	0.27± 0.05	0.29± 0.05	0.54± 0.07
PS-2012	0.64 ± 0.06	0.18 ± 0.05	0.16 ± 0.04	0.43 ± 0.08
PS-2013	0.63 ± 0.06	0.12 ± 0.04	0.12 ± 0.04	0.38 ± 0.08
PS-2014	0.65 ± 0.06	0.23 ± 0.05	0.20 ± 0.05	0.45 ± 0.08
PS-2015	0.56 ± 0.06	0.20 ± 0.05	0.17 ± 0.04	0.44 ± 0.09
PS-2016	0.53 ± 0.06	0.18 ± 0.05	0.18 ± 0.05	0.38 ± 0.07

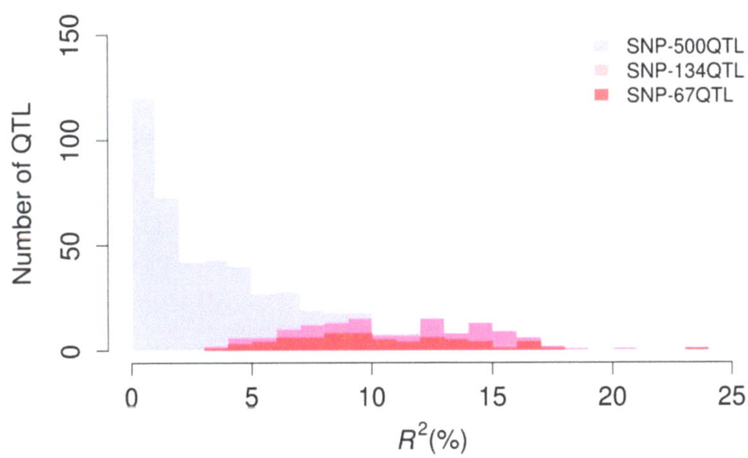

Figure 2. Distribution of R2 (%) (phenotypic variation explained by individual QTL) in the three QTL marker sets.

2.3 Accuracy of Genomic Prediction in Relation to Marker Sets and Pasmo Severity Datasets

Genomic prediction models were constructed using RR-BLUP with pairwise combinations of the four marker sets and the six PS datasets. Statistical models for the 24 combinations were generated and evaluated for their accuracy (r) and relative efficiency (RE) using a five-fold random cross-validation scheme (Table 3). RE represents the relative efficiency of GP over direct phenotypic selection which depends on the heritability of a selective trait. Direct phenotypic selection for a trait was considered to have a baseline efficiency of 1. Thus, RE values greater than 1 indicate GP models more efficient than direct phenotypic selection in one selection cycle [25–27]. Analysis of variance (ANOVA) indicated that r and RE both significantly differed among the four marker sets and the six PS datasets; there was also a significant interaction effect between marker sets and PS datasets (Table

S1). Owing to the significant marker × phenotype dataset interaction, multiple comparisons of the 24 combinations were performed. For all marker sets, the PS-mean models significantly outperformed those based on individual year datasets (Table 3). The SNP-500QTL marker set models generated significantly higher r and RE values than any other marker sets (Figure 3). Interestingly, the SNP-67QTL derived models produced slightly but significantly higher values of r and RE than SNP-134QTL models. The highest r and RE values were obtained for models combining the SNP-500QTL and PS-mean datasets (Table 3, Figure 3). Intriguingly, the SNP-52347 models yielded the lowest r and RE values despite including all QTL markers (Table 3, Figure 3); both BL and Bayesian ridge regression (BRR) corroborated this finding (Figure S1). No significant differences in r and RE values were observed among the three statistical models: RR-BLUP, BL and BRR (Figure S1).

2.4 Sample Size of Training Populations versus Genomic Prediction Accuracy

To find an optimal size for the TP, the relationship between TP size and prediction accuracy was analysed. TPs of various sizes from 18 to 351, corresponding to 5% to 95% of the total 370 accessions, were used to build models with the SNP-500QTL marker set and the PS-mean phenotypic dataset. The prediction accuracy significantly increased for TP sizes up to 100, followed by smaller accuracy gains with every additional TP size increments (Figure 4). A GP accuracy >0.9 was obtained once the TP size reached 185 (Figure 4).

Table 3. Accuracy (r) and relative efficiency (RE) values of the 24 combinations representing the four marker sets and six pasmo severity (PS) datasets using RR-BLUP obtained using a random five-fold cross-validation.

Marker Set	PS Dataset	r (x ± s) 1	RE (x ± s) 1
SNP-500QTL	PS-mean	0.92 ± 0.02a	1.84 ± 0.04a
	PS-2012	0.84± 0.03b	1.68 ± 0.06b
	PS-2013	0.81 ± 0.04c	1.62 ± 0.07c
	PS-2014	0.82 ± 0.04c	1.63± 0.07c
	PS-2015	0.76 ± 0.05d	1.52 ± 0.09d
	PS-2016	0.76 ± 0.05d	1.52 ± 0.11d
SNP-134QTL	PS-mean	0.75 ± 0.06e	1.49 ± 0.11e
	PS-2012	0.68 ± 0.06f	1.36 ± 0.11f
	PS-2013	0.60± 0.07ij	1.19 ± 0.14ij
	PS-2014	0.60 ± 0.07i	1.21 ± 0.14i
	PS-2015	0.47 ± 0.09o	0.94 ± 0.18o
	PS-2016	0.56 ± 0.09l	1.12 ± 0.17l

SNP-67QTL	PS-mean	0.76 ± 0.05d	1.53 ± 0.1d
	PS-2012	0.67 ± 0.06g	1.35 ± 0.11g
	PS-2013	0.60 ± 0.07ij	1.20 ± 0.14ij
	PS-2014	0.60 ± 0.07ij	1.20 ± 0.14ij
	PS-2015	0.50 ± 0.09n	1.00 ± 0.17n
	PS-2016	0.59 ± 0.08k	1.17 ± 0.17k
SNP-52347	PS-mean	0.67 ± 0.07g	1.33 ± 0.14g
	PS-2012	0.63 ± 0.06h	1.27 ± 0.12h
	PS-2013	0.59 ± 0.07jk	1.19 ± 0.14jk
	PS-2014	0.53 ± 0.08m	1.06 ± 0.17m
	PS-2015	0.38 ± 0.09q	0.77 ± 0.17q
	PS-2016	0.46 ± 0.09p	0.93 ± 0.18p

1 Different letters represent multiple test significance among the 24 combinations at the 0.05 probability level.

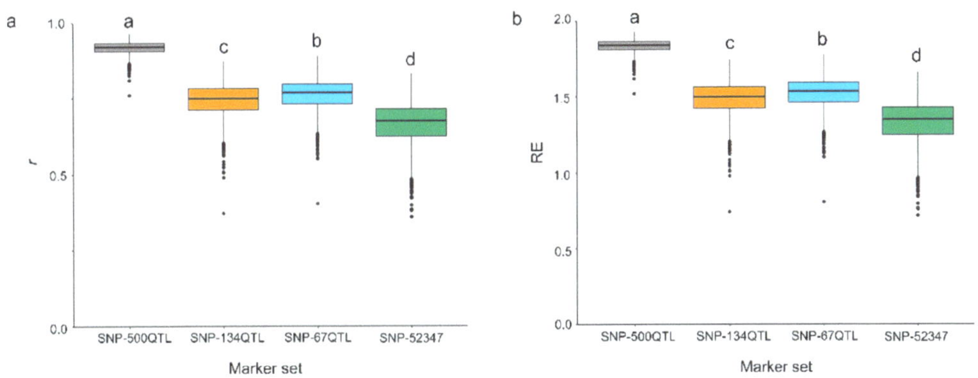

Figure 3. Accuracy (r) (a) and relative efficiency (RE) (b) of RR-BLUP prediction models built with combinations of four marker sets using the five-year average PS dataset (PS-mean) and random five-fold cross-validations. Letters above box plots indicated statistical significance ($p < 0.05$) for r and RE among marker sets.

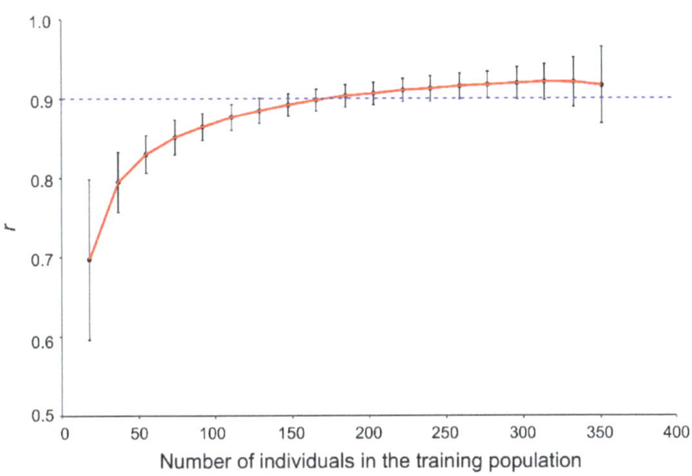

Figure 4. Relationship between the genomic prediction accuracy (r) and the size of the training population based on the SNP-500QTL marker set, the PS-

mean dataset and the RR-BLUP models. The dash line represents a prediction accuracy of 0.9.

2.5 Prediction Models of Pasmo Resistance

All 370 accessions were used as a training population to build a prediction model using the SNP-500QTL genotypic dataset and the PS-mean phenotypic dataset because this combination outperformed all other models. The model was then employed to predict PS in each year (Table 4). Prediction accuracies (r) ranging from 0.71 to 0.81 and RE values of 1.42 to 1.62 were obtained when predicting PS for individual years (Table 4).

A prediction accuracy as high as 0.98 and a RE value of 1.96 were obtained when the model was used to predict PS-means of the 370 accessions (Table 4). A linear relationship was observed between the observed (y) and predicted PS (x): y = 1.0522x - 0.3267 (R2 = 0.96) (Figure 5a). Based on this equation, the average prediction interval between the two red dashed lines, representing the 95% confidence interval, was only less than 1 (an average of 0.97) on the PS ratings (Figure 5a).

NPQTL in the 370 accessions for the 500 QTL set was tallied. Significant linear correlation between PS-mean and NPQTL (r = 0.86 or R2 = 0.73) was observed (Figure 5b). This correlation was less than but close to the accuracy of the GP model with SNP-500QTL and higher than the GP models using other marker sets (Table 3). However, the single linear regression equation (y = -0.0262x + 11.934) of the observed PS (y) to NPQTL (x) had a large standard deviation for each prediction value, with an average prediction interval width of 2.70, nearly three times the average prediction interval width of the GP model; that is, the NPQTL model had a higher prediction error than the GP model.

Table 4. Accuracy (r) and relative efficiency (RE) of genomic prediction for pasmo severity in different years using the RR-BLUP model built with the SNP-500QTL marker set and the PS-mean phenotypic data using all 370 accessions as training data set.

PS Dataset for Prediction	r	RE
PS-mean	0.98	1.96
PS-2012	0.73	1.46
PS-2013	0.71	1.42
PS-2014	0.81	1.62
PS-2015	0.71	1.43
PS-2016	0.77	1.55

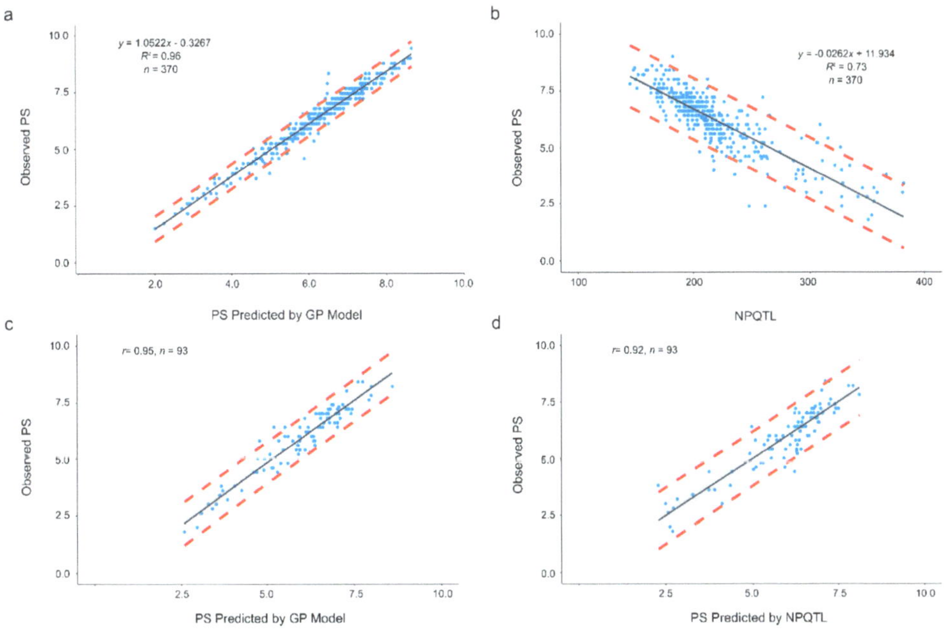

Figure 5. Relationship of observed pasmo severity (PS) with PS predicted by a GP model (a,c) or with PS predicted by the number of QTL with positive-effect alleles (NPQTL) (b,d). (a) Linear regression of observed PS (y) to predicted PS (x) using the genomic prediction model built with the PS-mean dataset and the SNP-500QTL marker set of all 370 accessions as training data set. (b) Linear regression of observed PS (y) to NPQTL (x) in the 370 flax accessions. (c) Relationship of observed PS of 93 randomly chosen accessions with the PS predicted by the genomic model constructed with the SNP-500QTL marker set and PS-mean dataset when a random subset of 277 accessions was used as training population. (d) Relationship of observed PS of 93 randomly chosen accessions with the PS predicted by NPQTL (Figure S2) The red dashed lines represent upper and lower boundaries of the 95% prediction intervals, that is, it is expected that the value of a sample lies within that prediction interval in 95% of the samples. The grey band represents the 95% confidence interval, that is, 95% of those intervals include the true value of the population mean.

2.6 A Case Study of Genomic Prediction

To assess GP prediction accuracy, a training-testing partition was generated with random assignment of breeding lines to either training or testing subsets. Considering the different improvement status of accessions in the population (cultivars, breeding lines, landraces or unknown types) and different levels of resistance, we randomly chose 20% of the 370 accessions in the population, that is, 93 accessions (52 cultivars, 21 breeding lines, 3 landraces and 17 unknown types) as validation dataset, that is, a five-fold random cross-validation set. To predict the PS of these 93 accessions, a RR-BLUP model using the SNP-500QTL set and the

PS-mean of the remaining 277 accessions as TP set was built to predict PS. The predicted results are shown in Figure 5c and Table S2. The prediction accuracy was as high as 0.95 (r between observed and predicted PS). Similarly, a linear regression model of observed PS (y) to NPQTL (x) of the 277 accessions (the same TP as GP) produced y = -0.026x + 11.902 (Figure S2), which was similar to the regression equation previously obtained with the complete accession set (Figure 5b). Using this prediction model, predicted PS and intervals were calculated (Figure 5d, Table S2). The prediction accuracy of 0.92 for NPQTL was slightly inferior to that of the GP model. The observed PS values all fell within prediction intervals (Table S2).

3. DISCUSSION

Cross-validation remains the most popular method to evaluate GP accuracy [14,28]. Our RR-BLUP model prediction accuracy of 0.92 for PR is the highest of all published GP models for plant disease resistance traits [14]. This model is especially valuable because PR has low heritability and high inheritance complexity [3,4]. The QTL markers, multi-year phenotypic data and the genetic diversity and size of the population likely contributed positively to this high prediction accuracy [29].

3.1 All Detected QTL Used as Markers in Genomic Prediction

Three sets of QTL markers (SNP-500QTL, SNP-134QTL and SNP-67QTL) and a genome-wide SNP marker set (SNP-52347) were evaluated here. GP models built using SNP-500QTL consistently outperformed models derived with any of the other three marker sets (Table 3, Figure S1), lending credence to the robustness and reliability of the QTL identified using multiple single-locus and multi-locus GWAS statistical methods [4]. Most GWAS aim to detect large-effect QTL, such as the SNP-67QTL set. While potentially useful in MAS, these tend to explain a reduced portion of the phenotypic variation compared to more comprehensive models (Table 2). Consequently, the GP models built with such marker sets have lower GP accuracies. Therefore, using all potential QTL associated with the selective trait to build GP models is advantageous because it greatly improves prediction accuracy. Prediction accuracies of models obtained with SNP-134QTL and SNP-67QTL data sets were comparable (Table 3 and they explained a similar proportion of the phenotypic variation for PS, confirming the redundancy or overlap between the two datasets. Removal of redundant QTL from SNP-134QTL to produce SNP-67QTL produced slightly higher accuracy models (Figure 3). Simplifying GP models by removal of redundant and unrelated markers will ease the practical implementation of GP in breeding programs.

3.2 Superior Performance of Genomic Prediction Combined with GWAS

Surprisingly, the GP models built using SNP-52347 generated a lower prediction accuracy than the models with SNP-500QTL (Table 3, Figure S1), regardless of the statistical methods (Figure S1). Similarly, SNP-52347 explained a lower percentage of the phenotypic variation for PS than SNP-500QTL (Table 2). Besides interaction between SNPs, introduction of noise from genome-wide markers [30], the low prediction accuracy may also be owing to some of the erroneously called SNPs and imputation of missing SNP data. SNP-500QTL includes all or nearly all QTL potentially associated with PS; additional markers, not only failed to increase but actually reduced the prediction accuracy, further emphasizing the effectiveness of the QTL identification methodology adopted in our previously published GWAS study [4]. Similar findings were found for FHB in wheat where deoxynivalenol (DON) concentration QTL-linked markers significantly improve prediction accuracy compared to random genome-wide markers [30]. Markers linked to QTL underlying important traits are deemed more useful for prediction strategies because genome-wide markers may introduce noise, thereby reducing accuracy [30]. Using QTL for GP models may be beneficial to balance genetic backgrounds along with maximum gain of breeding value [31]. Genome-wide prediction models based on ~5000 SNPs from de novo GWAS for tropical rice improvement were as effective for prediction as the full marker set of 108,005 SNPs, indicating that the relationship between marker number and prediction accuracy is neither strict nor linear [32]. To sum up, combined applications of the QTL discovered via GWAS and the accelerated breeding cycles through GP facilitate the full use of genome-wide markers in crop disease resistance breeding [10,33]. Removal of redundant markers has the potential to alleviate the effect of the "large p, small n" issue.

3.3 Accuracy of GP Modelling by Environment, Training Population and Statistical Methods

G × E interactions, which affects the accuracy of trait assessment, are common for plant traits. A strong G × E interaction was observed in flax PR [4]. As a consequence, different PS QTL were identified for individual years and for the 5-year average [4]; similarly, GP efficiencies differed when individual yearly and average PS data sets were used as training sets (Table 3). The highest accuracies were obtained when the 5-year mean phenotypic data was used as training data (Table 4), suggesting that the average phenotypic data across multiple environments should be used for GP model construction. Because phenotypic values of genotypes in each year had one replication, the average phenotypic data across multiple years is actually equivalent to the best linear unbiased prediction values (BLUPs) or the best linear unbiased estimators (BLUEs). Therefore, the means across multiple environments estimate or reflect the true breeding values of a trait.

Some studies report that prediction accuracy of GP is highly affected by the size of the TP. In general, the prediction accuracy increases with TP size [21,28,29,34–36]. In the GP of seed weight in soybean, for example, prediction accuracy was sensitive to changes in TP size, which may have led to changes of relatedness between training and validation sets [21]. Lorenzana and Bernardo observed that, in an Arabidopsis family, prediction accuracy improved by 0.10 when TP size increased from 48 to 96, by an additional 0.07 when TP size was increased to 192 and by a further 0.05 with a TP size of 332 [37]. Here GP accuracy >0.9 was observed when the TP size reached 185 which slightly increased to 0.921 with a TP size of 314 (Figure 4). Large TPs provide the statistical power needed to improve prediction accuracy [38], especially for traits with low heritability [34,39]. When TP size is sufficiently large, even low heritability traits can be accurately predicted [28,40], including the low heritability PS studied therein. Diversity of the population also affect prediction accuracy [21,29,34,41–43]. A diverse TP may contain more QTL associated with selective traits and increase the correlation of the TP with validation populations (VPs) or test/prediction populations (PPs), resulting in a subsequent increase in prediction accuracy. Although some breeding lines [11,30,44] and bi-parental derived lines [25,41,45,46] are used for TPs, many studies have opted for a more diverse TP germplasm [29,41–43]. Our core collection TP preserves the variation present in the world collection of 3378 accessions maintained by Plant Gene Resources of Canada (PGRC) and represents a broad range of geographical origins, different improvement statuses (landraces, historical and modern cultivars, breeding lines) and two morphotypes (linseed and fibre types) [1,3]. This collection also contains most parents of modern Canadian flax cultivars [25]. Therefore, diverse phenotypic and genetic variabilities within the flax core collection render it useful as a resource for breeding and as a TP for GP model construction.

A variety of statistical methods have been proposed to estimate marker effects for GP. In general, GP methods are based on additive genetic models and their accuracies may vary depending on genetic architecture of target traits. According to the assumptions for statistical distributions of the marker effects, two groups of GP models have been proposed. The first group of models, such as RR-BLUP, genomic BLUP (GBLUP) and BRR, assume that all markers have some effects on the target trait and the same variance, that is, all makers contribute to the variation of the trait. The second group of models, including BayesA, BayesB, BayesC and BL, assume a specific variance for each marker. Some of these models such as BayesB, BayesC and BL, also allow variable (marker) selection when some of markers have very small or no effects. Based on these assumptions, the first group of models are expected to be useful for complex quantitative traits that have a polygenic architecture, while the second group of models are suitable for traits that controlled by a small number of genes or QTL with large effects. Several studies have shown better performance of BayesB for traits controlled by a few of genes with large effect [47–50]. Some simulation studies have also shown that BayesB outperformed GBLUP that is equivalent to RR-BLUP, when the number of QTL underlying a trait are small [47,51]. However, BayesB, RR-BLUP and other models had a similar prediction accuracy under the infinitesimal model [51] or for some complex traits [19,49]. In this study, no difference among RR-BLUP, BRR and BL was observed (Figure S1), primarily because flax pasmo resistance is a complex and polygenic trait and most of QTL associated with it had similar and small effects (Figure 2). RR-BLUP is most commonly used because of some superior features [11,14,42,52–54]. For example, RR-BLUP successfully recognized complex patterns with additive effects and delivered good GP in wheat disease resistance [55]. RR-BLUP also has a clear-cut computational efficiency compared with any other statistical models [11,54,56,57]. Here the RR-BLUP model with the 500 QTL markers and the 5-year mean PS produced high prediction accuracy and is therefore recommended for the prediction of PR in flax.

3.4 Pasmo Severity Prediction Using Number of Positive-Effect QTL

A highly significant correlation ($r = 0.86$ or $R2 = 0.73$) between NPQTL and PS (Figure 5b) provides an alternative approach to directly predict PS phenotypes. The prediction accuracy using the linear regression equation of PS to NPQTL was inferior to the GP model (Figure 5) because the QTL effects were variable (Figure 2), whereas the linear regression equation considered only the number of QTL but not their individual effects. However, NPQTL is advantageous because it can be readily calculated based on the genotyping by sequencing (GBS) or other genotyping data for the QTL markers [14] and the prediction accuracy based on the NPQTL is comparable to most GP models. Thus, the NPQTL-based prediction equation provides a simple alternative model for PS prediction.

3.5 Breeding Application of Genomic Prediction

Plant breeding is to pyramid favourite alleles from distinct parents using different approaches such as conventional crossing, mutation or transgenic methods to develop new varieties. However, most traits of agronomic importance are genetically controlled by polygenes and have a low heritability such as seed yield and horizontal resistance to diseases. Conventional phenotype selection for these traits is usually inefficient because assessment for them must be performed in multiple environments to obtain breeding values of individuals and thus it is very costly, time consuming and inaccurate; and also because of difficulty of evaluation in fields, greenhouses or laboratories. GS or GP provides an efficient approach to increase selection efficiency by not only increasing selection accuracy but also shortening breeding cycles [58]. In this study, we demonstrate a good example of GP for flax pasmo resistance that is environment-sensitive, costly and difficult for field evaluation. As high as 0.92 of prediction accuracy was obtained for PR, corresponding to 1.84 of relative efficiency over the direct phenotypic selection (Table 3), demonstrating efficiency of GP for low heritability traits. Because the training population underlying the GP models is a diverse germplasm collection that contains more than 90 breeding lines and 245 varieties from different breeding programs [3], the GP models developed in this study are expected to be used for germplasm evaluation, parent selection and individual selection of segregation populations for PR.

4. MATERIALS AND METHODS

4.1 Population

A total of 370 diverse flax accessions from the core collection [1] were used to evaluate different GP models. This subset of the core collection collected from 38 countries in 12 geographic regions has been used to identify the QTL associated with PS used in our PS models [4].

4.2 Pasmo Resistance Data

All flax accessions were assessed for PS in the same pasmo nursery from 2012 to 2016 at the Morden Research and Development Centre, Agriculture and Agri-Food Canada (AAFC), Morden, Manitoba, Canada [4]. A type-2 modified augmented design (MAD2) [59,60] was used for the field trials [3]. Accessions were seeded during the second or third week of May every year. Approximately 200 g of pasmo-infested chopped straw from the previous growing season was spread between rows as inoculum when plants were approximately 30-cm tall. A misting system was operated for 5 min every half hour for 4 weeks, except on rainy days, to ensure conidia dispersal and disease infection and development. Field assessments were conducted at the early (P1) and late flowering stages (P2, 7–10 days after P1), the green boll stage (P3, 7–10 days after P2) and the early brown boll stage (P4, 7–10 days after P3). In 2014 and 2015, only the first three field assessments were conducted because early maturity of the plants did not allow for a fourth rating. The PS observed at green boll stage or maturity was used for GP as previously described [4]. PS was assessed on leaves and stems of all plants in a single row plot using a 0–9 scale (0 = no sign of infection and 9 = > 90% leaf and stem area infected) [4]. Six sets of PS, including five individual year datasets and the 5-year average, were used for GP modelling. The function "chart.Correlation" of the R package PerformanceAnalytics (v1.5.2, https://cran.r project.org/web/packages/PerformanceAnalytics /index.html) was used to analyse correlations between different PS datasets and draw histograms and scatter plots.

4.3 Genomic Data

A total of 258,873 SNPs were obtained from the 370 accessions after pruning by removing redundant SNPs [4]. The missing data of SNPs (on average 14.13% of a missing data rate) were imputed using Beagle v.4.2 with default parameters

[61]. Our previous GWAS analyses of PS in flax were conducted separately for combinations of the five individual year and the 5-year average datasets with ten statistical methods [4]. The statistical methods for GWAS included three single locus models (GLM [62], MLM [63] and GEMMA [64]) and seven multi-locus models (FarmCPU [65], mrMLM [66], FASTmrEMMA [67], ISIS EM-BLASSO [68], pLARmEB [69], pKWmEB [70], FASTmrMLM [71]). For GLM, MLM and FarmCPU, the first six principal components (PCs), accounting for 33.04% of the total variation, were chosen as covariates to measure population structure, while Frappe (http://med.stanford.edu/tanglab/software/frappe.html) was used to estimate the population structure of the 370 accessions for other six multi-locus models. GEMMA does not require a Q matrix. The threshold of significant associations for all three single-locus methods (GLM, MLM and GEMMA) and the multi-locus method FarmCPU was determined by a critical p value ($\alpha = 0.05$) subjected to Bonferroni correction, that is, the corrected p value = 1.93 × 10−7 (0.05/258,873 SNPs), while a log of odds (LOD) score of three was used to detect robust association signals for the remaining six multi-locus models. The R package MVP (https://github.com/XiaoleiLiuBio/MVP) was used for GWAS analyses for the GLM, MLM and FarmCPU, the GEMMA software (https://github.com/genetics-statistics/GEMMA) for GEMMA and the R package mrMLM (https://cran.r-project.org/web/packages/mrMLM/index.html) for the additional six multi-locus models. The details of GWAS analyses were described in Reference [4]. A total of 500 non-redundant QTL for PS were identified from 370 diverse flax accessions, including 134 QTL that statistically stable in all five years and 67 QTL with relatively stable and large effects [4]. These three QTL datasets (500 unique QTL, 134 statistically stable QTL and 67 stable and large-effect QTL) were used for GP model construction. In addition, we performed Pearson's $\chi 2$ test with Yate's continuity correction to detect all SNPs significantly associated with PS using a 10–5 probability level. The three QTL sets and the genome-wide SNP set were used to construct the GP models. Thus, GP models with the 24 combinations of the four marker sets and the six phenotypic datasets were built and compared.

4.4 Genomic Prediction Models

Three statistical methods RR-BLUP [9,17,20], Bayesian LASSO (BL) [20,25,33] and Bayesian ridge regression (BRR) [25,72] were used to build GP models for PS. These predictive models estimate marker effects by modelling markers as random effects. No fixed effects were fitted in the models. The statistical models and their computation procedures are described in detail elsewhere [40,73]. The R package rrBLUP [56] was used to fit the RR-BLUP model and the R package BLR [74] was used to fit the BL and BRR models. The parameters used to fit BL and BRR were determined based on suggestions of de los Campos et al. [74]. Broad-sense heritability (0.25) of PS estimated in the population [3] was used. When preparing

QTL marker data for model construction, the positive-effect allele of the tag SNP of a QTL was coded '1' and the alternative allele '-1'. Similarly for the SNP marker set, the reference allele of an SNP was coded '1' and the alternative allele '-1'. Missing data were coded '0'. The EM algorithm implemented in the R package rrBLUP [56] was used to impute the missing marker data because missing marker data were not allowed in the model construction.

4.5 Evaluation of Prediction Models

Two validation methods were used to evaluate prediction models generated from combinations of statistical models, marker sets and PS datasets. The first method was a five-fold random cross-validation. The 370 flax accessions were randomly partitioned into five subsets. For a given partition, each subset was in turn used as validation or test data and the remaining four subsets made the training dataset. This partitioning was repeated 500 times. In this manner, a total of 2500 training data sets were created to build GP models and estimate marker effects. These were used to predict the breeding values of the individuals in the corresponding 2500 test/validation datasets. The accuracy of the genomic predictions (r) was defined by the Pearson's simple correlation coefficient between the genetic values predicted by GP and the observed phenotypic values. The relative efficiency of genomic prediction over phenotypic selection (RE) was estimated using |r|/H2 [26,27], where H2 refers to the broad-sense heritability of PS, estimated to be 0.25 [3]. RE was used as a criterion to compare the response to one cycle of genome-wide selection versus one cycle of phenotypic selection. Means of r and RE of the 500 samplings for each marker set, GP model and PS dataset were used to describe the prediction accuracy of GP and the efficiency of one GP cycle relative to one phenotypic selection cycle, respectively. A combined analysis of variance with Tukey multiple pairwise comparisons was conducted to assess the statistical significance of differences in r and RE among various marker and PS datasets using R. For the case study, we randomly chose 20% of the 370 accessions as a validation dataset, while the remaining 277 accessions were utilized as a training dataset to construct a GP model for the genomic prediction of unknown germplasm. The second cross-validation method entailed comparisons among various PS datasets, wherein each of the six comprehensive PS phenotypic datasets served as training datasets to construct GP models, which were subsequently applied to both the respective dataset and the other five phenotypic datasets. The identical collection of markers was employed for training and validation across all 370 accessions. This method evaluates the applicability of models constructed from single-year phenotypic data to forecast phenotypes assessed in various years.

5. PHENOTYPIC VARIATION EXPLAINED BY MARKERS

The phenotypic variation explained by all markers in various marker sets, denoted $h2_{SNP}$, was estimated for all PS datasets based on the mixed linear model [75] implemented in the GCTA software [76]. The detailed calculation is described in Reference [77].

6. CONCISE SUMMARY

Utilizing a diverse global flax core collection comprising 370 accessions as a training and testing cohort, alongside 500 QTL identified through GWAS, we established a highly efficient GP model with a prediction accuracy reaching 0.92 for pasmo, a trait characterized by low heritability and high inheritance complexity, based on 5-year average PS data and the RR-BLUP statistical model. This represents the best documented accuracy value among all GP models for plant disease resistance attributes and is equivalent to previously published findings. We created a linear regression prediction model utilizing NPQTL, which achieved a high prediction accuracy of 0.86. The GP model and the NPQTL-based regression equation were tested and considered suitable for evaluating flax germplasm, including parent selection for PR. Utilizing every potential QTL linked to a target trait would be advantageous, since the elimination of a significant number of irrelevant markers would enhance the development of extremely precise GP models.

REFERENCES

1. Diederichsen, A.; Kusters, P.M.; Kessler, D.; Bainas, Z.; Gugel, R.K. Assembling a core collection from the flax world collection maintained by Plant Gene Resources of Canada. Genet. Resour. Crop Evol. 2012, 60, 1479–1485. [CrossRef]
2. Vera, C.L.; Irvine, R.B.; Duguid, S.D.; Rashid, K.Y.; Clarke, F.R.; Slaski, J.J. Pasmo disease and lodging in flax as affected by pyraclostrobin fungicide, N fertility and year. Can. J. Plant Sci. 2014, 94, 119–126. [CrossRef]
3. You, F.M.; Jia, G.; Xiao, J.; Duguid, S.D.; Rashid, K.Y.; Booker, H.M.; Cloutier, S. Genetic variability of 27 traits in a core collection of flax (Linum usitatissimum L.). Front. Plant Sci. 2017, 8, 1636. [CrossRef]
4. He, L.; Xiao, J.; Rashid, K.Y.; Yao, Z.; Li, P.; Jia, G.; Wang, X.; Cloutier, S.; You, F.M. Genome-wide association studies for pasmo resistance in flax (Linum usitatissimum L.). Front. Plant Sci. 2019, 9, 1982. [CrossRef]
5. Diederichsen, A.; Rozhmina, T.A.; Kudrjavceva, L.P. Variation patterns within 153 flax (Linum usitatissimum L.) genebank accessions based on evaluation for resistance to fusarium wilt, anthracnose and pasmo. Plant Genet. Resour. 2008, 6, 22–32. [CrossRef]
6. Collard, B.C.Y.; Mackill, D.J. Marker-assisted selection: An approach for precision plant breeding in the twenty-first century. Philos. Trans. R. Soc. Lond. B Biol. Sci. 2008, 363, 557–572. [CrossRef]
7. Heslot, N.; Jannink, J.L.; Sorrells, M.E. Perspectives for genomic selection applications and research in plants. Crop Sci. 2015, 55, 1–12. [CrossRef]

8. Xu, Y.; Crouch, J.H. Marker-assisted selection in plant breeding: From publications to practice. Crop Sci. 2008, 48, 391–407. [CrossRef]
9. Meuwissen, T.H.; Hayes, B.J.; Goddard, M.E. Prediction of total genetic value using genome-wide dense marker maps. Genetics 2001, 157, 1819–1829.
10. Lipka, A.E.; Kandianis, C.B.; Hudson, M.E.; Yu, J.M.; Drnevich, J.; Bradbury, P.J.; Gore, M.A. From association to prediction: Statistical methods for the dissection and selection of complex traits in plants. Curr. Opin. Plant Biol. 2015, 24, 110–118. [CrossRef]
11. Arruda, M.P.; Brown, P.J.; Lipka, A.E.; Krill, A.M.; Thurber, C.; Kolb, F.L. Genomic selection for predicting. Fusarium head blight resistance in a wheat breeding program. Plant Genome 2015, 8. [CrossRef]
12. Daetwyler, H.D.; Bansal, U.K.; Bariana, H.S.; Hayden, M.J.; Hayes, B.J. Genomic prediction for rust resistance in diverse wheat landraces. Theor. Appl. Genet. 2014, 127, 1795–1803. [CrossRef]
13. Technow, F.; Burger, A.; Melchinger, A.E. Genomic prediction of northern corn leaf blight resistance in maize with combined or separated training sets for heterotic groups. G3 Genes Genomes Genet. 2013, 3, 197–203. [CrossRef]
14. Poland, J.; Rutkoski, J. Advances and challenges in genomic selection for disease resistance. Annu. Rev. Phytopathol. 2016, 54, 79–98. [CrossRef]
15. Gianola, D. Priors in whole-genome regression: The Bayesian alphabet returns. Genetics 2013, 194, 573–596. [CrossRef]
16. Desta, Z.A.; Ortiz, R. Genomic selection: Genome-wide prediction in plant improvement. Trends Plant Sci. 2014, 19, 592–601. [CrossRef]
17. Whittaker, J.C.; Thompson, R.; Denham, M.C. Marker-assisted selection using ridge regression. Genet. Res. 2000, 75, 249–252. [CrossRef]
18. Tibshirani, R. Regression shrinkage and selection via the Lasso. J. R. Stat. Soc. Ser. B Methodol. 1996, 58, 267–288. [CrossRef]
19. Jiang, Y.; Zhao, Y.; Rodemann, B.; Plieske, J.; Kollers, S.; Korzun, V.; Ebmeyer, E.; Argillier, O.; Hinze, M.; Ling, J.; et al. Potential and limits to unravel the genetic architecture and predict the variation of Fusarium head blight resistance in European winter wheat (Triticum aestivum L.). Heredity 2015, 114, 318–326. [CrossRef]
20. Spindel, J.; Begum, H.; Akdemir, D.; Virk, P.; Collard, B.; Redona, E.; Atlin, G.; Jannink, J.L.; McCouch, S.R. Genomic selection and association mapping in rice (Oryza sativa): Effect of trait genetic architecture, training population composition, marker number and statistical model on accuracy of rice genomic selection in elite, tropical rice breeding lines. PLoS Genet. 2015, 11, e1004982.

21. Zhang, J.; Song, Q.; Cregan, P.B.; Jiang, G.L. Genome-wide association study, genomic prediction and marker-assisted selection for seed weight in soybean (Glycine max). Theor. Appl. Genet. 2016, 129, 117–130. [CrossRef]

22. Li, Y.; Ruperao, P.; Batley, J.; Edwards, D.; Khan, T.; Colmer, T.D.; Pang, J.; Siddique, K.H.M.; Sutton, T. Investigating drought tolerance in chickpea using genome-wide association mapping and genomic selection based on whole-genome resequencing data. Front. Plant Sci. 2018, 9, 190. [CrossRef]

23. Yu, J.; Buckler, E.S. Genetic association mapping and genome organization of maize. Curr. Opin. Biotechnol. 2006, 17, 155–160. [CrossRef]

24. Arojju, S.K.; Conaghan, P.; Barth, S.; Milbourne, D.; Casler, M.D.; Hodkinson, T.R.; Michel, T.; Byrne, S.L. Genomic prediction of crown rust resistance in Lolium perenne. BMC Genet. 2018, 19, 35. [CrossRef]

25. You, F.M.; Booker, H.M.; Duguid, S.D.; Jia, G.; Cloutier, S. Accuracy of genomic selection in biparental populations of flax (Linum usitatissimum L.). Crop J. 2016, 4, 290–303. [CrossRef]

26. Dekkers, J.C. Prediction of response to marker-assisted and genomic selection using selection index theory. J. Anim. Breed. Genet. 2007, 124, 331–341. [CrossRef]

27. Ziyomo, C.; Bernardo, R. Drought tolerance in maize: Indirect selection through secondary traits versus genomewide selection. Crop Sci. 2013, 53, 1269–1275. [CrossRef]

28. Crossa, J.; Perez-Rodriguez, P.; Cuevas, J.; Montesinos-Lopez, O.; Jarquin, D.; de Los Campos, G.; Burgueno, J.; Gonzalez-Camacho, J.M.; Perez-Elizalde, S.; Beyene, Y.; et al. Genomic selection in plant breeding: Methods, models, and perspectives. Trends. Plant Sci. 2017, 22, 961–975. [CrossRef]

29. Gowda, M.; Das, B.; Makumbi, D.; Babu, R.; Semagn, K.; Mahuku, G.; Olsen, M.S.; Bright, J.M.; Beyene, Y.; Prasanna, B.M. Genome-wide association and genomic prediction of resistance to maize lethal necrosis disease in tropical maize germplasm. Theor. Appl. Genet. 2015, 128, 1957–1968. [CrossRef]

30. Rutkoski, J.; Benson, J.; Jia, Y.; Brown-Guedira, G.; Jannink, J.-L.; Sorrells, M. Evaluation of genomic prediction methods for Fusarium head blight resistance in wheat. Plant Genome 2012, 5, 51–61. [CrossRef]

31. Deshmukh, R.; Sonah, H.; Patil, G.; Chen, W.; Prince, S.; Mutava, R.; Vuong, T.; Valliyodan, B.; Nguyen, H.T. Integrating omic approaches for abiotic stress tolerance in soybean. Front. Plant Sci. 2014, 5, 244. [CrossRef]

32. Spindel, J.E.; Begum, H.; Akdemir, D.; Collard, B.; Redona, E.; Jannink, J.L.; McCouch, S. Genome-wide prediction models that incorporate de novo GWAS

are a powerful new tool for tropical rice improvement. Heredity 2016, 116, 395–408. [CrossRef]

33. Kayondo, S.I.; Pino del Carpio, D.; Lozano, R.; Ozimati, A.; Wolfe, M.; Baguma, Y.; Gracen, V.; Offei, S.; Ferguson, M.; Kawuki, R.; et al. Genome-wide association mapping and genomic prediction for CBSD resistance in Manihot esculenta. Sci. Rep. 2018, 8, 1549. [CrossRef]
34. Wang, X.; Xu, Y.; Hu, Z.L.; Xu, C.W. Genomic selection methods for crop improvement: Current status and prospects. Crop J. 2018, 6, 330–340. [CrossRef]
35. Jarquin, D.; Kocak, K.; Posadas, L.; Hyma, K.; Jedlicka, J.; Graef, G.; Lorenz, A. Genotyping by sequencing for genomic prediction in a soybean breeding population. BMC Genom. 2014, 15, 740. [CrossRef]
36. Asoro, F.G.; Newell, M.A.; Beavis, W.D.; Scott, M.P.; Jannink, J.-L. Accuracy and training population design for genomic selection on quantitative traits in elite North American oats. Plant Genome 2011, 4, 132–144. [CrossRef]
37. Lorenzana, R.E.; Bernardo, R. Accuracy of genotypic value predictions for marker-based selection in biparental plant populations. Theor. Appl. Genet. 2009, 120, 151–161. [CrossRef]
38. Goddard, M. Genomic selection: Prediction of accuracy and maximisation of long term response. Genetica 2009, 136, 245–257. [CrossRef]
39. Nielsen, H.M.; Sonesson, A.K.; Yazdi, H.; Meuwissen, T.H.E. Comparison of accuracy of genome-wide and BLUP breeding value estimates in sib based aquaculture breeding schemes. Aquaculture 2009, 289, 259–264. [CrossRef]
40. Lorenz, A.J.; Chao, S.; Asoro, F.G.; Heffner, E.L.; Hayashi, T.; Iwata, H.; Smith, K.P.; Sorrells, M.E.; Jannink, J.L. Genomic selection in plant breeding. Adv. Agron. 2011, 110, 77–123.
41. Cuevas, J.; Crossa, J.; Montesinos-Lopez, O.A.; Burgueno, J.; Perez-Rodriguez, P.; de los Campos, G. Bayesian genomic prediction with genotype x environment interaction kernel models. G3 Genes Genomes Genet. 2017, 7, 41–53.
42. Dong, H.; Wang, R.; Yuan, Y.; Anderson, J.; Pumphrey, M.; Zhang, Z.; Chen, J. Evaluation of the potential for genomic selection to improve spring wheat resistance to Fusarium head blight in the Pacific Northwest. Front. Plant Sci. 2018, 9, 911. [CrossRef]
43. Isidro, J.; Jannink, J.L.; Akdemir, D.; Poland, J.; Heslot, N.; Sorrells, M.E. Training set optimization under population structure in genomic selection. Theor. Appl. Genet. 2015, 128, 145–158. [CrossRef] [PubMed]
44. Rutkoski, J.E.; Poland, J.A.; Singh, R.P.; Huerta-Espino, J.; Bhavani, S.; Barbier, H.; Rouse, M.N.; Jannink, J.-L.; Sorrells, M.E. Genomic selection for quantitative adult plant stem rust resistance in wheat. Plant Genome 2014, 7. [CrossRef]

45. McElroy, M.S.; Navarro, A.J.R.; Mustiga, G.; Stack, C.; Gezan, S.; Pena, G.; Sarabia, W.; Saquicela, D.; Sotomayor, I.; Douglas, G.M.; et al. Prediction of cacao (Theobroma cacao) resistance to Moniliophthora spp. diseases via genome-wide association analysis and genomic selection. Front. Plant Sci. 2018, 9, 343. [CrossRef]

46. Enciso-Rodriguez, F.; Douches, D.; Lopez-Cruz, M.; Coombs, J.; de Los Campos, G. Genomic selection for late blight and common scab resistance in tetraploid potato (Solanum tuberosum). G3 Genes Genomes Genet. 2018, 8, 2471–2481. [CrossRef]

47. Daetwyler, H.D.; Pong-Wong, R.; Villanueva, B.; Woolliams, J.A. The impact of genetic architecture on genome-wide evaluation methods. Genetics 2010, 185, 1021–1031. [CrossRef]

48. Jannink, J.L.; Lorenz, A.J.; Iwata, H. Genomic selection in plant breeding: From theory to practice. Brief Funct. Genom. 2010, 9, 166–177. [CrossRef]

49. Thavamanikumar, S.; Dolferus, R.; Thumma, B.R. Comparison of genomic selection models to predict flowering time and spike grain number in two hexaploid wheat doubled haploid populations. G3 Genes Genomes Genet. 2015, 5, 1991–1998. [CrossRef]

50. VanRaden, P.M.; Van Tassell, C.P.; Wiggans, G.R.; Sonstegard, T.S.; Schnabel, R.D.; Taylor, J.F.; Schenkel, F.S. Invited review: Reliability of genomic predictions for North American Holstein bulls. J. Dairy Sci. 2009, 92, 16–24. [CrossRef]

51. Clark, S.A.; Hickey, J.M.; van der Werf, J.H. Different models of genetic variation and their effect on genomic evaluation. Genet. Sel. Evol. 2011, 43, 18. [CrossRef]

52. Rutkoski, J.; Singh, R.P.; Huerta-Espino, J.; Bhavani, S.; Poland, J.; Jannink, J.L.; Sorrells, M.E. Genetic gain from phenotypic and genomic selection for quantitative resistance to stem rust of wheat. Plant Genome 2015, 8. [CrossRef]

53. Gonzalez-Camacho, J.M.; Ornella, L.; Perez-Rodriguez, P.; Gianola, D.; Dreisigacker, S.; Crossa, J. Applications of machine learning methods to genomic selection in breeding wheat for rust resistance. Plant Genome 2018, 11. [CrossRef]

54. Liabeuf, D.; Sim, S.C.; Francis, D.M. Comparison of marker-based genomic estimated breeding values and phenotypic evaluation for selection of bacterial spot resistance in tomato. Phytopathology 2018, 108, 392–401. [CrossRef]

55. Ornella, L.; Singh, S.; Perez, P.; Burgueño, J.; Singh, R.; Tapia, E.; Bhavani, S.; Dreisigacker, S.; Braun, H.-J.; Mathews, K.; et al. Genomic prediction of genetic values for resistance to wheat rusts. Plant Genome 2012, 5. [CrossRef]

56. Endelman, J.B. Ridge regression and other kernels for genomic selection with R package rrBLUP. Plant Genome 2011, 4, 250–255. [CrossRef]
57. Piepho, H.P. Ridge regression and extensions for genomewide selection in maize. Crop Sci. 2009, 49, 1165–1176. [CrossRef]
58. Bassi, F.M.; Bentley, A.R.; Charmet, G.; Ortiz, R.; Crossa, J. Breeding schemes for the implementation of genomic selection in wheat (Triticum spp.). Plant Sci. 2016, 242, 23–36. [CrossRef]
59. Lin, C.S.; Poushinsky, G. A modified augmented design (type 2) for rectangular plots. Can. J. Plant Sci. 1985, 65, 743–749. [CrossRef]
60. You, F.M.; Duguid, S.D.; Thambugala, D.; Cloutier, S. Statistical analysis and field evaluation of the type 2 modified augmented design (MAD) in phenotyping of flax (Linum usitatissimum) germplasms in multiple environments. Aust. J. Crop Sci. 2013, 7, 1789–1800.
61. Browning, S.R.; Browning, B.L. Rapid and accurate haplotype phasing and missing-data inference for whole-genome association studies by use of localized haplotype clustering. Am. J. Hum. Genet. 2007, 81, 1084–1097. [CrossRef] [PubMed]
62. Price, A.L.; Patterson, N.J.; Plenge, R.M.; Weinblatt, M.E.; Shadick, N.A.; Reich, D. Principal components analysis corrects for stratification in genome-wide association studies. Nat. Genet. 2006, 38, 904–909. [CrossRef] [PubMed]
63. Yu, J.; Pressoir, G.; Briggs, W.H.; Vroh Bi, I.; Yamasaki, M.; Doebley, J.F.; McMullen, M.D.; Gaut, B.S.; Nielsen, D.M.; Holland, J.B.; et al. A unified mixed-model method for association mapping that accounts for multiple levels of relatedness. Nat. Genet. 2006, 38, 203–208. [CrossRef]
64. Zhou, X.; Stephens, M. Genome-wide efficient mixed-model analysis for association studies. Nat. Genet. 2012, 44, 821–824. [CrossRef] [PubMed]
65. Liu, X.; Huang, M.; Fan, B.; Buckler, E.S.; Zhang, Z. Iterative usage of fixed and random effect models for powerful and efficient genome-wide association studies. PLoS Genet. 2016, 12, e1005767. [CrossRef] [PubMed]
66. Wang, S.B.; Feng, J.Y.; Ren, W.L.; Huang, B.; Zhou, L.; Wen, Y.J.; Zhang, J.; Dunwell, J.M.; Xu, S.; Zhang, Y.M. Improving power and accuracy of genome-wide association studies via a multi-locus mixed linear model methodology. Sci. Rep. 2016, 6, 19444. [CrossRef] [PubMed]
67. Wen, Y.J.; Zhang, H.; Ni, Y.L.; Huang, B.; Zhang, J.; Feng, J.Y.; Wang, S.B.; Dunwell, J.M.; Zhang, Y.M.; Wu, R. Methodological implementation of mixed linear models in multi-locus genome-wide association studies. Brief. Bioinform. 2017, 19, 700–712. [CrossRef] [PubMed]

68. Tamba, C.L.; Ni, Y.L.; Zhang, Y.M. Iterative sure independence screening EM-Bayesian LASSO algorithm for multi-locus genome-wide association studies. PLoS Comput. Biol. 2017, 13, e1005357. [CrossRef] [PubMed]

69. Zhang, J.; Feng, J.Y.; Ni, Y.L.; Wen, Y.J.; Niu, Y.; Tamba, C.L.; Yue, C.; Song, Q.; Zhang, Y.M. pLARmEB: Integration of least angle regression with empirical Bayes for multilocus genome-wide association studies. Heredity 2017, 118, 517–524. [CrossRef]

70. Ren, W.L.; Wen, Y.J.; Dunwell, J.M.; Zhang, Y.M. pKWmEB: Integration of Kruskal-Wallis test with empirical Bayes under polygenic background control for multi-locus genome-wide association study. Heredity 2017, 120, 208–218. [CrossRef]

71. mrMLM. Available online: https://cran.r-project.org/web/packages/mrMLM/index.html (accessed on 25 August 2018).

72. De los Campos, G.; Naya, H.; Gianola, D.; Crossa, J.; Legarra, A.; Manfredi, E.; Weigel, K.; Cotes, J.M. Predicting quantitative traits with regression models for dense molecular markers and pedigree. Genetics 2009, 182, 375–385. [CrossRef] [PubMed]

73. De los Campos, G.; Hickey, J.M.; Pong-Wong, R.; Daetwyler, H.D.; Calus, M.P. Whole-genome regression and prediction methods applied to plant and animal breeding. Genetics 2013, 193, 327–345. [CrossRef] [PubMed]

74. de Los Campos, G.; Perez, P.; Vazquez, A.I.; Crossa, J. Genome-enabled prediction using the BLR (Bayesian Linear Regression) R-package. Methods Mol. Biol. 2013, 1019, 299–320. [PubMed]

75. Yang, J.; Benyamin, B.; McEvoy, B.P.; Gordon, S.; Henders, A.K.; Nyholt, D.R.; Madden, P.A.; Heath, A.C.; Martin, N.G.; Montgomery, G.W.; et al. Common SNPs explain a large proportion of the heritability for human height. Nat. Genet. 2010, 42, 565–569. [CrossRef] [PubMed]

76. Yang, J.; Lee, S.H.; Goddard, M.E.; Visscher, P.M. GCTA: A tool for genome-wide complex trait analysis. Am. J. Hum. Genet. 2011, 88, 76–82. [CrossRef]

77. You, F.M.; Xiao, J.; Li, P.; Yao, Z.; Jia, G.; He, L.; Kumar, S.; Soto-Cerda, B.; Duguid, S.D.; Booker, H.M.; et al. Genome-wide association study and selection signatures detect genomic regions associated with seed yield and oil quality in flax. Int. J. Mol. Sci. 2018, 19, 2303. [CrossRef]

3 Chapter

GENOME-WIDE ASSOCIATION STUDY OF MUCILAGE AND HULL COMPOSITION IN FLAX (LINUM USITATISSIMUM L.) SEEDS.

1. INTRODUCTION: AN OVERVIEW

Flaxseed (Linum usitatissimum L.), one of the most ancient crops, has been utilized as sustenance for humans and livestock since antiquity [1]. The two primary morphotypes of cultivated L. usitatissimum are the oil morphotype (flaxseed) and the fiber morphotype (fiber flax). Flaxseed plants are diminutive, more branched, and possess larger seeds, with branches encompassing a greater part of the primary stem in contrast to fiber flax. Flaxseed presently possesses enhanced opportunities in the functional food sector due to increasing consumer demand for health-promoting foods [1]. Flaxseed is abundant in bioactive substances, including α-linolenic acid (omega-3) with cardiovascular advantages, lignans with anticancer characteristics, and both insoluble and soluble fiber (mucilage) that helps reduce cholesterol and insulin levels. [2].

Flaxseed mucilage is a heterogeneous polysaccharide composed of xylose, arabinose, glucose, galactose, rhamnose, and fructose [3] that can be purified into neutral and acidic polymers. Mucilage abounds in the seed coat, where it makes up to 8–10% of the seed weight [4]. Mucilage synthesis is tightly linked to seed coat development [5] and both tissues form the seed hull, a structure representing 37–48% of the seed weight [6,7]. These two fractions, rich in polysaccharides, are components of the flaxseed meal, primarily used as a protein rich livestock and poultry feed [6,8]. Absorption of flaxseed mealts advantageous 31–45% protein content [9] may be hindered by mucilage and cell wall polysaccharides. This is due to the swelling capacity of polysaccharides in the digestive tract of monogastric animals that causes concomitant growth depression and reduced feed efficiency [7,10]. In that context, reduction of mucilage (MC) and hull (HC) contents in flaxseed meal is desirable to achieve improved feeding value for livestock and poultry. Studies of flaxseed mucilage degradation are focused on chemical retting, enzyme retting, and steam explosion [11]. Reduction of the hull content in flaxseed and rapeseed meal has been achieved through dehulling methods [12] and the use of yellow-seeded genotypes [7,13]. Food and feed markets demand flaxseed cultivars differing in mucilage and hull content. It is, therefore, crucial to decipher the genetic factors underlying these complex traits in order to accelerate the development of market-specific flaxseed cultivars.

In the model plant Arabidopsis thaliana, the genes necessary for the synthesis, modification, and release of mucilage, as well as seed coat development, are well understood [5,14]. Putative flax orthologs of the RHAMNOSE SYNTHASE (AtRHM1), GALACTURONOSYLTRANSFERASE-LIKE3 (GATL3),GALACTURONOSYLTRANSFERASE 11 (GAUT11), XYLOGLUCAN ENDOTRANSGLUCOSYLASE/HYDROLASE 3(XTH3), and ALPHA-XYLOSIDASE-1 (AtBXL1), involved in mucilage production, have been identified using gene expression analysis during seed development [15]. Similarly, putative flax orthologs of the TRANSPARENT TESTA 3, 4, 5, and 7 (TT3, TT4, TT5, and TT7), FLAVONOL SYNTHASE (FLS) and BANYULS (BAN), involved in flavonoids synthesis during seed coat development, have also been identified [15].

Genetic variation for MC and HC in flaxseed has been assessed [4,7,16–18] but no quantitative trait loci (QTL) have been reported so far. QTL for Fusarium wilt resistance [19], powdery mildew [20], iodine value, palmitic, linoleic, and linolenic acids [21,22], and seed and flower color [23], were reported. QTL for seed protein, cell wall, straw weight, yield-related traits, and phenological traits, have also been reported using bi-parental mapping and association mapping [22,24,25]. Recently, genome-wide association studies (GWAS) have been conducted for agronomic and seed quality traits using thousands of single nucleotide polymorphism (SNP) loci [26,27]. GWAS mines the natural sequence diversity within a species and captures historical recombination events. It is therefore a suitable approach to discover loci that control complex traits, leading to a higher mapping resolution to facilitate the identification of candidate genes [28]. Thus, the suite of genomic tools available for flaxseed genetic studies [21,22,26,27,29–32] make genomic evaluation of MC and HC feasible.

The objective of this research was to identify QTL and candidate genes contributing to mucilage content and hull content that could be capitalized upon to assist in breeding flaxseed cultivars with different mucilage content and with reduced hull content. Improving these traits will increase seed value of this important cash crop.

2. RESULTS

2.1 Phenotypic Evaluation

Two hundred flax accessions from the Canadian flax core collection were planted in two environments. Evaluation of MC and HC showed a normal distribution across the two environments according to the Shapiro–Wilk normality test and normality plots (Table S1, Figure S1). Variance component analysis indicated significant effects of genotype, environment, and genotype × environment interaction, according to the Wald statistic ($p < 0.001$). The phenotypic variation for MC in Vilcún location 2014 (CAR2014) ranged from 23.52 to 103.57 mg g-1 with an average of 58.67 mg g-1. A lower variation was observed for MC in Huichahue location 2015 (HU2015), which ranged from 18.88 to 91.90 mg g-1 with an average of 55.04. mg g-1 HC variation ranged from 35.56 to 48.59% in CAR2014 and from 35.73 to 48.59% in HU2015 (Figure 1, Table S1). MC and HC were significantly positively correlated in CAR2014 and HU2015 with coefficients of 0.28 and 0.29, respectively. Narrow sense heritability (h2) for MC attained 70.7 and 73.8% in CAR2014 and HU2015, respectively. Lower h2 of 51.4 and 46.2% for HC at CAR2014 and HU2015 were observed. MC did not differ statistically between flax morphotypes nor seed color classes, according to the Kruskal–Wallis non-parametric test. The average MC was 55.33 and 56.63 mg g-1 for the fiber and oil morphotypes, respectively ($p = 0.651$) (Figure S2a). The average MC registered values of 56.63 and 59.22 ($p = 0.517$) for the brown and yellow seeded classes, correspondingly. The average HC did not differ statistically between flax morphotypes (fiber = 43.41%, oil = 42.79%; $p = 0.373$). On the other hand, yellow-seeded genotypes averaged 2.66% less HC than brown seeded accessions ($p = 3.2 \times 10\text{-}5$) (Figure S2b).

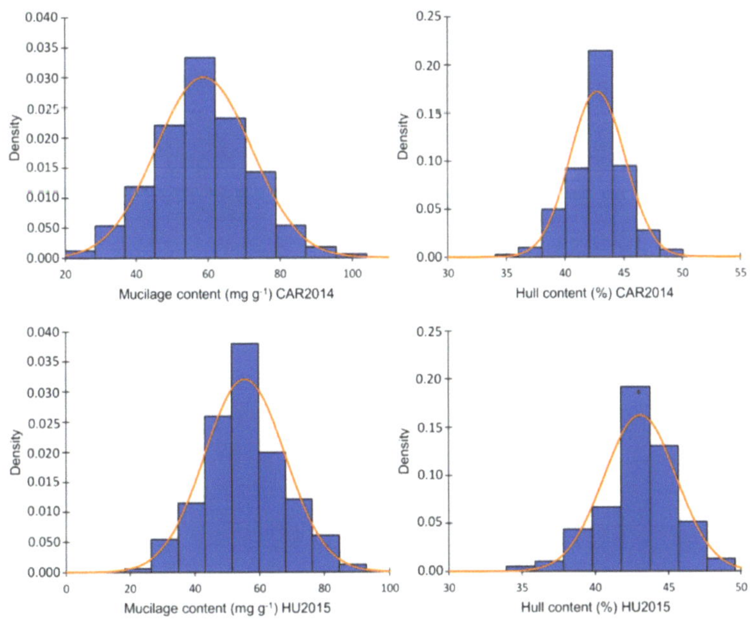

Figure 1. Mucilage (MC) and hull (HC) contents distribution in the association panel in two environments: CAR2014 = Vilcún location 2014, HU2015 = Huichahue location 2015. Values represent the mean of three biological replicates for each trait.

2.2 Population Structure and Linkage Disequilibrium

The dendrogram based on 771,914 SNPs and the STRUCTURE plot grouped the 200 individuals into two major clusters arbitrarily assigned as "red" and "blue" (Figure 2a,b). In the K against ΔK plot, a break in the slope was clearly observed at K = 2 (Figure 2b). The red cluster comprised almost exclusively genotypes belonging to the oil morphotype, while the blue cluster included both flax morphotypes. The coefficient of population differentiation (FST = 0.08) indicated a weak population structure between the two clusters.

The genome-wide linkage disequilibrium (LD) decayed to r2 = 0.1 at a physical distance of ~100 kb (Figure 2c). Intrachromosomal LD decayed to r2 = 0.1 at a distance between marker pairs ranging from ~40 kb on chromosome 8 to ~210 kb on chromosome 13 (Figure S3). A highly significant positive correlation (r = 0.75, p = 0.0012) between marker density and the intrachromosomal LD blocks was observed (Figure 3). For example, chromosomes 4 and 8 with the smallest number of markers, and chromosomes 6 and 13 with the largest, displayed the fastest and slowest LD decays, respectively (Figure 3 and Figure S3). The fast LD decays observed in this association panel are indicative of its advantageous potential for reducing QTL intervals and fine mapping of candidate genes for MC and HC.

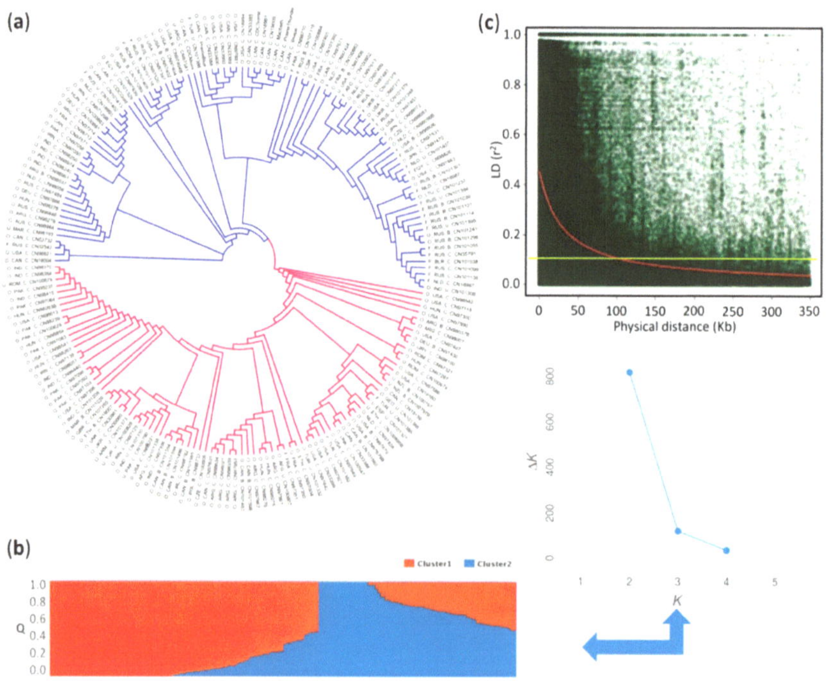

Figure 2. Population structure and genome-wide linkage disequilibrium decay. (a) Neighbor-joining (NJ) tree for 200 flax accessions based on 779,914 single nucleotide polymorphisms (SNPs); (b) Model-based population structure of 200

flax accessions belonging to two clusters predefined by the STRUCTURE software. Each accession is represented by a vertical bar. The color subsections within each vertical bar indicate membership coefficient (Q) to different clusters; (c) Genome-wide linkage disequilibrium decay of r2 values (red line), against physical distance (kb) using the Hill and Weir (1988) function in L. usitatissimum. Yellow line indicates the cutoff value (r2 = 0.1) used to determine the genome-wide linkage disequilibrium (LD) block size.

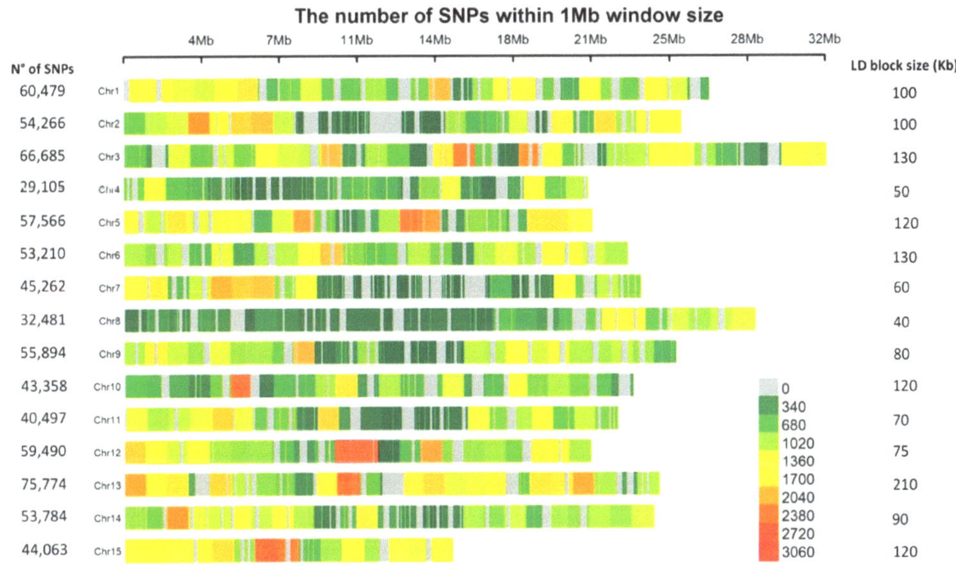

Figure 3. Single nucleotide polymorphism (SNP) density plot across the L. usitatissimum genome. Numbers of SNPs and LD blocks are also indicated for each of the 15 chromosomes.

2.3 Genome-Wide Association Analysis

Three GWA models were tested, including GLM-Q, GLM-PCA, and MLM-K. According to the quantile-qualtile (Q–Q) plot results, the GLM-Q model showed a strong skew toward significance for every trait, indicating that the Q matrix was insufficient to account for population structure and cryptic relatedness. Conversely, the MLM-K, which only used the kinship matrix, led to an overcorrection of these confounding factors, particularly for HC. The GLM-PCA was tested with 5 and 10 PCA covariates for HC and MC, accounting for 30.1 and 37.1% of the variation, respectively. Both GLM-5PCA and GLM-10PCA models performed well in controlling the rate of false positives, providing suitable statistical power to identify significant marker–trait associations for MC and HC). Therefore, the GLM-PCA model was applied for GWA in this study.

GWA analysis identified 12 and 17 significant associations for MC in CAR2014 and HU2015, respectively ($p < -\log10 (P) = 6.88$), and markers Lu5-3808878, Lu7-

13225294, and Lu11-2498303 were significant in both environments (Table 1, Figure S5). Various significant SNP markers fell into the same LD blocks. For example, five other significant markers surrounded the peak SNP Lu5-3808878 (Figure S5), thus, they were considered the same QTL. Following this criterion, seven QTL were delineated on chromosomes 2, 3, 5, 7, and 11. The peak SNPs of these QTL accounted for 11.8 to 17.3% of phenotypic variation, and the combined three consistent QTL accounted for 43.6% of the MC variation (Table 1).

Table 1. Genome-wide significant peak SNPs for mucilage content (MC) and hull content (HC).

Trait	Marker	Chromosome	Allele	MAF 1	-log10 (P)		R2 (%)	
					CAR2014	HU2015	CAR2014	HU2015
MC	Lu2-22298066	2	T/C	0.07	8.69	3.41ns 2	17.32	ns 2
	Lu3-25559600	3	G/T	0.06	7.45	4.13ns 2	13.42	ns 2
	Lu3-26033342	3	C/G	0.07	7.68	4.23ns 2	13.25	ns 2
	Lu3-7398487	3	C/T	0.41	4.96ns 2	7.02	ns 2	11.82
	Lu5-3808878	5	G/A	0.10	8.03	10.21	14.97	16.52
	Lu7-13225294	7	G/A	0.34	8.10	6.91	16.46	12.05
	Lu11-2498303	11	C/G	0.16	7.05	7.47	14.25	13.18
HC	Lu7-6577527	7	A/C	0.13	6.90	7.36	14.66	15.79
	Lu10-21552161	10	G/A	0.09	6.90	6.16ns 2	16.32	ns 2
	Lu12-5267706	12	C/T	0.06	5.91ns 2	6.92	ns 2	13.83
	Lu13-2803224	13	T/C	0.06	7.83	8.45	17.43	18.20

1 MAF: minor allele frequency; 2 ns: not significant at the threshold value -log10 lo (P) = 6.88.

A total of three and four significant associations were detected for HC in CAR2014 and HU2015, respectively ($p < -\log 10$ (P) = 6.88). Markers Lu7-6577527 and Lu13-2803224 were significant in both environments (Table 1). The four QTL identified on chromosomes 7, 10, 12, and 13 explained between 13.8% and 17.8% of the HC variation. The two consistent QTL Lu7-6577527 and Lu13-2803224, accounted for a combined 33% of the HC variation (Table 1).

The peak SNPs effect for MC and HC were all significant according to the non-parametric Kruskal–Wallis test ($p < 0.05$), except for Lu3-26033342 associated with MC (Figure 4a,b and Figure S6). Accessions with a thymine (T) allele at Lu2-22298066 displayed, on average, an increase of 15.3 and 9.4 mg g−1 in MC, compared to accessions with a cytosine (C) allele in CAR2014 and HU2015, respectively (Figure 4a). Similarly, accessions with a "T" allele at Lu3-7398487 had, on average, 6.64 and 8.4 mg g−1 higher MC compared to accessions with a "C" allele in CAR2014 and HU2015, correspondingly.

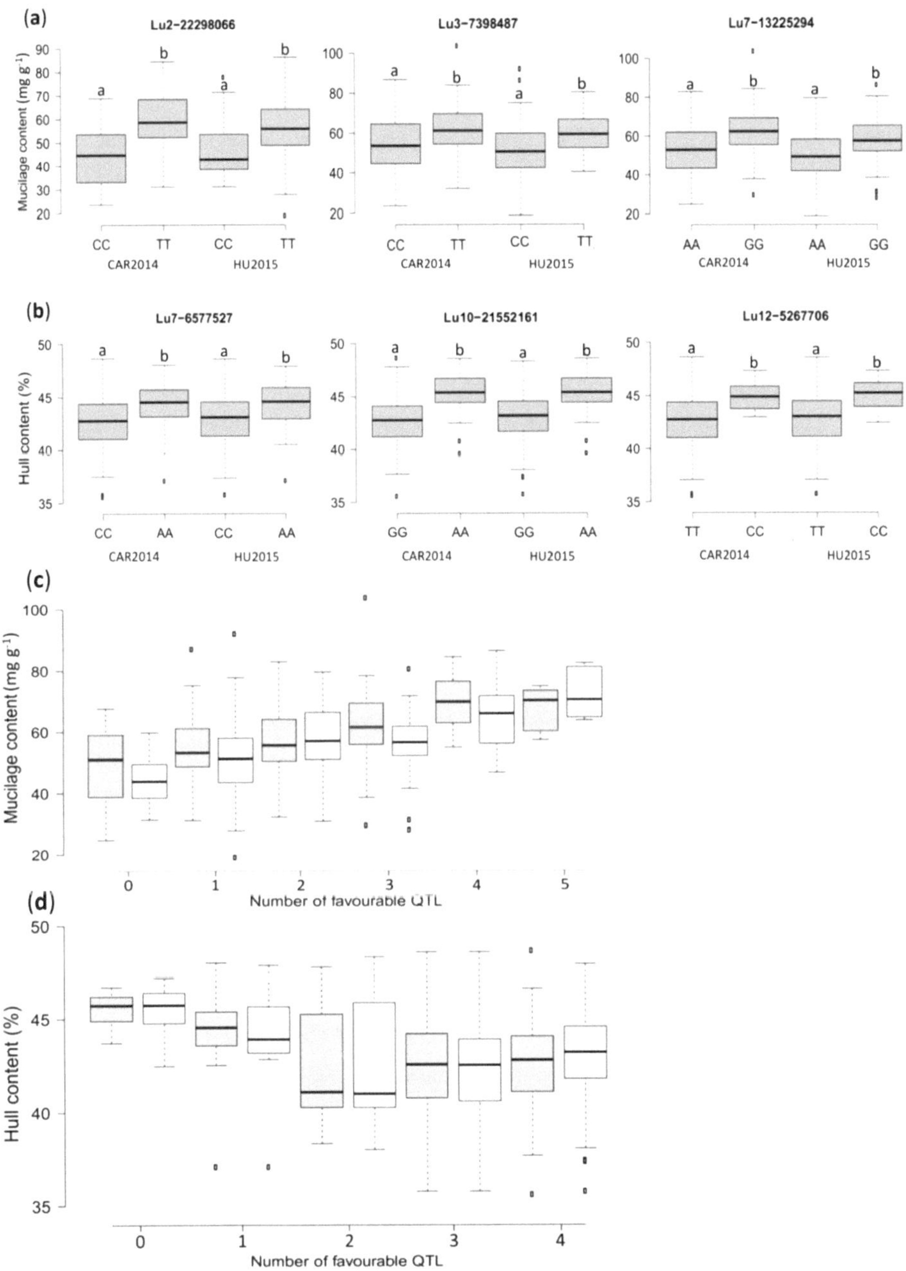

Figure 4. Box plots illustrating the phenotypic differences between flaxseed accessions carrying different alleles of the significant SNPs, and combined phenotypic effects of favorable QTL in the association panel. (a) Mucilage content (MC); (b) Hull content (HC). CAR2014 = Vilcún location 2014, HU2015 = Huichahue location 2015. Different letters indicate significant statistical differences according to the Kruskal-Wallis non-parametric test ($p < 0.05$); (c) QTL effect for MC; (d) QTL effect for HC. Grey and white boxplots represent the CAR2014 and HU2015 locations, respectively.

Accessions with a guanine (G) allele at Lu7-13225294 had, on average, 8.56 and 7.71 mg g-1 more mucilage compared to accessions carrying an adenine (A) allele in CAR2014 and HU2015, respectively (Figure 4a). The allelic effect for the other four peak SNPs is illustrated in Figure S6. The allelic effect of peak SNPs for HC revealed that accessions harboring a "C" allele at Lu7-6577527 had, on average, 1.4 and 1.3% less HC compared with "A" allele genotypes in CAR2014 and HU2015, correspondingly (Figure 4b). On average, HC was reduced by 1.4 and 1.3% (Lu7-6577527) to 2.6 and 2.7% (Lu13-2803224) in CAR2014 and HU2015, respectively (Figure 4b and Figure S6).

The combined QTL effect revealed that the MC of accessions harboring none of the favorable QTL alleles averaged 44.6 and 48.9 mg g−1, compared to 72.1 and 67.6 mg g−1, for those with five favorable alleles in CAR2014 and HU2015, respectively (Figure 4c). No accession had all seven favorable QTL alleles. The combined QTL effect for HC indicated that genotypes with none of the favorable QTL alleles averaged 45.5% and 45.3% HC compared to genotypes with four favorable QTL alleles, in which HC averaged 42.7% and 42.9% in CAR2014 and HU2015, respectively (Figure 4d).

2.4 Identification of Candidate Genes

The LD blocks harboring the peak SNPs were mined for genes relevant to MC and HC using the L. usitatissimum v.1.0 reference genome. A total of 204 and 118 candidate genes were identified for MC and HC, respectively (Table 2 and Table S2). Several genes ascribed to carbohydrate metabolism, seed mucilage synthesis, modification, and release, and cell wall synthesis and modification were identified at the MC QTL loci (Table 2). Five particularly promising candidate genes were identified. The SNP marker Lu3-26033342 was located 58.92 and 49.60 kb from Lus1007101 and Lus10007083 that encode the ortholog of A. thaliana's TRANSPARENT TESTA 8 (TT8) and SUBTILISIN-LIKE SERINE PROTEASE (SBT1.7) (Figure S5a, Table 2). In another independent QTL on chromosome 3, the SNP marker Lu3-25559600 was located 64.41 and 67.02 kb from Lus10009311 and Lus10009288 that encode the ortholog of A. thaliana's GALACTUROSYL TRANSFERASE-LIKE 5 (GATL5) and MUCILAGE-MODIFIED 4 (MUM4). On chromosome 5, Lu5-3508878 is located 100.78 kb from Lus10008285, an ortholog of another A. thaliana gene implicated in mucilage transcriptional regulation, NAC-REGULATED SEED MORPHOLOGY 1 (NARS1) (Figure S5a, Table 2).

Table 2. Candidate genes within LD blocks harboring peak SNPs associated with MC and HC.

Trait	Marker	Gene ID	Scaffold	A. Thaliana	Gene Bank	Identity (%)	E-Value	Distance from Peak SNP (kb)
MC	Lu3-25559600	Lus10009311	318	GATL5	at1g02720	27	7 ×10−27	64.41
		Lus10009288	318	MUM4	at1g53500	26	4 × 10−23	67.02
		Lus10009287	318	PME36	at3g60730	61	2 × 10−110	70.33
		Lus10009313	318	SBT1.7	at5g67360	45	0.0	75.66
	Lu3-26033342	Lus10007101	772	TT8	at4g09820	38	5×10^{-15}	58.92
		Lus10007083	772	SBT1.7	at5g67360	39	1×10^{-152}	49.60
	Lu5-3808878	Lus10008285	489	NARS1	at3g15510	52	9×10^{-45}	100.78
HC	Lu7-6577527	Lus10035456	151	AGL62	at5g60440	43	6×10^{-39}	11.40
	Lu12-5267706	Lus10018306	163	GH17	at2g39640	34	9×10^{-86}	39.93
	Lu13-2803224	Lus10026902	651	DBR1	at4g31770	68	0.0	96.87
		Lus10026926	651	UGT79B1	at5g54060	25	2×10^{-32}	238.19

Genes related to embryo, endosperm, and seed coat development, cell wall biogenesis/degradation, anthocyanin biosynthesis, and seed dormancy, were found at QTL loci associated with HC (Table 2 and Table S2). Among the relevant candidate genes, Lus10035456 encodes the ortholog of A. thaliana's AGAMOUS-LIKE MADS-BOX PROTEIN AGL62 (AGL62) and is located 11.40 kb from the SNP marker Lu7-6577527 (Figure S5b, Table 2). On chromosome 12, Lu12-5267706 was situated 39.93 kb from Lus10018306 that encodes the ortholog of A. thaliana's GLYCOSYL HYDROLASE FAMILY 17 (GH17). Two other interesting candidate genes, Lus10026902 and Lus10026926, were situated 96.87 and 238.19 k, respectively, from the SNP marker Lu13-2803224. Lus10026902 and Lus10026926 encode the ortholog of A. thaliana's LARIAT DEBRANCHING ENZYME (DBR1) and UDP-GLUCOSE FLAVONOL 3-O-GLUCOSYLTRANSFERASE (UGT79B1), respectively.

3. DISCUSSION

3.1 Phenotypic Variation of Mucilage and Hull Contents

Flaxseed mucilage and seed hull possess valuable nutritional and rheological attributes [33,34] but are also known to affect animal performance [7]. The presence of mucilage and fiber components (i.e., acid detergent lignin) in flaxseed meal reduces the energy uptake in both monogastric and ruminant animals [35]. Therefore, knowledge about the phenotypic variation and genetic control of seed mucilage content (MC) and hull content (HC) is pivotal to better design breeding strategies aiming to improve the overall food and feed value of flaxseed. The broad phenotypic variation of MC and HC in the association panel and the degree of additivity of the genetic components hint at the potential for improving flaxseed for either high or low MC and reduced HC through marker-assisted selection.

Kaewmanne et al. [4] reported MC ranging from 1.8 to 2.9% in seven Italian flaxseed cultivars, while Oomah et al. [16] found that MC ranged from 3.6 to 8.0% in 109 flaxseed accessions. We found a slightly wider range from 2 to 10% in our diversity panel. Little information exists for HC variation in large collections of flaxseed. In general, HC ranges from 22–27% to 36–48% were reported in mechanically treated and hand-dissected seeds, respectively [7,36], which is much higher than canola at 18.6% and soybean at 16.1% [6]. Reduction of HC can be achieved through the use of yellow-seeded cultivars, known to contain higher oil content and less HC than their brown-seeded counterparts [7,37]. Indeed, the yellow-seeded accessions displayed a lower HC compared to the brown-seeded genotypes. Nevertheless, caution should be exercised in adopting yellow-seeded flaxseed cultivars for reduced HC flaxseed because their susceptibility to natural splitting and mechanical cracking of the seed coat can negatively affect seed quality [38]. Consequently, breeding and seed tests to mechanical damage during harvesting should be conducted together in order to identify the ideal HC that would ensure seed mechanical resistance. All considered, our association panel harbored abundant phenotypic variation for dissecting the genetic landscape of MC and HC.

3.2 Population Structure and Linkage Disequilibrium

When the main factors accounting for population subdivision correlate with a trait under study (i.e., geographic distribution and flowering time), then marker–trait associations will undergo a more accentuated inflation of observed p-values as effect of the structure confounding factor [39]. In flaxseed, population structure has been assessed in varying numbers of accessions, where geographic origin and flax morphotype seemed to have been the main factors underlying population subdivisions [40–42]. In our association panel, the "red" and "blue" clusters were slightly differentiated (FST = 0.08), with a weak morphotypic effect on dendrogram topology, possibly due to the small number of fiber types (n = 33) compared to the larger number of oilseed type accessions (n = 153).

Linkage disequilibrium (LD) is the main factor influencing marker density requirement and mapping resolution in GWAS. Mating system and genetic diversity influence LD decay. LD decays more rapidly in outcrossing plant species than in self-pollinated plants [43] and, similarly, in wild relatives and landraces compared to modern cultivars [44]. Here, we observed a rapid LD decay for most of the chromosomes, comparable to some maize commercial elite inbred lines [45] and faster than winter-type Brassica napus (480 to 1283 kb, r2 = 0.1) [46]. Therefore, the 200 flaxseed accessions of our diversity panel are expected to contain plentiful allelic diversity, as suggested by the generally short LD blocks for the 15 chromosomes, thereby assisting the search for candidate genes through efficient narrowing of the putative QTL regions.

3.3 Genome-Wide Association Analysis

Several general (GLM) and mixed (MLM) linear models have been proposed to control both population structure and cryptic relatedness [47–49]. In flax, MLM has been the preferred association model for multiple traits [24–26,42]. The "red" and "blue" clusters were weakly differentiated, and MC and HC between flax morphotypes was not statistically significant (Figure S2a,b), in contrast to a report comparing indica and japonica rice types assessed for 34 traits [39]. Hence, the genetic architecture of MC and HC seem to be only weakly correlated with population and family structures, and GLM-PCA was sufficient to control the rate of false positive associations.

The discovery of QTL for agronomic and economically important traits in crops is of great importance for marker-assisted breeding. This is the first report of QTL for MC in flax, likely because this trait has not been a breeding priority in the most important breeding programs of the world [18]. In the present study, GWAS identified seven QTL for MC, and their effects clearly suggest the promise of marker-assisted selection for modifying MC.

Chromosome 3ts multiple MC QTL harbored candidate genes orthologous to Arabidopsis TT8 gene, which is part of a transcription factor complex that, along with GLABRA2 (GL2), regulates MUM4 gene expression [50]. MUM4 is required to produce rhamnose, a key substrate for mucilage biosynthesis [50], and chromosome 3 Lus10009311 is its flax ortholog. In Arabidopsis, GATL5 encodes a glycosyltransferase involved in rhamnogalacturonan I (RG I) backbone synthesis [51]. The presence of a L. usitatissimum ortholog Lus10009311 in a LD block, with a peak SNP for MC, corresponds to the expected role of RG I synthesis. Arabidopsis gene SBT1.7 triggers the activation of cell wall-modifying enzymes necessary for mucilage release upon imbibition [52]. In line with the expected seed coat mucilage dynamics, we identified two orthologous copies of this gene in two independent QTL (Table 2). Arabidopsis PECTIN METHYLESTERASE INHIBITOR 6 (PMEI6) mutants were defective in seed coat mucilage release [53]. An ortholog of the Arabidopsis gene, PECTIN METHYLESTERASE 36 (PME36), another family member, was located at one of the MC QTL loci identified herein. While PME36 has not been shown to be involved in mucilage release, it might participate indirectly because it exerts a similar role to that of PMEI6 in pectin synthesis and cell wall modification [54].

Oil content is an economically important but genetically complex trait. MC is negatively correlated with oil content, therefore, reducing MC should facilitate increasing oil content. Indeed, reduced accumulation of mucilage accompanied by increased oil content was observed in Arabidopsis MUM4 or GL2 mutants [55]. We observed a significant negative correlation ($r = -0.15$, $p = 0.03$) between MC and oil content in the association panel (data not shown). This is perhaps due to increased carbon allocation to the embryo in reduced or no seed coat mucilage synthesis in low MC accessions as proposed in Arabidopsis [55].

Increasing seed oil content and reducing the fiber fraction of the meal have been important goals in oil crop breeding. In B. napus and L. usitatissium, seed coat thickness or HC are negatively correlated with seed oil and protein content, as well as seed color [56–58]. QTL for seed coat color to indirectly increase oil content and minimize HC have been identified in B. napus and soybean [37,59,60]. In flax, a pleiotropic QTL controlling yellow seed and white flower color was recently dissected at the molecular level, but its effect on HC has not been addressed [23]. Here, we identified four QTL whose effects reduced HC by 2.6%, on average. Chromosome 7 harbored Lus10035456, which resembles the A. thaliana transcription factor AGL62. AGL62 mutants initiated embryo and endosperm formation, but failed to form a seed coat [61]. Light seed color and low HC are thought to coincide because the biochemical pathways leading to lignin and pigment synthesis share common precursors [59]. In Arabidopsis, the core components of seed coat pigments are proanthocyanidins (PAs) [62]. Chromosome 12 encompassed three candidate genes including the ortholog of Arabidopsis O-GLYCOSYL HYDROLASES FAMILY 17 gene. GH17 is coexpressed with TT12, AHA10, and BAN, that might process glycosylated flavan-3-ol monomers, leading to accumulation of PAs in the seed coat [63]. In black seed soybean, a UDP-GLUCOSE:FLAVONOID 3-O-GLUCOSYLTRANSFERASE (UGT78K1), was isolated from the seed coat, a key enzyme that catalyzes the final step in anthocyanin biosynthesis [64]. Chromosome 13 contained Lus10026926, an ortholog of the A. thaliana UGT79B1, a gene also involved in anthocyanin biosynthesis. Yellow seed color stems from the blocked biosynthesis of PAs that impart the brown color to the seed coat [65]. The flaxseed meal derived from brown-seeded cultivars contains PAs that negatively affect protein digestion [66], hence low PA meal is preferred in animal ration. Additional advantages of modifying the seed color and reducing MC and HC include higher limpidity of the crude oil from the removal of gum-like residues and dark pigments, higher protein content and better feeding value of flaxseed meal for livestock and poultry [7].

Few accessions combined favorable alleles for reduced MC and HC. It should be possible to combine these attributes in a single genotype through the pyramiding of the respective favorable alleles owing to the fact that the significant QTL for both traits did not co-locate in the flax genome.

The development of yellow-seeded cultivars with low HC and either low or high MC for different industrial uses is an opportunity to increase market share and value.

4. MATERIALS AND METHODS

4.1 Plant Material, Field Trials, Phenotyping, and Statistical Analyses

A total of 200 L. usitatissimum accessions from the Canadian flax core collection [67] were selected for this study based on their geographic distribution and genetic diversity (Table S3). The 200 genotypes were planted in 2014 and 2015 at the Agriaquaculture Nutritional Genomic Center (CGNA) experimental stations located in Vilcún (CAR2014) and Huichahue (HU2015), La Araucania region, Chile, using a completely randomized design (CRD) with three biological replicates. Genotypes were arranged in rows and columns in order to take into account spatial heterogeneity.

The seed mucilage content (MC) was determined in three biological replicates following the procedure described by Kaewmanee et al. [4] with minor modifications. A total of 2 g of seeds were incubated in 20 mL of water at 100 °C for 15 min in 50 mL Falcon tubes. Next, the tubes were shaken for 30 min at 250 rpm. The soluble extract was recovered by centrifugation at 6132 relative centrifugal force (RCF) for 30 min, and the mucilage fraction was precipitated by incubating in 30 mL of ethanol (95%) overnight at 4 ◦C. The seeds were recovered, and the extraction procedure was carried out twice more to maximize mucilage recovery. The mucilage pellet was weighed and expressed as milligrams of mucilage per gram of seed (mg g-1).

HC was determined in three biological replicates by separating the hull from the embryos using a dissecting needle and tweezers from 50 seeds after imbibition in water for 24 h. Both fractions were dried at 90 ◦C for 4 h before their dry weights were measured. HC was expressed as (hull dry weight/(hull dry weight + embryo dry weight)) × 100, averaged from 50 seeds.

$h^2 = \sigma^2{}_a/\sigma^2{}_a + \sigma^2{}_e,$ Variation of phenotypic data was analyzed individually for each environment using a restricted maximum likelihood (REML) analysis. Spatial correction in row and column directions was used with different variance–covariance structures. Spatial models were compared with Akaike information criterion (AIC) and Bayesian information criterion (BIC), and the most appropriate model in each environment was used to obtain a best linear unbiased estimate (BLUEs) for mucilage and hull contents in GenStat v.16 [68]. Descriptive statistics and Shapiro–Wilk normality test were conducted in the R package MVN [69]. Narrow sense heritability (h2) was estimated using variance components from TASSEL v.5.2.31 [70]. Trait h2 estimates were computed using the equation: , where σ 2a is the additive genetic variance and σ 2e is the residual error variance [70].

4.2 Whole Genome Resequencing and SNP Calling

Genotyping by sequencing (GBS) methodology was adopted to genotype the 407 accessions from the Canadian flax core collection. The 407 individuals were grown in pots in a greenhouse with a 16 h light and 8 h dark cycle. Young leaf tissues from single plants were collected for DNA extraction using the DNeasy 96 Plant kit (Qiagen, Mississauga, ON, Canada) according to manufacturer's instructions. Genomic DNA was quantified, sheared, size-selected, and libraries were constructed for each genotype by the Michael Smith Genome Sciences Centre of the BC Cancer Agency, Genome British Columbia (Vancouver, BC, Canada) which also sequenced the libraries generating 100 bp paired-end reads on the Illumina HiSeq 2000 platform (Illumina Inc., San Diego, CA, USA). A total of 26.875 billion 100 bp pair-end reads were generated, corresponding to 6587 MB sequences and 15.5× genome equivalents of the reference genome (~370 MB) [32,71], on average, per individual.

All reads from each individual of the population was aligned to the flax reference sequence (the flax pseudomolecules v2.0) [72] using BWA (v0.6.1) [73] with default parameters. The alignment file for each individual was used as input for SNP discovery using the software package SAMtools [74]. SAMtools converts the sequence alignment files from sequence alignment/map (SAM) to sorted binary alignment/map (BAM) format and call all potential variants (SNP and indels) into the pileup files. Then, all variants were further filtered to get a set of high-quality SNPs. SNPs were filtered by the following criteria: (1) candidate SNP loci must be more than 10 bp away from each other; each candidate SNP must have three mapped reads on the region; and (3) all the singleton SNPs were excluded. All steps of this procedure were implemented in an annotation-based genome-wide SNP discovery (AGSNP) pipeline [75,76]. The coordinates of all SNPs were extracted from the chromosome-based flax pseudomolecules v2.0 [72] totaling 1.7 million SNPs. Two hundred accessions, as previously mentioned, were selected for this study.

4.3 Population Structure, LD, Genome-Wide Association Study, and Candidate Genes

Population structure was estimated with 259 neutral SSR loci [41] distributed across flax's 15 chromosomes. The software STRUCTURE v.2.3.4 [47] was employed with predefined numbers of genetic clustering (K) from 1–5, using 50,000 burn-in iterations, followed by 100,000 MCMC across five independent runs for each K values. The number of clusters (K) was calculated with the Evanno method [77] implemented in the R package POPHELPER v.1.1.10 [78]. A total of 771,914 SNPs, filtered from the 1.7 million SNPs by removing those with a minor allele frequency <0.05 and >10% missing data, were used to produce a dendrogram using the neighbor-joining (NJ) algorithm implemented in TASSEL v.5.2.31 [70]. Genome-wide linkage disequilibrium (LD) and intrachromosomal LD between pairs of SNPs using the 771,914 filtered SNPs were estimated using squared allele frequency correlations (r2) in TASSEL v.5.2.31 [70]. LD values were plotted against physical distance to determine the LD decay using the Hill and Weir [79] function. A cut off value of r2 = 0.1 was set to estimate the average LD blocks [41].

GWAS was performed in TASSEL v.5.2.31 [70] using the 771,914 filtered SNPs. Three models were evaluated, including GLM-Q, GLM-PCA, and MLM-K. The Q matrix generated from STRUCTURE was used as a cofactor to adjust for population stratification (GLM-Q). A GLM-PCA was assessed, including up to ten principal component covariates. The ten PCAs were generated in TASSEL v.5.2.31 [70] with 105,038 SNPs (MAF > 0.05 and at least 95% present among the 200 genotypes). For the MLM-K, a kinship matrix was created in TASSEL v.5.2.31 [70] with the set of 105,038 SNPs, and used as covariate to account for cryptic relatedness. A quantile–quantile (Q–Q) plot was displayed using the R package qqman [80] to evaluate the fitness and efficiency of the different models. The final Manhattan plots were also displayed using the qqman package [80]. The Bonferroni correction (0.1/771,914 = -log (P) = 6.88) was used as threshold for the significance of marker–trait associations.

To identify candidate genes associated with significant SNPs, the Jbrowse feature of Phytozome v.12.1 (http://phytozome.jgi.doe.gov/pz/portal.html) was used to examine the L. usitatissimum v.1.0 genome [71] for genes relevant to MC and HC in flaxseed. As mentioned above, a cut-off value of r2 = 0.1 was set to estimate the average LD block for each chromosome. The defined physical distance was used to pinpoint candidate genes on either side of the most significant SNPs. A plausible candidate gene was defined by the following criteria: (a) the gene had a function known to be related to the trait evaluated based on gene ontology term descriptions in Phytozome; (b) BLASTX searches from the Arabidopsis genome returned orthologous protein sequences with functions associated to the phenotypes of interest.

5. CONCISE SUMMARY

We conducted a GWAS utilizing 771,914 SNPs, discovering seven and four QTL for MC and HC, respectively. Chromosome 3 contained three QTL that included promising candidate genes for MC. Three of the QTL linked to HC encompassed credible candidate genes pertinent to seed coat and anthocyanin production. These advantageous QTL alleles will facilitate the development of market-specific flaxseed cultivars with diminished HC while preserving elevated MC for food and low MC for feed. The utilization of the found SNP markers in molecular-assisted breeding for MC and HC, two complicated traits that require labor-intensive and time-consuming phenotyping, may facilitate the swift incorporation of advantageous alleles into well-adapted elite flaxseed cultivars, therefore reducing the breeding cycle. Our findings, along with prior gene expression research, suggest that the genetic regulation of mucilage and hull content in flax may include conserved processes analogous to those in Arabidopsis. Additional confirmation of candidate genes, such as LuTT8, LuSBT1.7, LuMUM4, and LuAGL62, via gene expression analysis or gene editing, will be essential to substantiate the aforementioned notion. The characterization of genes associated with the QTL will enhance understanding of the intricate dynamics of cell wall processes involved in seed mucilage and seed coat production in flaxseed..

REFERENCES

1. Rabetafika, H.N.; van Remoortel, V.; Danthine, S.; Paquot, M.; Beckler, C. Flaxseed proteins: Food uses and health benefits. Int. J. Food Sci. Technol. 2011, 46, 221–228. [CrossRef]

2. Kristensen, M.; Jensen, M.G.; Aarestrup, J.; Petersen, K.; Søndergaard, L.; Mikkelsen, M.S.; Astrup, A. Flaxseed dietary fibers lower cholesterol and increase fecal fat excretion, but magnitude of effect depend on food type. Nutr. Metab. 2012, 9, 1–8. [CrossRef] [PubMed]

3. Hunt, K.; Jones, J.K.N. The structure of linseed mucilage: Part II. Can. J. Chem. 1962, 40, 1266–1279. [CrossRef]
4. Kaewmanee, T.; Bagnasco, L.; Benjakul, S.; Lanteri, S.; Morelli, C.F.; Speranza, G.; Cosulich, M.E. Characterisation of mucilages extracted from seven Italian cultivars of flax. Food Chem. 2014, 148, 60–69. [CrossRef] [PubMed]
5. Haughn, G.; Chaudhury, A. Genetic analysis of seed coat development in Arabidopsis. Trends Plant Sci. 2005, 10, 472–477. [CrossRef] [PubMed]
6. Bhatty, R.S.; Cherdkiatgumchai, P. Compositional analysis of laboratory-prepared and commercial samples of linseed meal and of hull isolated from flax. J. Am. Oil Chem. Soc. 1990, 67, 79–84. [CrossRef]
7. Gajardo, II.A.; Quian, R.; Soto-Cerda, B. Agronomic and quality assessment of linseed advanced breeding lines varying in seed mucilage content and their use for food and feed. Crop Sci. 2017, 57, 2979–2990. [CrossRef]
8. Cherian, G.; Quezada, N. Egg quality, fatty acid composition and immunoglobulin Y content in eggs from laying hens fed full fat camelina or flax seed. J. Anim. Sci. Biotechnol. 2016, 7, 15. [CrossRef] [PubMed]
9. Oomah, B.D.; Mazza, G. Processing of flaxseed meal: Effect of solvent extraction on physicochemical characteristics. LWT-Food Sci. Technol. 1993, 26, 312–317. [CrossRef]
10. Sosulski, F.W.; Bakal, A. Isolated proteins from rapeseed, flax and sunflower meals. Can. Inst. Food Sci. Technol. J. 1969, 2, 28–32. [CrossRef]
11. Kessler, R.W.; Kohler, R. New strategies for exploiting flax and hemp. Chemtech 1996, 26, 34–42.
12. Sosulski, F.; Zadernowski, R. Fractionation of rapeseed meal into hour and hull components. J. Am. Oil Chem. Soc. 1981, 58, 96–98. [CrossRef]
13. Daun, J.K.; DeClercq, D.R. Quality of yellow and dark seeds in Brassica campentris canola varieties Candle and Tobin. J. Am. Oil Chem. Soc. 1988, 65, 122–126. [CrossRef]
14. Francoz, E.; Ranocha, P.; Burlat, V.; Dunand, C. Arabidopsis seed mucilage secretory cells: Regulation and dynamics. Trends Plant Sci. 2015, 8, 515–524. [CrossRef] [PubMed]
15. Venglat, P.; Xiang, D.; Qiu, S.; Stone, S.L.; Tibiche, C.; Cram, D.; Alting-Mees, M.; Nowak, J.; Cloutier, S.; Deyholos, M.; et al. Gene expression analysis of flax seed development. BMC Plant Biol. 2011, 11, 74. [CrossRef] [PubMed]
16. Oomah, B.D.; Kenaschuk, E.O.; Cui, W.; Mazza, G. Variation in the composition of water-soluble polysaccharides in flaxseed. J. Agric. Food Chem. 1995, 43, 1484–1488. [CrossRef]

17. Oomah, B.D.; Mazza, G. Effect of dehulling on chemical composition and physical properties of flaxseed. LWT-Food Sci. Technol. 1997, 30, 135–140. [CrossRef]
18. Diederichsen, A.; Raney, J.P.; Duguid, S.D. Variation of mucilage in flax seed and its relationship with other seed characters. Crop Sci. 2006, 46, 365–371. [CrossRef]
19. Spielmeyer, W.; Green, A.G.; Bittisnish, D.; Mendham, N.; Lagudah, E.S. Identification of quantitative trait loci contributing to Fusarium wilt resistance on an AFLP linkage map of flax (Linum usitatissimum). Theor. Appl. Genet. 1998, 97, 633–641. [CrossRef]
20. Asgarinia, P.; Cloutier, S.; Duguid, S.; Rashid, K.; Mirlohi, A.F.; Banik, M.; Saeidi, G. Mapping quantitative trait loci for powdery mildew resistance in flax (Linum usitatissimum L.). Crop Sci. 2013, 53, 2462–2472. [CrossRef]
21. Cloutier, S.; Ragupathy, R.; Niu, Z.; Duguid, S. SSR-based linkage map of flax (Linum usitatissimum L.) and mapping of QTL underlying fatty acid composition traits. Mol. Breed. 2011, 28, 437–451. [CrossRef]
22. Kumar, S.; You, F.M.; Duguid, S.; Booker, H.; Rowland, G.; Cloutier, S. QTL for fatty acid composition and yield in linseed (Linum usitatissimum L.). Theor. Appl. Genet. 2015, 128, 965–984. [CrossRef] [PubMed]
23. Sudarshan, G.P.; Kulkarni, M.; Akhov, L.; Ashe, P.; Shaterian, H.; Cloutier, S.; Rowland, G.; Wei, Y.; Selvaraj, G. QTL mapping and molecular characterization of the classical D locus controlling seed and flower color in Linum usitatissimum (flax). Sci. Rep. 2017, 7, 15751. [CrossRef] [PubMed]
24. Soto-Cerda, B.J.; Duguid, S.; Booker, H.; Rowland, G.; Diederichsen, A.; Cloutier, S. Genomic regions underlying agronomic traits in linseed (Linum usitatissimum L.) as revealed by association mapping. J. Integr. Plant Biol. 2014, 56, 75–87. [CrossRef] [PubMed]
25. Soto-Cerda, B.J.; Duguid, S.; Booker, H.; Rowland, G.; Diederichsen, A.; Cloutier, S. Association mapping of seed quality traits using the Canadian flax (Linum usitatissimum L.) core collection. Theor. Appl. Genet. 2014, 127, 881–896. [CrossRef] [PubMed]
26. Xie, D.; Dai, Z.; Yang, Z.; Sun, J.; Zhao, D.; Yang, X.; Zhang, L.; Tang, Q.; Su, J. Genome-wide association study identifying candidate genes influencing important agronomic traits of flax (Linum usitatissimum L.) using SLAF-seq. Front. Plant Sci. 2018, 8, 2232. [CrossRef] [PubMed]
27. You, F.M.; Xiao, J.; Li, P.; Yao, Z.; Jia, G.; He, L.; Kumar, S.; Soto-Cerda, B.; Duguid, S.D.; Booker, H.M.; et al. Genome-Wide Association Study and Selection

Signatures Detect Genomic Regions Associated with Seed Yield and Oil Quality in Flax. Int. J. Mol. Sci. 2018, 19, 2303. [CrossRef] [PubMed]

28. Ersoz, E.S.; Yu, J.; Buckler, E.S. Applications of linkage disequilibrium and association mapping in maize. In Molecular Genetic Approaches to Maize Improvement; Kriz, A., Larkins, B., Eds.; Springer: Berlin, Germany, 2009; pp. 173–195.
29. Cloutier, S.; Niu, Z.; Datla, R.; Duguid, S. Development and analysis of EST-SSRs for flax (Linum usitatissimum L.). Theor. Appl. Genet. 2009, 119, 53–63. [CrossRef] [PubMed]
30. Cloutier, S.; Miranda, E.; Ward, K.; Radovanovic, N.; Reimer, E.; Walichnowski, A.; Datla, R.; Rowland, G.; Duguid, S.; Ragupathy, R. Simple sequence repeat marker development from bacterial artificial chromosome end sequences and expressed sequence tags of flax (Linum usitatissimum L.). Theor. Appl. Genet. 2012, 125, 685–694. [CrossRef] [PubMed]
31. Cloutier, S.; Ragupathy, R.; Miranda, E.; Radovanovic, N.; Reimer, E.; Walichnowski, A.; Ward, K.; Rowland, G.; Duguid, S.; Banik, M. Integrated consensus genetic and physical maps of flax (Linum usitatissimum L.). Theor. Appl. Genet. 2012, 125, 1783–1795. [CrossRef] [PubMed]
32. Ragupathy, R.; Rathinavelu, R.; Cloutier, S. Physical mapping and BAC-end sequence analysis provide initial insights into the flax (Linum usitatissimum L.) genome. BMC Genom. 2011, 12, 217. [CrossRef] [PubMed]
33. Altunkaya, A. Dermal Lubricant and Moisturizer. WO 2006/075236 A1, 20 July 2006.
34. Anttila, M.; Kankaanpää-Anttila, B.; Sepponen, M.; Timonen, H.; Autio, K. Improving of Texture of Dairy Products. WO 2008/000913 A1, 3 January 2008.
35. Kracht, W.; Dänicke, S.; Kluge, H.; Keller, K.; Matzke, W.; Henning, U.; Schumann, W. Effect of dehulling of rapeseed on feed value and nutrient digestibility of rape products in pigs. Arch. Anim. Nutr. 2004, 58, 389–404. [CrossRef] [PubMed]
36. Oomah, B.D.; Mazza, G. Fractionation of flaxseed with a batch dehuller. Ind. Crop Prod. 1998, 9, 19–27. [CrossRef]
37. Yan, X.Y.; Li, J.N.; Fu, F.Y.; Jin, M.Y.; Chen, L.; Liu, L.Z. Co-location of seed oil content, seed hull content and seed coat color QTL in three different environments in Brassica napus L. Euphytica 2009, 170, 355–364. [CrossRef]
38. Saedi, G.; Rowland, G.G. Seed color and linolenic acid effects on agronomic traits in flax. Can. J. Plant Sci. 1999, 79, 521–526. [CrossRef]

39. Zhao, K.; Tung, C.W.; Eizenga, G.C.; Wright, M.H.; Ali, M.L.; Price, A.H.; Norton, G.J.; Islam, M.R.; Reynolds, A.; Mezey, J.; et al. Genome-wide association mapping reveals a rich genetic architecture of complex traits in Oryza sativa. Nat. Commun. 2011, 2, 467. [CrossRef] [PubMed]

40. Smýkal, P.; Bacˇová-Kerteszová, N.; Kalendar, R.; Corander, J.; Schulman, A.H.; Pavelek, M. Genetic diversity of cultivated flax (Linum usitatissimum L.) germplasm assessed by retrotransposon-based markers. Theor. Appl. Genet. 2011, 122, 1385–1397. [CrossRef] [PubMed]

41. Soto-Cerda, B.J.; Diederichsen, A.; Ragupathy, R.; Cloutier, S. Genetic characterization of a core collection of flax (Linum usitatissimum L.) suitable for association mapping studies and evidence of divergent selection between fiber and linseed types. BMC Plant Biol. 2013, 13, 78. [CrossRef] [PubMed]

42. Chandrawati, N.S.; Kumar, R.; Kumar, S.; Singh, P.K.; Yadav, V.K.; Ranade, S.A.; Yadav, H.K. Genetic diversity, population structure and association analysis in linseed (Linum usitatissimum L.). Physiol. Mol. Biol. Plants 2017, 23, 207–219. [CrossRef] [PubMed]

43. Abdurakhmonov, I.; Abdukarimov, A. Application of association mapping to understanding the genetic diversity of plant germplasm resources. Int. J. Plant Genom. 2008, 2008, 574297. [CrossRef] [PubMed]

44. Xu, J.; Ranc, N.; Munos, S.; Rolland, S.; Bouchet, J.P.; Desplant, N.; Le Paslier, M.C.; Liang, Y.; Brunel, D.; Causse, M. Phenotypic diversity and association mapping for fruit quality traits in cultivated tomato and related species. Theor. Appl. Genet. 2013, 126, 567–581. [CrossRef] [PubMed]

45. Jung, M.; Ching, A.; Bhattramakki, D.; Dolan, M.; Tingey, S.; Morgante, M.; Rafalski, A. Linkage disequilibrium and sequence diversity in a 500-kbp region around the adh1 locus in elite maize germplasm. Theor. Appl. Genet. 2004, 109, 681–689. [CrossRef] [PubMed]

46. Hatzig, S.V.; Frisch, M.; Breuer, F.; Nesi, N.; Ducournau, S.; Wagner, M.H.; Leckband, G.; Abbadi, A.; Snowdon, R.J. Genome-wide association mapping unravels the genetic control of seed germination and vigor in Brassica napus. Front. Plant Sci. 2015, 6, 221. [CrossRef] [PubMed]

47. Pritchard, J.K.; Stephens, M.; Rosenberg, N.A.; Donnelly, P. Association mapping in structured populations. Am. J. Hum. Genet. 2000, 67, 170–181. [CrossRef] [PubMed]

48. Price, A.L.; Patterson, N.J.; Plenge, R.M.; Weinblatt, M.E.; Shadick, N.A.; Reich, D. Principal components analysis corrects for stratification in genome-wide association studies. Nat. Genet. 2006, 38, 904–909. [CrossRef] [PubMed]

49. Yu, J.; Pressoir, G.; Briggs, W.; Vroh, B.; Yamasaki, M.; Doebley, J.; McMullen, M.D.; Gaut, B.S.; Nielsen, D.M.; Holland, J.B.; et al. A unified mixed-model method for association mapping that accounts for multiple levels of relatedness. Nat. Genet. 2006, 38, 203–208. [CrossRef] [PubMed]

50. Western, T.L.; Young, D.S.; Dean, G.H.; Tan, W.L.; Samuels, A.L.; Haughn, G.W. MUCILAGE-MODIFIED4 encondes a putative pectin biosynthetic enzyme developmentally regulated by APETALA2, TRANSPARENT TESTA GLABRA1, and GLABRA2 in the Arabidopsis seed coat. Plant Physiol. 2004, 134, 296–306. [CrossRef] [PubMed]

51. Kong, Y.; Zhou, G.; Abdeen, A.A.; Schafhauser, J.; Richardson, B.; Atmodjo, M.A.; Jung, J.; Wicker, L.; Mohnen, D.; Western, T.; Hahn, M.G. GALACTURONOSYLTRANSFERASE-LIKE5 is involved in the production of Arabidopsis seed coat mucilage. Plant Physiol. 2013, 163, 1203–1217. [CrossRef] [PubMed]

52. Rautengarten, C.; Usadel, B.; Neumetzler, L.; Hartmann, J.; Büssis, D.; Altmann, T. A subtilisin-like serine protease essential for mucilage release from Arabidopsis seed coats. Plant J. 2008, 54, 466–480. [CrossRef] [PubMed]

53. Saez-Aguayo, S.; Ralet, M.C.; Berger, A.; Botran, L.; Ropartz, D.; Marion-Poll, A.; North, H.M. PECTIN METHYLESTERASE INHIBITOR6 promotes Arabidopsis mucilage release by limiting methylesterification of homogalacturonan in seed coat epidermal cells. Plant Cell 2013, 25, 308–323. [CrossRef] [PubMed]

54. Louvct, R.; Cavel, E.; Gutierrez, L.; Guénin, S.; Roger, D.; Gillet, F.; Guerineau, F.; Pelloux, J. Comprehensive expression profiling of the pectin methylesterase gene family during silique development in Arabidopsis thaliana. Planta 2006, 224, 782–791. [CrossRef] [PubMed]

55. Shi, L.; Katavic, V.; Yu, Y.; Kunst, L.; Haughn, G. Arabidopsis glabra2 mutant seeds deficient in mucilage biosynthesis produce more oil. Plant J. 2012, 69, 37–46. [CrossRef] [PubMed]

56. Wang, R.; Li, J.N.; Chen, L.; Tang, Z.L.; Zhang, X.K. Genetic correlation analysis for main characters in yellow-seeded rapeseed lines (Brassica napus L.). Chin. J. Oil Crop Sci. 2003, 25, 8–11.

57. Khan, N.A.; Booker, H.; Yu, P. Molecular structures and metabolic characteristics of protein in brown and yellow flaxseed with altered nutrient traits. J. Agri. Food Chem. 2014, 62, 6556–6564. [CrossRef] [PubMed]

58. Qu, C.; Hasan, M.; Lu, K.; Liu, L.; Zhang, K.; Fu, F.; Wang, M.; Liu, S.; Bu, H.; Wang, R.; et al. Identification of QTL for seed coat colour and oil content in Brassica napus by association mapping using SSR markers. Can. J. Plant Sci. 2015, 95, 387–395. [CrossRef]

59. Badani, A.G.; Snowdon, R.J.; Wittkop, B.; Lipsa, F.D.; Baetzel, R.; Horn, R.; De Haro, A.; Font, R.; Lühs, W.; Friedt, W. Colocalization of a partially dominant gene for yellow seed colour with a major QTL influencing acid detergent fibre (ADF) content in different crosses of oilseed rape (Brassica napus). Genome 2006, 49, 1499–1509. [CrossRef] [PubMed]
60. Zhang, D.; Sun, L.; Li, S.; Wang, W.; Ding, Y.; Swarm, S.A.; Li, L.; Wang, X.; Tang, X.; Zhang, Z.; et al. Elevation of soybean seed oil content through selection for seed coat shininess. Nat. Plants 2018, 4, 30–35. [CrossRef] [PubMed]
61. Roszak, P.; Köhler, C. Polycomb group proteins are required to couple seed coat initiation to fertilization. Proc. Natl. Acad. Sci. USA 2011, 108, 20826–20831. [CrossRef] [PubMed]
62. Mizzotti, C.; Ezquer, I.; Paolo, D.; Rueda-Romero, P.; Guerra, R.F.; Battaglia, R.; Rogachev, I.; Aharoni, A.; Kater, M.M.; Caporali, E.; et al. SEEDSTICK is a master regulator of development and metabolism in the Arabidopsis seed coat. PLoS Genet. 2014, 10, e1004856. [CrossRef] [PubMed]
63. Marinova, K.; Pourcel, L.; Weder, B.; Schwarz, M.; Barron, D.; Routaboul, J.M.; Debeaujon, I.; Klein, M. The Arabidopsis MATE transporter TT12 acts as a vacuolar flavonoid/H+-antiporter active in proanthocyanidin-accumulating cells of the seed coat. Plant Cell 2007, 19, 2023–2038. [CrossRef] [PubMed]
64. Kovinich, N.; Saleem, A.; Arnason, J.T.; Miki, B. Functional characterization of a UDP-glucose:flavonoid 3-O-glucosyltransferase from the seed coat of black soybean (Glycine max (L.) Merr.). Phytochemistry 2010, 71, 1253–1263. [CrossRef] [PubMed]
65. Xu, W.; Dubos, C.; Lepiniec, L. Transcriptional control of flavonoid biosynthesis by MYB–bHLH–WDR complexes. Trends Plant Sci. 2015, 20, 176–185. [CrossRef] [PubMed]
66. Mole, S.; Waterman, P.G. Tannic acid and proteolytic enzymes: Enzyme inhibition or substrate deprivation? Phytochemistry 1986, 26, 99–102. [CrossRef]
67. Diederichsen, A.; Kusters, P.M.; Kessler, D.; Bainas, Z.; Gugel, R.K. Assembling a core collection from the flax world collection maintained by Plant Gene Resources of Canada. Genet. Resour. Crop Evol. 2013, 60, 1479–1485. [CrossRef]
68. VSN International. Genstat for Windows, 18th ed.; VSN International: Hemel Hempstead, UK, 2015; Available online: http://www.Genstat.co.uk (accessed on 10 May 2015).
69. Korkmaz, S.; Goksuluk, D.; Zararsiz, G. MVN: An R Package for Assessing Multivariate Normality. R J. 2014, 6, 151–162.

70. Bradbury, P.J.; Zhang, Z.; Kroon, D.E.; Casstevens, T.M.; Ramdoss, Y.; Buckler, E.S. TASSEL: Software for association mapping of complex traits in diverse samples. Bioinformatics 2007, 23, 2633–2635. [CrossRef] [PubMed]

71. Wang, Z.; Hobson, N.; Galindo, L.; Zhu, S.; Shi, D.; McDill, J.; Yang, L.; Hawkins, S.; Neutelings, G.; Datla, R.; et al. The genome of flax (Linum usitatissimum) assembled de novo from short shotgun sequence reads. Plant J. 2012, 72, 461–473. [CrossRef] [PubMed]

72. You, F.M.; Xiao, J.; Li, P.; Yao, Z.; Jia, G.; He, L.; Zhu, T.; Luo, M.C.; Wang, X.; Deyholos, M.K.; et al. Chromosome-scale pseudomolecules refined by optical, physical and genetic maps in flax. Plant J. 2018, 95, 371–384. [CrossRef] [PubMed]

73. Li, H.; Durbin, R. Fast and accurate short read alignment with Burrows-Wheeler Transform. Bioinformatics 2009, 25, 1754–1760. [CrossRef] [PubMed]

74. Li, H.; Handsaker, B.; Wysoker, A.; Fennell, T.; Ruan, J.; Homer, N.; Marth, G.; Abecasis, G.; Durbin, R. The sequence alignment/map format and SAMtools. Bioinformatics 2009, 25, 2078–2079. [CrossRef] [PubMed]

75. Kumar, S.; You, F.M.; Cloutier, S. Genome wide SNP discovery in flax through next generation sequencing of reduced representation libraries. BMC Genom. 2012, 13, 684. [CrossRef] [PubMed]

76. You, F.M.; Deal, K.R.; Wang, J.; Britton, M.T.; Fass, J.N.; Lin, D.; Dandekar, A.M.; Leslie, C.A.; Aradhya, M.; Luo, M.C.; et al. Genome-wide SNP discovery in walnut with an AGSNP pipeline updated for SNP discovery in allogamous organisms. BMC Genom. 2012, 13, 354. [CrossRef] [PubMed]

77. Evanno, G.; Regnaut, S.; Goudet, J. Detecting the number of clusters of individuals using the software STRUCTURE: A simulation study. Mol. Ecol. 2005, 14, 2611–2620. [CrossRef] [PubMed]

78. Francis, M.R. POPHELPER: An R package and web app to analyse and visualise population structure. Mol. Ecol. Resour. 2017, 1, 27–32. [CrossRef] [PubMed]

79. Hill, W.G.; Weir, B.S. Variances and covariances of squared linkage disequilibria in finite populations. Theor. Popul. Biol. 1998, 33, 54–78. [CrossRef]

80. Turner, S.D. QQMAN: An R package for visualizing GWAS results using Q-Q and manhattan plots. biorXiv 2014. [CrossRef]

Chapter 4

COMPUTATIONAL AND EXPERIMENTAL INSTRUMENTS FOR MONITORING GENOME-WIDE VARIATIONS IN PLANT MRNA TRANSLATION EFFICIENCY: BENEFITS, CONSTRAINTS, AND REMEDIES

1. INTRODUCTION: AN OVERVIEW

In plants, as in other eukaryotes, genetic information is executed by a sequential sequence of biological activities, with transcription and translation as the principal operations. Contemporary experimental omics technologies for genomic monitoring of plant gene expression facilitate the tracing of genetic information from the genome to the proteome and subsequently to the metabolome. Novel experimental methodologies, such as RNA-Seq and DNA microarrays, have elucidated numerous fundamental mechanisms governing transcription regulation in plants, which represents the initial phase of gene expression and is the most accessible for experimental investigation. Transcriptome studies, which involve the qualitative and quantitative assessment of gene expression across the entire genome, have provided compelling evidence of dynamic alterations in the transcriptomes of diverse plant species during growth, development, and in response to environmental factors. Comparative omics studies in plants unequivocally reveal a minimal association between transcription levels (the abundance of specific mRNAs) and translation levels (the quantities of matching proteins in the proteome). It is noteworthy that variations in transcript levels do not invariably result in alterations in the corresponding protein levels [1]. This indicates a complex nature of the mechanisms facilitating genome decoding, which encompass not only differential transcription but also differential translation.

Translation is a complex biological process with numerous players, including mRNAs, tRNAs, ribosomes, and manifold protein factors. Undoubtedly, each is important for efficient translation. The mRNAs themselves comprise different regions, namely, the 5' untranslated region (5'UTR) and coding region (CDS) and 3' untranslated region (3'UTR), which modulate translation at a number of "checkpoints": translation initiation, elongation, and termination. In the current view, numerous regulatory elements may be concealed in the nucleotide contexts of these mRNA regions and each of them individually or in combination can determine the development of any transcript in translational process [2].

The paradox of misfit between the levels of mRNAs and their protein products observable in different plant species at all stages of their growth and development as well as upon the impact of various environmental factors focuses the attention of researchers on two key problems, namely detection of the specific sets of differentially-translated transcripts, i.e., the sets of transcripts that are efficiently translated under certain conditions, and the sets of transcripts with repressed or unchanged translation under the same conditions and (ii) clarification of the particular regions or specific structural features of the mRNA nucleotide composition that mediate this differential translational control.

This review focuses on the experimental methods for genome-wide analysis of translational control, computational algorithms to search and analyze various regulatory contexts within mRNAs, and approaches for subsequent experimental verification of their correlation with mRNA translation in plants. Currently, we cannot refer to deficiency in publications comprehensively reporting the basic protocols of various methods for genome-wide analyses of translational control in general, including the methods applicable to plant objects. However, reviews that consider and discuss the three key components of the general strategy for identification of regulatory contexts in mRNA that may play a key role in differential translation are still absent in the scientific literature. Our goals here are (i) to consider the experimental approaches aiming to clarify differential translation on a plant genome-wide scale; (ii) to summarize the current data on the computational algorithms used for detection of the specific structural and functional features of key determinants within plant mRNAs and their interrelation with the translation efficiency; (iii) to highlight the methods for experimental verification of existed data and theoretical predictions of the intrinsic features of plant mRNAs important for their differential translation; and (iv) to discuss the ways of decoding the specific structural features of plant mRNA that mediate differential translational control by combining computational and experimental approaches. This review delineates the essential and pivotal phases for each approach within this overarching strategy, the domains of their application, and the principal outcomes derived from plant subjects, as well as their contribution to our understanding of the intricate mechanisms of translation in plants.

2.EXPERIMENTALAPPROACHESTODETERMINEDIFFERENTIALLY-TRANSLATED MRNAS IN PLANTS

Initially, proteomics methods were used to identify the correlation between the observed fluctuations in the expression of a transcript and the actual level of peptides in plants [3]. However, the proteomics approaches have certain limitations in the case of a spatiotemporal study of a large pool of translated mRNAs and are mainly applied for assessing translation of the known peptides and proteins. Moreover, the methods of proteomics are laborious and expensive, while preparation of the specimens, quantification of proteome, and subsequent peptide sequencing require specialized technical experience [4]. Advances in high-throughput technologies, such as microarrays and deep sequencing, have made it possible to develop the new experimental approaches to studying mRNA translation efficiency on a global scale. Three basic experimental approaches are currently used for these purposes: (a) polysome profiling; (b) translating ribosome affinity purification (TRAP); and ribosome profiling or Ribo-Seq. These approaches are based on (i) the production of the mRNA pool with the ribosomes arrested on them; (ii) separation of actively-translated mRNAs (polysomal mRNAs and mRNAs bound to several ribosomes), moderately translated mRNAs (monosomal mRNAs and mRNAs bound to one ribosome), and untranslated mRNAs (steady-state mRNAs that are not bound to ribosomes); and (iii) subsequent quantitative assessment of an individual transcript or an mRNA population represented in polysomal complexes relative to the total amount of the transcript in the assayed plant specimens. Note that polysomes are several ribosomes performing translation from one mRNA and this process is regulated for individual mRNAs.

2.1 Profiling Polysomes

The translational status of the mRNA pool on a genome-wide scale can be estimated using polysome profiling. The basic protocols for the polysome profiling in plants are described in several publications [5]. Simple protocols have been additionally designed and verified for individual plant species, including Arabidopsis thaliana, Nicotiana benthamiana, Solanum lycopersicum, and Oryza sativa, as well as for individual plant tissues [5]. This method is based on the separation of the polysomal mRNAs, monosomal mRNAs, and steady-state mRNAs using sucrose density gradient centrifugation, referred to as polysome fractionation assays (Figure 1). Then, the transcripts (mRNAs) associated with each mRNA pool are analyzed by hybridization on microarrays or undergo RNA sequencing. Assembly, mapping, and in silico analysis of the sequencing data for different pools (polysomes, monosomes, and steady-state mRNAs) provide the researcher with the initial lists of the transcripts with different translation activities [6,7].

According to the experimental data, the results of polysome profiling can be used for a quantitative estimation of mRNA translation efficiency both at different plant growth and developmental stages and under the effect of adverse environmental factors [6,8] or for assessment of quantitative changes in the translational status of individual mRNAs [9]. As a rule, the polysome score (PS) or polysome ratio (PR) are used for this purpose; they are computed as the relative abundances of RNAs in polysomes versus RNAs in nonpolysomes or versus total mRNA. Where total mRNA is the total mRNA level in polysomal and nonpolysomal fractions, respectively [6,8,9].

The polysome profiling appeared rather efficient in the studies on differential translation regulation of specific plant mRNAs under the influence of several abiotic environmental factors [2,6,10]. For example, it has been convincingly demonstrated that the main part of the transcripts under stress displays different degrees of translation repression; moreover, a specific set of transcripts that avoids such repression and retain their transcriptional activity was detected. Below are several examples that in our view, illustrate the abilities of this method in clarifying the mechanisms of translation control in plants. In particular, it is shown that the shares of individual mRNA species in A. thaliana polysomal fractions under controlled growth upon a moderate dehydration stress vary from 5% to 95% and that this stress causes a decrease in the ribosome load for over 60% of all mRNAs [2]. The results of genome-wide assay of the relative amounts of individual mRNAs in polysomal versus nonpolysomal fractions under heat shock in the A. thaliana cell culture gave the set of genes with different translational responses, i.e., the genes that either considerably increased or considerably decreased the amounts of their mRNAs in polysomal fractions [10]. These results formed the background for further identification of the new cis-regulatory elements in 5'UTRs that influenced differential translation in response to heat shock in A. thaliana [8].

In another study, polysome profiling was used for a global assessment of the translation efficiency of mRNA pools during the growth and development of A. thaliana leaves. It was demonstrated that the degree of association of each mRNA with the polysomal fraction was different and considerably (from a strong repression to activation at a constant level) changed throughout these processes. Analysis of the functional categories of the mRNAs associated with polysomal fraction showed that the translation control, being of physiological significance during plant growth and development, was especially pronounced in the mRNAs associated with signaling and protein synthesis. In general, these results emphasize the importance of the dynamic changes in mRNA translation during plant growth and development and suggest that mRNA translation may be controlled via complex mechanisms underlying the response to each factor [6].

Although polysome profiling has been successfully used for a global study of plant mRNA translation efficiency, this method still has some limitations [11]. One of these, it cannot precisely determine the ribosome density, i.e., the number of ribosomes per mRNA, because the mRNA–ribosome complexes from the same differential centrifugation fractions may contain a different number of ribosomes. Moreover, polysome profiling fails to determine the actual ribosome distribution along the transcript, i.e., it is impossible to determine a mRNA region (5'UTR, CDS, or 3'UTR) in which reside the arrested ribosomes. This is very important since it allows for assessing of the translation stage (initiation, elongation, or termination) associated with differential translation of an individual transcript. As a consequence, this makes it not possible to specifically search for the regulatory determinants in particular mRNA regions important for an efficient translation.

Nonetheless, these limitations of the polysome profiling technique do not diminish its tremendous potential for the study of the fine mechanisms of translation in plants on a global scale. This method not only makes it possible to determine the correlations between the observed translational and transcriptional fluctuations under normal conditions and under stress factors, but also provides researchers with general information useful for further insights into the rules of mRNA decoding, i.e., allows defining the pools of transcripts with different translation efficiency and to find regulatory contexts of mRNAs or their combinations important for translation efficiency using computational analysis (this will be considered below in more detail). According to the available experimental data, polysome profiling is, as a rule, applicable to the search for actively-translated mRNAs and the subsequent analysis, although the understanding of the mechanisms associated with the repression of translation in a certain pool of transcripts is of the same importance; perhaps, researchers will focus on this area in future. It should be also emphasized that most studies utilizing polysome profiling performed so far, involve the plants with annotated genomes. However, the use of this method is not limited to the plant species with annotated genomes and can be extended to other plant species, including those genomes that have not been yet determined or those already sequenced but poorly annotated.

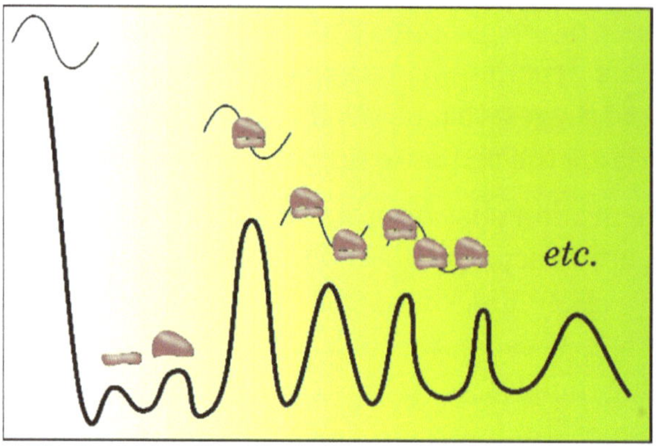

Figure 1. Polysome profiling in a sucrose density gradient. Separation of the transcripts depending on the ribosome loading: the first peak corresponds to the mRNAs unbound to ribosomes; second and third peaks, to the ribosome small and large subunits, respectively; and the fourth and subsequent peaks, to the mRNAs with different ribosome loadings.

2.2 Translating Ribosome Affinity Purification (TRAP) and TRAP-Seq

The experimental approach referred to as translating ribosome affinity purification (TRAP) is a modification of the traditional polysome profiling procedure and was for the first time described for A. thaliana [12]. This method utilizes the plant transgenic lines that express an epitope-tagged variant of ribosomal protein L18 (usually referred to as RPL18). As a rule, these plant transgenic lines express FLAG epitope-tagged RPL18 in the N-terminal region [12,13]. The cell lysates of transgenic plants are produced under the conditions that stabilize the ribosomes on RNA and block translation. The transcripts bound to the ribosomes that carry the labeled RPL18 are selectively separated using the absorption on anti-FLAG-M2 agarose. This enables ribosome capture from crude cell extracts by a single-stage immune precipitation (Figure 2) and, as a rule, allows the pool of RNAs (designated as TRAP RNA) that are actively translated to be obtained.

This method is described and discussed in detail in several papers [3,13–15]. Note that both the traditional polysome profiling approach and TRAP give analogous proportions of the small and large polysomes (i.e., ribosome profiles) [13]. Both approaches also have similar limitations on their application, namely, in the assessment of the number of ribosomes per mRNA length and the distribution of ribosomes along the transcript (see above). However, note that a wide use of the experimental TRAP approach is also limited by the available plant transgenic lines but, nonetheless, the use of transgenic lines gives certain advantages as compared with the traditional polysome profiling. This advantage consists in the possibility of not only constitutive, but also tissue-specific RPL18 expression by using different tissue-specific promoters [14]. Thanks to the tissue-specific RPL18 expression, TRAP is applicable to profiling of actively-translated RNAs in different populations of plant cells, namely, in (i) different root cells (epidermis, cortex, or endodermis); (ii) companion phloem cells, meristem cells, and leaf mesophyll cells; and (iii) microspores, pollen, and other plant tissues and cell types [14]. For example, the use of APETALA1, APETALA3, and AGAMOUS for expression of FLAG-RPL18 in early flower development allowed for the discovery of new levels of the expression control in developing flowers associated with differential translation [16]. A systemic analysis of the mRNAs in different specimens relative to the pollen grains within buds and in vitro-germinated pollen tubes has been performed with the help of the A. thaliana transgenic lines expressing epitope-tagged RPL18 under the control of ProLAT52 promoter, which allowed for the identification of a cohort of the transcripts that regulate late stages of pollination in flowering plants; this paves the way for better understanding of the pollen-based mechanisms that promote fertilization [15]. It should be emphasized that the in vivo proteomic studies of pollen tubes are extremely complicated because of the difficulties with pollen collection; the selective immune purification of the transcripts associated with the polysomes in pollen tubes in this case assisted in identification of the genes important for the in vivo pollen biology. Thus, the TRAP approach has an important advantage for efficient isolation of the population of mRNA complexes from particular cell types.

The sensitive moment when using TRAP approach is during the selection of the transgenic line that expresses FLAG-RPL18, which is extremely important for a successful analysis of the tissue-specific responses. A position effect associated with the T-DNA integration site in the genome of transgenic plants is known. In this regard, the new transgenic lines intended for this research should be selected bearing in mind the presence of known tissue-specific genes in the corresponding tissues or cell types. This will ensure selection of the most appropriate line for further analysis.

According to the current opinion, not only stable plant transformants, but also a transient expression of FLAG epitope-tagged RPL18 can be used for identification of the differentially-translated mRNA pools in plant genomes, for example, utilizing the agroinfiltration of Medicago truncatula hairy root cultures or of N. benthamiana leaves by Agrobacterium rhizogenes.

The FLAG tag may be also added to other proteins in order to determine their role in translation. For example, the expression of tagged oligouridylate binding protein 1 (UBP1) with subsequent immune purification of the mRNA–protein complexes (mRNPs) clarified the role of this protein in the dynamic and reversible aggregation of translationally repressed mRNAs in hypoxia [17]. In particular, UBP1 constitutively binds a subpopulation of the mRNAs with the 3'UTRs enriched for uracil under normoxic conditions. In hypoxia, UBP1 is associated with non-uracil-rich mRNAs, which increases its aggregation in microscopically-visible cytoplasmic foci, referred to as UBP1 stress granules (SGs). This UBP1–mRNA association leads to a global decrease in the protein synthesis. The translation limitation for the transcripts associated into SGs reduces the energy spending, thereby determining the priority in synthesis of the proteins that enhance plant survival in stress. The UBP1 SGs rapidly disaggregate during reoxygenation, which coincides with the mRNA return to polysomes. In this process, the mRNAs that are significantly induced and translated in hypoxia to a considerable degree manage to avoid UBP1 sequestration. Thus, it has been shown that the SG-nucleating RNA-binding UBP1 is a component of the mechanism that post-translationally reprograms plant gene expression, thereby enhancing plant survival in hypoxia [17].

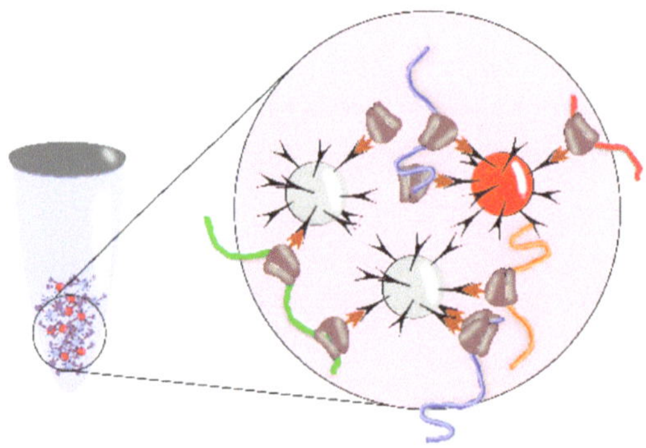

Figure 2. Polysome profiling using translating ribosome affinity purification (TRAP). General principle of selective separation on anti-FLAG-M2 agarose of the transcripts bound to ribosomes carrying the epitope-tagged variant of ribosomal protein. Brown arrows denote the FLAG epitope in ribosomal protein and black icons denote the anti-FLAG on agarose beads.

2.3 Ribosome Profiling, or Ribo-Seq

Ribosome Profiling (RP), or Ribo-Seq, elaborated by Ingolia, Newman, and Weissman in 2009 [18], is based on the isolation and sequencing of the mRNA fragments protected by ribosome. This gives a "snapshot" of the ribosome positions along mRNA on a genome-wide scale, i.e., gives the possibility to determine both the number and positions of the ribosomes in the mRNA coding region in vivo (Figure 3).

As a rule, many studies use the RP experimental protocol, which comprises five interrelated stages: (i) preparation of RNA specimens with the arrested ribosomes; (ii) controlled hydrolysis of these specimens by RNase to generate small RNA fragments associated with a ribosome (referred to as footprints); (iii) their subsequent isolation; (iv) preparation of purified footprints with a size of 28–30 nucleotides; and (v) construction of the library and its high-throughput sequencing, as a rule, with the help of short-read sequencers. The deep sequencing reads of the footprints are analyzed using bioinformatics methods and the translation efficiency is derived by normalizing the number of reads of the footprints to the number of reads of the total transcriptome by RNA-Seq.

As is mentioned above, the first experimental protocol for ribosome profiling was described in 2009 [18] and has been constantly developed, in particular, for its application to different organisms [18,19], including plants [20,21] and plant organelles—chloroplasts [20] and mitochondria [22]. The individual protocols differ in the particular details providing optimization of each of the five interrelated stages, including the differences in tissue and cell processing; pH and composition of the buffer for cell lysis; prepurification of polysomes before RNase hydrolysis (done or omitted); type of RNase used for generating monosomes [23]; and the methods used to purify the monosome fractions and construct sequencing libraries. Ribosome affinity purification (TRAP method), including the tissue-specific purification, can be also used as the starting point for ribosome profiling [20].

In general, the RP results allow for determination of the precise positions of the translating ribosomes on mRNA with an unprecedented resolution, to a single nucleotide. The specialized software for analysis, interpretation, and visualization of RP data is currently available (for detailed review, see [24]). By assessing the relative number and location of ribosomes on mRNA, the researcher can estimate the general translation pattern i.e., to assess the translation efficiency, which is calculated as the ratio of translation (the data on the number of footprint reads in individual mRNA) to transcription (RNA-Seq data at the level of individual mRNA) (Figure 4). Note that it is possible not only to directly quantify the mRNAs that will be translated into proteins, but also to detect the new types of contexts in the plant mRNAs associated with translation, for example, uORFs (upstream ORFs) and frameshifts; to precisely determine the translation initiation site (TIS) of the main ORF; and to find new translated ORFs, including those residing in intergenic RNAs or putative noncoding short RNAs (ncRNAs) (Figure 4) [3,24,25]. The researcher gets these additional options thanks to the fact that the 80S ribosomes associate only with the portion of the transcript that will be most likely decoded into the protein product. The 80S ribosome and transcript will associate not only in CDS, but also in 5'UTRs if they contain an uORF, i.e., short translated reading frame located upstream of the main ORF (CDS), which may have an important role in translation regulation. Another most important aspect that can be studied in terms of the RP experimental data is assessment of the dynamics of ribosome movement along individual mRNAs and the rate at which certain codons are translated. This is possible because three nucleotide bases in the sequenced footprints are reflected in a periodic mode as a consequence of the ribosome movement along the mRNA coding region, since the ribosome moves along the overall coding sequence in a codon-wise manner, the 5' region of ribosome footprints tend to be mapped at the same position of each codon.

Find below several examples which in our view illustrate the distinctive capabilities of RP in clarification of the fine mechanisms underlying the translational control in plants, such as the detection of new ORFs, including those annotated as noncoding RNAs and pseudogenes. In particular, the study of translation regulation under normoxic and sublethal hypoxic conditions (hypoxia) in A. thaliana shoots with the help of RP not only detected an inhibitory effect of the uORF on the translation of downstream protein coding regions in normoxia, which was further modulated by hypoxia, but also determined the alternatively spliced mRNAs as well as the fact that ribosomes were associated with certain noncoding RNAs [21]. An RP study of the maize shoots under drought showed a statistically significant change in the translation efficiency of 931 genes, which according to further analysis of the transcripts was associated with the nucleotide composition of the sequence, including GC content, length of coding sequences, and normalized minimum free energy. In addition, potential translation of 3036 open reading frames (uORFs) in 2558 genes was detected; the authors believe that these uORFs are able to influence the translation efficiency of the downstream main open reading frames (ORFs) [26]. In another study, the Ribo-Seq data detected 27 and 37 translated sORFs (short ORFs) among the annotated noncoding ncRNAs and pseudogenes of A. thaliana, respectively [27]. Moreover, 187 translated uORFs were identified with a high degree of reliability. In addition, the events of translation from the start codons other than AUG were identified in the dataset among both annotated genes and uORFs. They also demonstrated that 15 of the 19 detected single-exon sORFs had homologs in various flowering plants, which suggests their functional significance [27].

Lukoszek et al. [28] used RNA-Seq and Ribo-Seq to assess reprogramming of the A. thaliana global gene expression during a long-term heat shock (3 h at 37 ◦C) at both the transcriptional and translational levels. They have shown that translation is globally impaired in the early period of the heat impact (15 to 45 min), while the stress response appears mainly at the expense of transcriptional programs. In this process, a long-term stress impact (3 h) activated translational programs, which eventually form the adaptive response. The transcripts regulated via translation display a number of common characteristics, namely, the presence of relatively conserved A/G-rich motifs in their 5'UTRs or 3'UTRs that are similar to the sequences identified as protein-binding nucleotide motifs. Another specific feature widespread among the genes upregulated in heat stress is that they are less inclined to form secondary structures, which is likely to ensure their binding with ribosomes and to enhance translation. In addition, several transcripts prevalently induced by heat contain a putative G2 quadruplex in their 5'UTRs. Note that an increased number of reads for RP footprints in quadruplex structures correlates with an expanded expression of the downstream CDSs. This suggests an important role of these structures in translation activation of the downstream ORF according to yet unknown mechanism [28]. Ribosome profiling has been used to analyze translation of the chloroplast transcripts in maize shoots in response to changes in light conditions. According to the experimental data, all chloroplast mRNAs except for psbA maintain similar numbers of ribosomes after short-term changes in light conditions but nonetheless are more efficiently translated in the light. On the other hand, the psbA mRNA displays a sharp increase in the ribosomes over several minutes after the plants are transferred to light and restores a low ribosome loading during 1 h in the dark, which correlates with the need to replace the damaged psbA in photosystem II. These results emphasize the unique translational response of psbA in mature chloroplasts, indicate the particular light-regulated steps in the context of photosystem II activity maintenance, and provide the background for the study into the mechanisms underlying both the psbA-specific and genome-wide effects of the light on the translation in chloroplasts [29].

The RP technology was also used to study several aspects in the translation of A. thaliana mitochondrial genome in a dynamic mode. As has been shown, the mitochondrial mRNAs are differentially-translated; in this process, the translational levels of the transcripts encoding the subunits of mitochondrial protein complexes, in particular, complex V, proportionally correlate with the stoichiometry of respiratory chain subunits. In general, the mitochondrial translation is shown to be controlled at the level of individual mRNAs and is directly involved in the activity regulation of plant mitochondria [22].

Note that Ribo-Seq technology is currently at a relatively early stage of its development, which leads to some experimental difficulties and technical artifacts influencing the Ribo-Seq data interpretation [18]. In particular, the RP results may display statistically significant differences associated with the modifications of one of the five stages in the basic protocol, such as the conditions of cell lysis, composition of buffer solutions, selection of nucleases and the absence of pronounced specificity to the sequences to be cleaved, and construction of the library; even more so as these details in many cases are not analyzed in a systematic manner [19,23]. This suggests the need to systematically study the effects of the corresponding experimental parameters of the used RP protocols [19].

The RP technique also has its limitations. According to the current scientific consensus, the basic limitation of RP approach is a static position of ribosomes along the mRNA. This prevents distinguishing between the ribosomes involved in translation from the ribosome in a steady-state [19]. Thus, the methods used in the majority of studies involving RP can overestimate the translation efficiency because of the data related to monosomes, in which mRNA is also protected by a ribosome (Figure 5A) [18,23,26]. Underrepresentation of the transcript regions with ribosome stacking is also possible; this is associated with the stacked polysomes and may prevent hydrolysis in monosomes, because of inaccessibility to RNases (Figure 5B) [4,30]. Correspondingly, the recommendation for an additional direct measurement of the polysome-protected mRNA looks most reasonable to overcome this limitation of the RP approach [26].

Note that the Ribo-Seq technique now is mainly applicable to study translation of the organisms with annotated genomes since the deep sequencing data for footprints are represented by very short reads (28–30 nucleotides), which, as a rule, are analyzed by mapping onto the genome data (Figure 4).

However, the current limitations of this method can be bypassed and this experimental technology will remain a useful tool in the omics [30]. RP data with high resolution is a priceless resource for studying noncanonical start codons and alternative start sites and can be useful for characterizing translation of different isoforms of transcripts, identifying new translated ORFs and their quantifying, and, in general, for improving the genome annotation of poorer characterized organisms. Additional ribosome profiling can also be a proxy for the proteome or assist in proteomics studies [27,30].

Completing this section, note that the genome-wide profiling of the transcripts associated with ribosomes utilizing one of the above experimental approaches may highlight the new aspects in gene expression unvisualizable by an ordinary profiling of the total cellular mRNA. In Table 1, we attempted to consolidate the advantages, limitations, and areas of applicability of the discussed experimental approaches to the study of differential translation on a genome-wide scale. Undoubtedly, selection of the appropriate experimental approach depends on the particular aims set by a researcher.

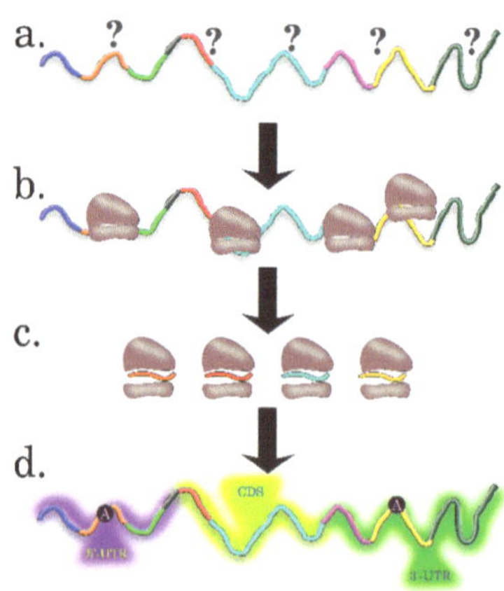

Figure 3. Scheme of application of ribosome profiling to functional characterization of mRNA regions.

(a) Scheme of an mRNA with unknown ribosome positions. (b) The mRNA with arrested ribosomes in the transcript regions potentially important for efficient translation. (c) Formation of the ribosome footprints by RNase hydrolysis. The resulting footprints characterize the translational functionality of a certain mRNA region. The footprints shown with different colors correspond to different mRNA regions. (d) Result of analysis of the precise positions of translating ribosomes along mRNA, where A is the identified alternative open reading frames in 5'UTRs or 3'UTRs and CDS (coding sequence) is the main reading frame of the transcript.

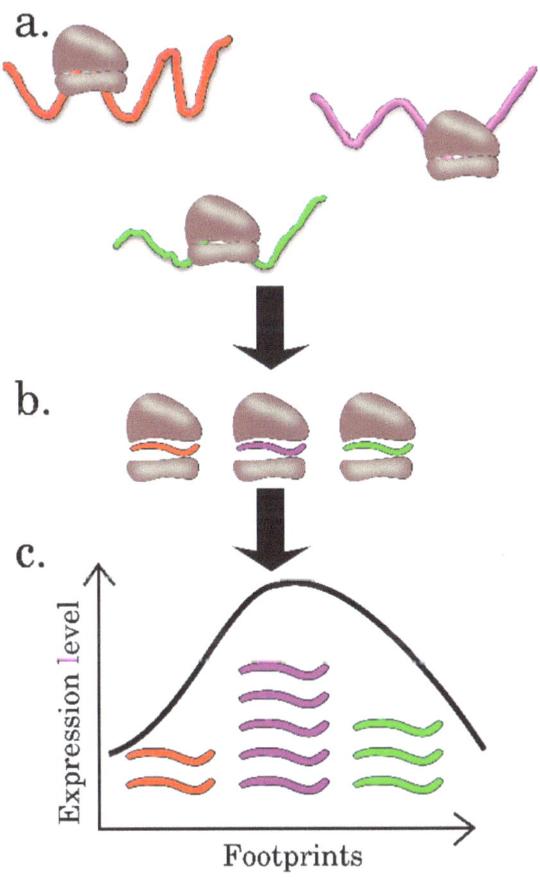

Figure 4. Principle of analysis, interpretation, and visualization of the ribosome profiling data.

(a) The ribosomes arrested on transcripts (b) form ribosome footprints after RNase hydrolysis.

(c) The footprints mapped onto genome can be associated with particular sequences to assess the relative amount and positions of ribosomes on the transcripts on a genome-wide scale.

Table 1. Comparative characterization of the experimental approaches producing pools of differentially-translated mRNAs.

Experimental Approach	Basic Protocol	Advantages	Limitations	References
Polysome Profiling	Separation of transcripts with Ribosome Profiling different ribosome loading by ultracentrifugation; supplemented by sequencing of different mRNA fractions, including transcriptome-wide analysis	Simplicity and possibility to analyze the plant species with annotated and unannotated genomes	Does not assess the number and location of ribosomes on each transcript	[4,5,21]
Translating Ribosome Affinity Purification	Separation of the transcripts with different ribosome loadings by absorption on anti-FLAG-M2 agarose; supplemented by mRNA sequencing, including transcriptome-wide analysis	Profiling of actively-translated RNAs from different plant tissues and particular cell types; Identification of only the mRNAs bound to ribosomes, which makes it possible to avoid the potential confusion with the transcripts associated with the other RNA-binding proteins	Cannot estimate the number and location of ribosomes on each transcript; Requires production and accurate selection of the plant transgenic lines that express an epitope-tagged variant of ribosomal protein L18	[3,12,14,15]
Ribosome Profiling	Isolation and sequencing of the ribosome-protected mRNA fragments; modification of the protocol is necessary for individual species	Identification of the ribosome number and location on each transcript; Detection of new translated ORFs and noncanonical translation start sites	Requires significant material, time and labor investments; A considerable amount of biological material is necessary; The transcript regions with stacked ribosomes may be underrepresented; Incorrect trace identification may result from the RNA interaction with RNA-binding proteins of an analogous size; Applicable only to the plants with a well-annotated genome	[4,18–20,24]

Note: The key advantages and limitations are shown for each approach; see the text for a comprehensive description.

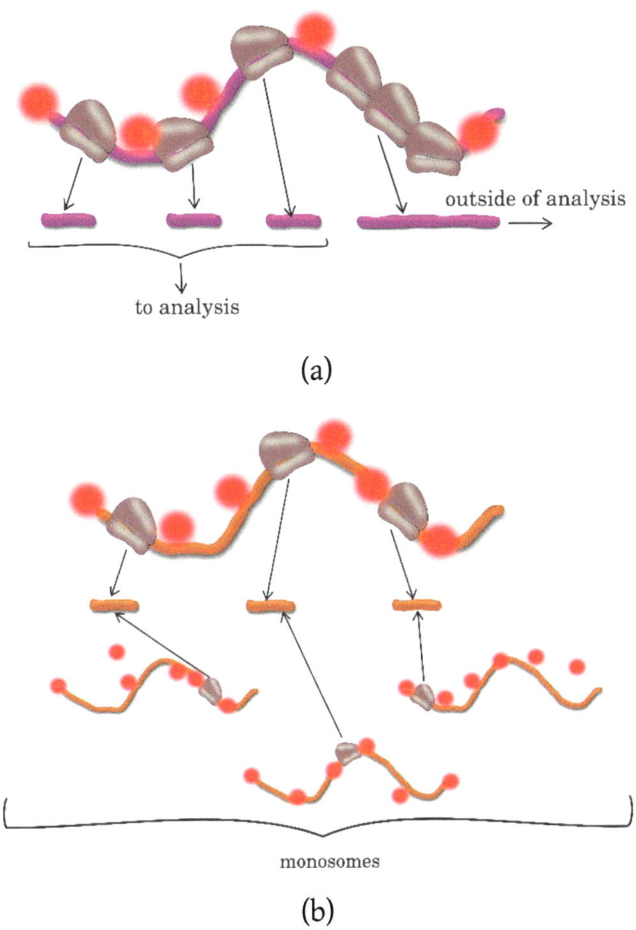

Figure 5. Limitations in the use of ribosome profiling: (a) overestimation of translation efficiency because of the footprints of monosomes, where mRNA is also protected by ribosome and (b) underrepresentation of the transcript region with stacked ribosomes; carefully stacked polysomes are inaccessible to RNases, thus cannot be digested into ribosome footprints of the tested size (28–30 nucleotides). Red spots denote the region attacked by RNase.

3.COMPUTATIONALALGORITHMSFORPREDICTINGTHEFEATURES OFPLANTMRNASIMPORTANTFORDIFFERENTIALTRANSLATION

The above described experimental approaches make it possible to detect the specific pools of transcripts with characteristic differential translation. Several computational resources are useful for identification of regions of specific structural features in mRNA nucleotide composition that can mediate differential translational control.

In this section, we summarized the resources and some computational algorithms that have been used to form the samples of target plant transcript sequences and to predict their peculiar characteristics, as well as their main functions and domains of application. Note that the resources and the corresponding software are rather numerous and, in fact, require a separate review. Here, we consider only those that have given the data on and predictions of regulatory contexts in transcripts with further experimental confirmation.

3.1 Preparing Datasets for Analysis

The key preparatory stage in the in silico predictions is a construction of the most representative sets of sequences for the transcript pools differing in their translation efficiency. Note that the researcher needs not only full-sized transcript sequences (cDNA), but also the sequences of individual regions of these transcripts, namely, coding (CDS) and untranslated (5'UTR and 3'UTR) regions, which, as mentioned above, can also contribute to translation efficiency. Currently, many internet resources have been elaborated that allow sets of such sequences to be downloaded, including the sequences for plants. In particular, TAIR is the information source for the model plant A. thaliana (https://www.arabidopsis.org/download/index-auto.jsp?dir=%2Fdownload_files% 2FSequences%2FTAIR10_blastsets) [31], which is widely used for loading 5'UTR, CDS, 3'UTR, and cDNA sequences using the tools "Download", "Sequences", and TAIR10 blastsets [32–34]. Another information resource containing CDS, cDNA, 5'UTR, and 3'UTR sequences of the representatives of six key kingdoms of the living organisms, including plants, is JetGene. JetGene is publicly available at http://jetgene.bioset.org/; its data are stored and updated at the Ensembl server [35]. The intuitively clear and friendly JetGene interface allows the cDNA, 5'UTR, CDS, and 3'UTR sequences to be extracted in a FASTA format, including the specific samples on user request. Note that only the sequences with complete information about the full-sized transcripts are in most cases selected for further analysis.

Once the sets of sequences (5'UTR, CDS, and 3'UTR) for the pools of differentially-translated transcripts are obtained, the researcher has to select for analysis the regions of transcripts and regulatory sequences that may be potentially involved in translation modulation. According to the current opinion, the complex multilevel information is encoded in the full-sized mRNA sequence (transcript) in general and in its individual parts—5'UTR, CDS, and 3'UTR (Figure 6). This gives researchers the grounds to include all these regions into in silico analysis to characterize the differentially-translated plant transcripts. Note that translation initiation is, as a rule, the stage limiting the translation rate and 5'UTR plays here the decisive role. The length, nucleotide composition, secondary structures, and regulatory elements of a smaller size, such as upstream start codons (uAUGs), uORFs, nucleotide motifs, and several other features in the 5'UTRs of transcripts, are closely examined in terms of their contributions to the translation efficiency. In this process, the probability to find the potential regulatory regions and contexts and to clarify how their properties influence the translation efficiency will be higher if more traits of this kind are involved in the initial in silico analysis.

The further aims of the researcher could be (i) to assess the variations in distribution of individual traits in the sequences from the examined transcript pools and to figure out the statistically significant differences that are positively correlated with the translation efficiency; (ii) to find and determine the statistically significant representation of the potential regulatory contexts in the transcripts with different translation efficiencies; and (iii) to identify the specific regulatory sequences if they are present in the examined pools.

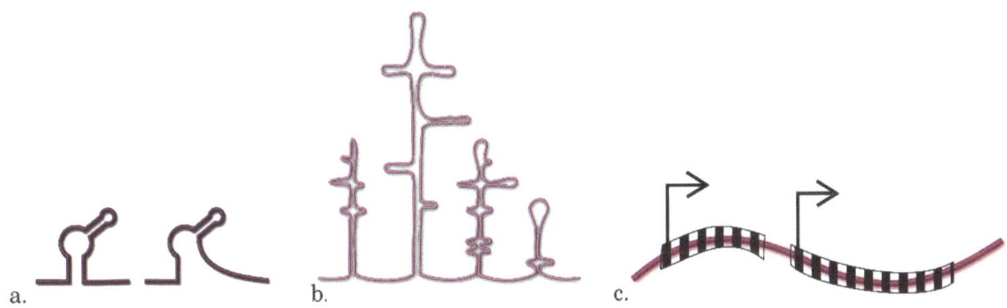

Figure 6. Examples of some mRNA cis-regulatory elements: (a) riboswitches; (b) internal ribosome entry sites (IRESs); and (c) alternative open reading frames.

3.2 Statistical Methods

The methods of mathematical statistics have been rather efficiently used for solution of the first task. As a rule, basic and extended statistical analyses are performable with the help of the available standard programs, such as Excel, STATA, and IBM SPSS Statistics 20 [26,32–34]. For example, the genome-wide monitoring of the changes in the translation efficiency of individual mRNAs in A. thaliana shoots after heat shock have demonstrated translation repression for the majority of mRNAs; however, some mRNAs still followed the differential translation pattern. Analysis of the differentially-translated mRNA sequences demonstrated that only some characteristics, such as the G + C content in 5'UTR and cDNA length, are putatively involved in the mechanisms providing discrimination of the mRNA loading with ribosomes and are associated with differential translation of a certain transcript cohort in response to a high temperature. In particular, the translationally active mRNAs have a low G + C content (on the average, 36%) versus the transcripts with repressed translation (42%). This selection mechanism also influences the differential polysomal loading of the transcripts associated with stress and, as a consequence, the efficiency of their translation [32].

In general, the methods of mathematical statistics have made it possible to (i) find the characteristics that are representative for the analyzed mRNA, (ii) discard the characteristics the effect of which can result from a bias to the group of particular genes, and (iii) determine the statistically significant differences displaying a positive correlation with the relative translation efficiency.

3.3 Methods for Identification of RNA Motifs

The methods that are used to identify potential regulatory motifs in mRNA sequences assess the statistical significance of represented potential regulatory motifs in the mRNAs with different translation efficiencies. These methods are in general the same or very similar to those for DNA motifs except for the methods addressing the DNA conformational properties. The latter utilize physical parameters of DNA double helix and can be plied both to prokaryotic [36] and eukaryotic [37] genomes. Having appropriate parameters to convert letter representation of RNA into numerical representation, the same methodology could be applied to analysis of mRNA. Conventional approaches are based on accounting for conserved nucleotides within a short motif. One of the most frequently used programs for the detection of motifs in the transcript pools with different translation efficiencies is MEME, which is based on the maximal likelihood optimization [38]. Ease of use and a wide set of the accompanying programs for visualization and further search are advantages of this program. The MEME suite comprises four main sections, namely, motif discovery, motif enrichment analysis, motif search, and motif comparison, altogether 14 different tools. This toolkit allows the researcher to both determine motifs de novo and to scan a dataset of sequences for the matches of the already known motifs. MEME shows a schematic arrangement of the found motifs on the initial sequence, constructs a graphical representation for them, and computes statistical significance (p-value) for these motifs.

In particular, MEME suite has allowed identification of a nine-nucleotide-long element present in both the 5'UTRs and 3'UTRs of numerous A. thaliana and Gynandropsis gynandra transcripts; the authors named it MEM2. Later, it was experimentally confirmed that the MEM2 motif residing in the 5'UTR was necessary for preferential protein accumulation in the mesophyll cells. It is assumed that this motif can be involved in the mechanism guiding preferential cellular accumulation of several enzymes necessary for C4 photosynthesis, which provides a more efficient carbon capture as compared with the ancestral C3 pathway [39]. The MEME suite has been used in a comparative analysis of the 5'UTR sequences for steady-state and polysomal A. thaliana mRNAs and allowed for discovery of two motifs (TAGGGTTT and AAAACCCT) present in many genes, which potentially suggests their contribution to the translation efficiency. Furthermore, it has been experimentally shown that only one of these motifs, TAGGGTTT, regulates gene expression at the level of translation [33].

However, the search for the motifs using this tool also has some limitations. Among the serious disadvantages of this program is the trend to find very long motifs (over 20 nucleotides), these motifs are present only on a small subgroup of sequences and/or frequently repeated motifs in one or just a few sequences. Although statistical significance (p-value) of such motifs is very high, the motifs themselves, as a rule, are rarely of any biological/practical interest and represent statistical artifacts.

Most likely, these limitations are the main reason why several studies of motifs failed to bring any positive results [8,32,33]. Correspondingly, other computational approaches were used for this search and their statistically significant representation in the transcripts with different translation efficiency, for example, by comparing the frequencies of mono-, di-, and trinucleotide sequences. Statistical tests, for example, the Kolmogorov–Smirnov or Fisher test, allow the detection of statistically significant differences in such nucleotide distributions. Moreover, the use of linear or logistic regression makes it possible to detect not only the individual contribution of each sequence, but also the effect of their combinations. In particular, partial least regression analysis has been applied to the detection of the short regions residing in the neighborhood of the 5'-proximal region of 5'UTR that can play an important role in differential translation in response to heat shock [8]. However, the linear or logistic regression methods are also not free from limitations. For example, it is not practically feasible to analyze motifs with a length of four nucleotides or longer, because their frequencies sharply decrease and, as a consequence, the computation of statistical characteristics becomes too complicated. In addition, these methods do not take into account the locations of motifs on sequences, which in terms of biology mean the equal contributions of the codons residing far from the translation start codon and in the immediate proximity.

3.4 Detection of Other Context Features of RNA

Statistical approaches are rather efficient when a set of potential characteristic features is determined for a pool of sequences and is used as a reference in the analysis. However, the specific cis-regulatory sequences in mRNAs that can modulate translation are identified using specialized approaches and/or resources for their prediction. The examples below illustrate the approaches to predict cis-regulatory sequences in the case studies of internal ribosome entry sites (IRESs) and upstream ORFs (uORFs), first and foremost, conserved peptide uORFs (CPuORFs).

IRES are the nucleotide sequences that mediate translation initiation of alternative reading frames (aORFs) under stress conditions, when the trivial cap-dependent translation mechanism is inhibited without the corresponding changes in gene transcription [40]. In general, the IRESs of plant mRNAs, unlike the IRESs of viruses, display considerable diversity in both nucleotide composition and structure [41]. Despite this diversity, characteristic functional modules are distinguishable in the IRESs, namely, (i) the presence of several start codons and their localization and (ii) the fact that some IRESs carry short conserved modules, which are recognized by the plant translational machinery and are directly involved in the immobilization of ribosome small subunit [42]. Polypurine blocks residing close to the start codon, which may be directly involved in the immobilization of ribosome small subunit, are an example of such conserved motifs [43].

The mRNAs potentially carrying IRESs can be selected by analyzing the experimental data obtained by polysome or ribosome profiling followed by deep sequencing and/or by mass spectrometry analysis. First and foremost, such mRNAs must retain a high level of their translational activity under the impact of adverse environmental factors and carry additional alternative start codons. The following strategy is appropriate for further selection and analysis of the mRNAs carrying IRESs. (i) Interspecific comparison of the transcript sequences of homologous genes, which allows for identification of the conserved region in the vicinity (30–50 nucleotides) of the alternative start codon followed by (ii) assessment of the context of the alternative start codon, the optimal neighborhood of which may suggest that translation can be potentially started from it. This strategy has been successfully implemented for predicting translation initiation of a short aORF with involvement of a polypurine block via internal ribosome entry [43]. Note that a commonly accepted confirmation for an IRES activity is still its ability to provide a coordinated translation of reporter genes within a bicistronic transcript (see below).

The advent of RP and high-throughput sequencing technologies made it possible to determine the translation start sites and to discover numerous mRNAs with aORFs that (i) may have a putatively inert sequence that acts as a mere translation barrier upstream of the main ORF or (ii) may encode short peptides referred to as CPuORFs [44]. The main difference between CPuORFs and the other ORFs is their length and, although there are no strictly defined frames, the ORFs shorter than 200–250 codons are regarded as short. In general, the search for CPuORFs is analogous to the approach for prediction of main ORFs and the strategy utilizing interspecific comparison of CPuORF sequences to identify the conserved regions is in most cases used for this search and estimation of the coding potential. This strategy is based on revealing the homology between such short peptide sequences and to a considerable degree depends on the range of the species selected for comparison. For example, it is quite possible that the fact of preservation of CPuORFs within the plant species selected for comparison is insufficient to reveal the conserved regions. In this case, the analyzed CPuORFs will not be identified although their sequence is sufficiently conserved among the other species. When comparing the

CPuORF sequences among closely related species, the observed similarity between short uORF peptide sequences may result from nucleotide sequence retention rather than conservation of these peptides. In order to overcome the problems associated with selection of the species for comparative analysis, a new method for CPuORF identification, BAIUCAS, was developed and tested [45]. BAIUCAS utilizes sets of EST (expressed sequence tag) data for thousands of plant species to search for homologs. The BAIUCAS algorithm consists of six successive stages: (i) exhaustive search for uORFs; (ii) search for homologs of CPuORF amino acid sequences over EST databases using tBLASTn; (iii) selection of CPuORFs based on the conservation of the stop codon position; (iv) selection of the CPuORFs conserved in a wide range of species, i.e., the CPuORFs with the conserved stop codons detected in each of several taxonomic categories; and (v) and (vi) are filtration stages, which excludes the "false" conserved CPuORFs [45]. Using this approach to the search for CPuORFs, 16 A. thaliana CPuORFs were identified; five of them are the new CPuORFs involved in the translation regulation of the main ORF, which has been experimentally confirmed [46].

The list of the computational approaches that have been so far successfully used in decoding the specific structural features in nucleotide composition of the plant mRNAs that mediate differential translation control is rather short. One of the possible efficient computational tools for analyzing the translatomes and predicting numerous regulatory codes in transcript sequences could be artificial neural networks (ANNs). This assumption relies on the facts that (i) most neural network architectures is theoretically able to approximate any function, i.e., it is potentially possible using ANNs to construct a model for almost any biological pattern, and (ii) the capabilities of the supercomputers have reached the appropriate level to model biological processes using neural networks. However, a positive result of analysis depends first on an adequately selected architecture of the network and second on the amount and composition of the training sample and a training strategy. Several recent reports confirm the utility of ANNs in deciphering the molecular mechanisms involved in decoding the eukaryotic genome. For example, the ANNs constructed based on RP data have been used to predict the yield of protein products [47]; to search for the motifs potentially able to influence translation [48]; to extract the biologically important information from omics data [49]; and to simulate the interaction between nucleic acids and different types of ligands (protein and peptides) [50]. The ANN potential is broad enough and it can be expected its broad application to the research in diverse and multilevel mechanism underlying translation in plants.

4. APPROACHES FOR EXPERIMENTAL VERIFICATION OF THE SYSTEMIC EXPERIMENTAL DATA AND THEORETICAL PREDICTIONS OF INTRINSIC FEATURES OF PLANT MRNAS IMPORTANT FOR DIFFERENTIAL TRANSLATION

As a rule, the experimental results on determination of the pools of differentially-translated plant mRNAs as well as the predictions of the regions and nucleotide contexts in differentially-translated transcripts require experimental confirmation of their functionality and contribution to the translation efficiency. In this section, we will consider the main experimental approaches that allow for convincing confirmation of which particular regions in individual mRNAs and/or features within plant transcripts are important for modulation of translation. It should be emphasized that methods for high-throughput experimental verification, i.e., concurrent analysis of a large pool of regulatory regions, are yet not available; thus, potential regulatory sequences are examined for each individual transcript [4].

4.1 Direct Measurments of Individual Transcripts

Several tools for determining the changes of individual transcripts at the level of translation are available, including (i) the systems for in vitro translation based on measuring the incorporation of labeled amino acids, such as FUNCAT (fluorescent noncanonical amino acid tagging), SILAC (stable isotope labeling by amino acids in cell culture), BONCAT (bioorthogonal noncanonical amino acid tagging), QuaNCAT (quantitative noncanonical amino acid tagging), and PUNCH-P (puromycin-associated nascent chain proteomics), or cell-free protein expression systems, such as the wheat-germ extract, which contain all factors necessary for translation of the target transcript; toeprinting, or the primer extension inhibition assay, utilizing reverse transcription to study the interaction of ribosomes with the target transcript; (iii) enzyme immunoassays, in particular, Western blot hybridization; and (iv) mass spectrometry-based methods for identifying the changes in the proteome or their combinations with in vitro translation methods, for example, PUNCH-P. The review by Mazzoni-Putman et al. [4] describes the principles, advantages, and limitations of these methods in detail.

However, these research methods are in most cases suitable for assessing the general changes in translation, require considerable time, amount of reagents, and specialized equipment; correspondingly, most research works now use the strategy of reporter systems for studying the structure–function characteristics of the target sequences. The strategy of reporter genes considerably enhances such research since it is much easier to record the protein product of a reporter gene as compared with a studied gene. It should be also emphasized that the reporter genes code for the proteins that display either unique specific features or unique enzyme activities, allowing their products to be easily isolated from the totality of intracellular and extracellular proteins. Thanks to these advantages of reporter systems over the other methods for studying the regulation of gene expression, they have been widely used for experimental verification of the regions and nucleotide contexts in differentially-translated transcripts. For studies of this type, expression cassettes are constructed that carry the reporter gene sequence with the expression controlled by a particular regulatory region or sequence selected by researcher (Figure 7). Researchers have at their disposal several reporter systems that have proved their efficiency in the studies of potential regulatory sequences or the nucleotide contexts that modulate translation in plant systems, in particular, β-glucuronidase (GUS); different variants of fluorescent proteins (for example, GFP and RFP); luciferases (Renilla luciferase, RLuc, and firefly luciferase, FLuc); and thermostable lichenase (LicBM) [51,52]. Commercial substrates and kits as well as the quantification methods for assessing the corresponding protein products are available for these reporter systems. The main approaches that have been applied in the studies of various regulatory regions or nucleotide contexts within transcripts with the use of reporter systems are illustrated below.

When experimentally confirming the role of the full-sized 5'UTRs in transcripts, these sequences are cloned upstream of the 5' region of a reporter gene (Figure 7). Quantitative estimate of the reporter gene protein product when using different target 5'UTRs versus the known translational enhancers of various origins makes it possible to assess their contribution to the translation efficiency [51]. For example, A. thaliana mRNAs that are stably translated under any growth and environmental conditions have been found by polysome profiling. Testing of the translation capability of mRNA 5'UTRs of candidate genes using the reporter gene strategy has convincingly demonstrated that the 5'UTR of 47 cold-regulated genes are an efficient translational enhancers, which enables a stable high level translation under any conditions of plant growth. This suggests the utility of this method for plant biotechnology, for example, when engineering plants producing biologically active substances or the plants resistant to some stress factors, including the schemes that involve genome editing technologies [9].

Recombinant 5'UTR sequences are designed for identification of the cis-regulatory elements in these mRNA regions, first and foremost, the motifs or specific nucleotide contexts potentially able to influence differential translation, using for this purpose a (i) combinatorial approach, (ii) site-specific mutagenesis, (iii) translation assessment of the second (3'-terminal) ORF of a bicistronic transcript, (iv) deletion analysis, (v) frameshift mutations, or a combination of these methods (Figure 7).

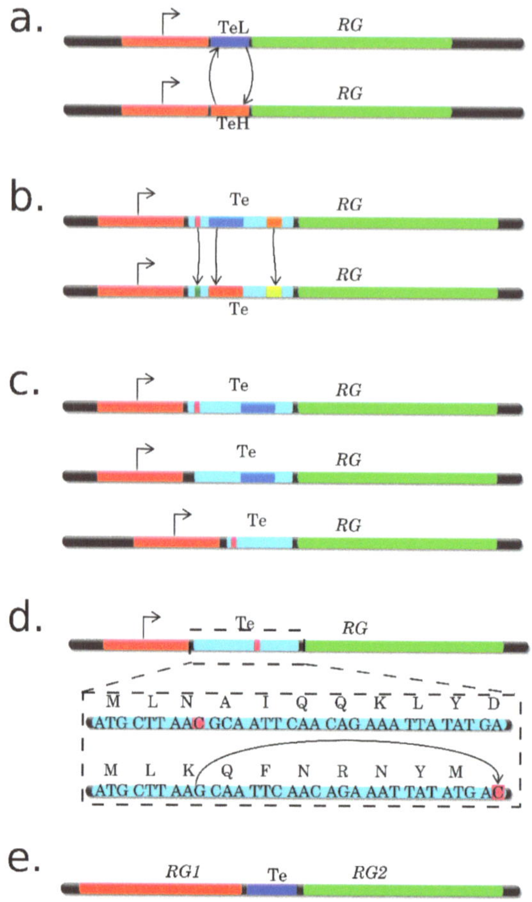

Figure 7. Approaches to experimental verification of the tested regulatory elements using the strategy of reporter systems.

(a) Combinatorial approach. Mutual substitutions of the predicted regulatory motifs TeL and TeH (denoted with arrows) are introduced into the pairs of 5'UTRs of the same length but considerably differing in the experimentally confirmed translation efficiencies. TeL and TeH are the tested elements characteristic of the transcripts with low and high translation efficiencies, respectively.

(b) Site-specific mutagenesis. The native regulatory sequence is above and the mutant regulatory sequence is below; different colors denote the region used for mutagenesis; direct arrows indicate the substituted regions in two sequences. (c) Deletion analysis. The native regulatory sequence is above; different colors denote the regions used for deletions: the regulatory region with deletion in the 5' region is in the middle (deletion of the pink region of the native sequence) and the regulatory region with deletion in the 3' region is below (deletion of the blue region of the native sequence).

(d) Frameshift analysis. The region for introducing frameshift is dashed; the native nucleotide and amino acid sequences are above and the mutant, below. The nucleotides colored red were frameshifted. Simultaneous introduction of deletions and insertions to positions -1 and +1 changes the amino acid composition of peptide sequence encoded by the alternative open reading frame preserving the presence of the overlapping peptide and its length. (e) Bicistronic construct for studying the functionality of IRESs. Two reporter genes are translated from the bicistronic construct in a coordinated manner; translation of one of them (RG2) is controlled by the tested element (Te) and the other (RG1) is translated according to the classical cap-dependent mechanism. In all panels, bent arrows denote the transcription start point; Te, tested element; and RG, reporter gene.

4.2 Site-Specific Substitutions

The combinatorial approach utilizes the fact that mutual substitutions of the predicted regulatory motifs are introduced into the pairs of 5'UTRs with the same lengths but with experimentally confirmed significant difference in translation efficiency (Figure 7a). Matsuura et al. [8] successfully used this approach to identify new cis-regulatory elements in 5'UTRs that influence the differential translation of A. thaliana transcripts in response to heat shock (HS). The genome-wide analysis of the changes in polysome loading of the transcripts in Arabidopsis cell culture allowed for selection of a set of genes with different translational responses to HS. The 5'UTR nucleotide sequences of the transcripts that change the level of reporter protein in the protoplasts affected by HS were used to predict the regulatory elements in 5'UTRs with the help of partial least square (PLS) method. These computational predictions suggested that two short regions residing in the vicinity of 5'-proximal region of the 5'UTR can play an important role in the relative activity of reporter protein and, thus, may be regarded as cis-regulatory region candidates. In order to experimentally confirm the predictions on the importance of these 5'UTR regions in differential translation control, a series of mutual substitutions of these regions in the pairs of 5'UTRs with equal length but different translation efficiencies were analyzed. Analysis of the reporter gives convincing evidence that the 5'-proximal region of the 5'UTR plays a key role and that certain specific determinants in 5'UTR mediate the differential translation in response to HS [8].

Site-specific mutagenesis makes it possible to introduce substitutions of one or a group of nucleotides within strictly defined regions of nucleic acid sequences. Comparison of the levels of a reporter protein translated when controlled by a native regulatory region (for example, 5'UTR) and the same regulatory regions but with the introduced mutations (Figure 7b) demonstrates how the modification of the primary sequences of nucleotide contexts and/or their secondary structures modulates the translation of specific mRNAs. This approach, along with other molecular methods (Western blotting, qPCR, polysome fractionation, and so on), has emerged to be most efficient when studying both the mechanism underlying formation of RNA G-quadruplex in the 5'UTR of the SUPPRESSOR OF MAX2 1-LIKE4/5 (SMXL4/5) mRNAs and the clarification of the role of a specialized structure, the regulator of phloem formation. In particular, a novel zinc finger protein, JUL, was identified; it specifically bound to consecutive guanine repeats in the 5'UTR of SMXL4/5 and induced RNA G-quadruplex. Moreover, convincing experimental data that both JUL1 and G-quadruplex are necessary for strong translation suppression rather than a single-stranded G-rich element have been obtained using the strategy of reporter systems. This suggests that the suppression of translation is caused by either JUL1-mediated formation of G-quadruplex or the G-quadruplex/JUL1 complex recruits an unknown translational suppressor [53].

Site-specific mutagenesis has emerged to be efficient for clarifying the role of the TAGGGTTT motif, overrepresented in the 5'UTRs of the transcripts regulated at the level of translation. A comparative study of two constructs, one with a native 5'UTR carrying this motif and the other with the 5'UTRs carrying mutations in this motif, has shown that the transcripts with the native 5'UTR are more efficiently translated provided that the number of transcript are equal. Thus, it is experimentally proved using reporter genes that the TAGGGTTT cis-element regulates expression of the gene particularly at the level of translation [33].

4.3 Analysis Using Deletions

Deletion analysis of 5'UTRs, implying construction of truncated variants of sequences and their subsequent fusion with a reporter gene, makes it possible to identify nucleotide contexts decisive for maintaining the structure of RNA that can be in two particular conformations, as for example, in riboswitches (Figure 7c). Note that an important specific feature of the riboswitches is their ability to both activate and inhibit translation from the controlled ORF, due to the presence of a specific regulatory region, the aptamer. This region with a particular secondary and sometimes tertiary structure, has the properties of a receptor for small molecules (ligands). In the over whelming majority of cases, riboswitches reside in 5'UTRs. A deletion analysis of the A. thaliana phytoene synthase (PSY) 5'UTR has shown that the long 5'UTR variants (403 and 330 nucleotides) with two predicted sequences of convertible RNA conformations, similar to riboswitches, inhibit translation of the reporter gene, in contrast to the short variant (252 nucleotides) of the PSY 5'UTR, lacking such hairpin structure. This allows the short 5'UTR variant to pass the translational control and rapidly elevate the protein levels [54].

4.4 Translation from Alternative ORFs

Translation of the alternative ORFs may be provided by a specific mRNA region, referred to as IRES (internal ribosome entry site). The functionality of an IRES can be experimentally confirmed by constructing bicistronic transcripts (Figure 7e) [40,41]. Note that the use of bicistronic constructs when studying the IRES functionality is determined, in particular, by the need to normalize a relative efficiency of the IRES-guided translation as compared with the cap-dependent translation. For example, functionality of the initially predicted IRES in the mRNA for a tobacco heat shock protein [55] was experimentally confirmed utilizing the ability of coordinated expression of reporter genes within a bicistronic transcript [56]. A structural analysis of the 5'UTR of a maize heat shock protein (Hsp101) mRNA has shown the presence of three stem loops towards the 5' end, which suggested the functioning of the 5'UTR structure as an IRES. This assumption was experimentally confirmed by comparing bicistronic reporter constructs; in particular, it has been shown that the overall hsp101 5'UTR sequence (150 nucleotides) acts as an IRES, since the deletion of 17 nucleotides from the 5' end decreases the translation efficiency of the reporter gene transcript by 87% as compared with the control sequence [57]. A functional analysis of the bicistronic constructs has revealed IRESs in the A. thaliana WUS mRNA, coding for the homeodomain transcription factor WUSCHEL (WUS), as well as the role of an additional protein, AtLa1 (an RNA-binding factor). As demonstrated, AtLa1 initiates an IRES-dependent WUS mRNA translation by binding 5'UTR (WUS is responsible for supporting the A. thaliana apical meristems under stress conditions) [58].

Alternative ORFs are among the most abundant regulatory elements in mRNAs; they are frequently present in the 5' leader regions of eukaryotic mRNAs (designated uORFs). Such uORFs may negatively modulate the translation efficiency of the downstream main ORF. According to the current estimate, approximately 20% of the A. thaliana protein-coding genes contain uORFs in their mRNA 5'UTRs [59]. Initially, such regulatory sequences are either predicted via searching for the conserved peptide uORFs (CPuORFs) [45] or experimentally detected, in particular, by ribosome profiling during, for example, plant cell response to a stress [43,60,61]. The highest interest of researchers is associated with the function of the regulatory sequences, such as CPuORFs, which are able to act in a sequence-dependent manner or cause ribosomal arrest, thereby modulating translation of the main ORF.

4.5 Frameshift Mutations

The approach of frameshift mutations utilizes concurrent introduction of deletions and insertions at -1 and +1 positions; this procedure changes only the amino acid composition of a peptide sequence coded for in CPuORFs but retains the presence and unchanged length of the overlapping CPuORFs (Figure 7d). This method has been used to analyze 16 recently predicted conserved CPuORFs of A. thaliana for assessing a sequence-dependent effect of each CPuORF on expression of the main ORF. A comparative analysis of the reporter protein activity of the variants when the translation is controlled by either native CPuORFs or the CPuORFs with introduced frameshift mutations has identified five novel CPuORFs that repress the expression of the main ORF in a sequence-dependent manner. Moreover, it has been convincingly demonstrated that the C-terminal regions of four of these CPuORF-encoded peptides play a crucial role in repressing the translation of the main ORF [46]. The functionality of three A. thaliana CPuORFs in arresting ribosomes during translation was tested in another study. This mechanism of CPuORF action was clarified using toeprinting analysis and the additional experimental evidence was obtained by constructing the following three types of reporter constructs. (i) With the CPuORF initiation codon removed from each reporter construct of the native CPuORF by replacing AUG with AAG; (ii) with frameshift only mutations, introduced to the CPuORF sequences; and (iii) with both removed initiation codon and frameshift mutations in CPuORF sequences. A comparative testing of all types of reporter constructs has shown that removal of the initiation codon from CPuORFs considerably increases the reporter gene expression; the frameshift mutations in CPuORFs also efficiently increase the reporter gene expression, although to a lower degree as compared with the removal of initiation codon; while the simultaneous presence of frameshift mutations and absence of the initiation codon have almost no effect on the reporter gene expression. These results clearly demonstrate that (i) the peptide sequences are partially responsible for strong repressive effects of these CPuORFs on the main ORF expression; (ii) repression of the main ORF expression (in this case, the ORF of reporter transcript) depends on CPuORF translation; and (iii) these CPuORFs induce ribosomal arrest and thereby considerably inhibit expression of the main ORF [62]. Thus, it is possible not only to insert regulatory sequences that control the reporter gene translation into the constructs carrying this reporter, but also to introduce manifold modifications, which allows their functional role in a key biological process, translation, to be assessed.

4.6 Characteristics and Reasoning for Selection of a Verification Method(s)

It should be emphasized that the selection of a method for assembling reporter constructs is of highest importance, since it is necessary to clone the target regulatory sequences with a reporter gene without the introduction of additional nucleotides, which may influence the translation modulation. The classical restriction–ligation cloning method does not allow generation of such constructs and requires several cloning stages. Several more economical and efficient technologies facilitating and accelerating the design of such constructs have been recently proposed. These technologies make it possible to produce seamless fusions of a "regulatory sequence–reporter gene". Most of them utilize the recombination between homologous sequences residing at the ends of the DNA fragments to be assembled [63]. For example, the Gateway®cloning system is based on the well-characterized site-specific recombinase system of the lambda phage for recombination of DNA segments. The DNA segments are flanked by the ATT recombination sites, which provide seamless assembly of almost any sequence [64]. However, this system also has certain limitations, namely, the need that the sequences overlap for at least 15 nucleotides at their ends. Correspondingly, the assembly of nonoverlapping DNA fragments requires additional terminal extensions or the use of bridge oligonucleotides [65]. Moreover, in our view, the approaches, such as Golden Gate system [66,67] or CPEC (circular polymerase extension cloning) strategy are more purposeful for designing the regulatory sequence–reporter gene constructs [68]. The Golden Gate method utilizes the IIS type restriction fragments to generate 4-nucleotide sticky ends, flanking each DNA portion, which then can be effectively ligated using T4 ligase. The assembly is performed in a single reaction and gives a seamless or nearly seamless target construct. This is determined by that the IIS type recognition sites are removed during ligation leaving only four nucleotide which positions may be determined by researcher [69,70].

The principle underlying the CPEC method utilizes the polymerase extension mechanism and one DNA polymerase for the in vitro assembly and cloning of sequences in any vector in a single-stage reaction. CPEC allows for an integrated, combinatorial, or multifragment assembly of sequences as well as for routine cloning procedures [68]. Thus, the Golden Gate and CPEC technologies have important advantages suggesting their utility in designing of the reporter constructs intended for studying and experimentally verifying the role of the regulatory regions in transcripts during translation.

As for the functional assessment of the constructs carrying a target regulatory sequence fused with a reporter gene, two basic experimental approaches are used: they utilized (i) a transient (temporary) and/or (ii) stable (constant) expression of reporter constructs in plants [4,71]. In the case of the transient expression of reporter constructs, transfection of the protoplasts derived from leaves of N. tabacum L. cv. BY2, lettuce, or A. thaliana are used as well as agroinfiltration of leaves (N. benthamiana, N. excelsior, N. tabacum var. xanth, or A. thaliana) or plant cell suspension culture (N. tabacum BY2 or A. thaliana T87) [4,71,72] (Supplement Table S1). In the current view, the transient expression of reporter constructs is regarded as a more efficient approach because of lower material and time expenditures. The protocols for generation of protoplasts and agroinfiltration have been elaborated for many model and non-model plants, widening the range of the used plant objects. However, there are some limitations in the application of transient expression associated with the variation of protoplasts in the transformation efficiency and with the delivery of constructs to plant cells in agroinfiltration [4,73].

A stable expression of reporter constructs in plants requires production of transgenic plants or transgenic plant cell suspension cultures. This makes it possible to solve increasingly more complex problems of translational control, such as translation regulation under different growth and stress conditions or long-term physiological effects of certain changes in a sequence that modulate translation. In particular, polysome fractionation of control and transgenic plants makes it possible to confirm that the transcript of a reporter gene controlled by a tested regulatory sequence is actually associated with the polysome fraction. Thus, it is possible to assess the translational status of the mRNA of a reporter gene fused with the tested regulatory sequence, including under the effect of stress factors [8].

When using the methods involving both stable and transient expressions, the choice of an adequate control is of a paramount importance to ascertain that the change in expression of the reporter protein is actually associated with the change in translation (rather than with transcription, protein stability, and so on) [4,51].

According to the available experimental data, the strategy of reporter systems is in demand for a wide range of studies into individual regulatory sequences or their nucleotide contexts. The use of this strategy gives the unique data on the functional role of target sequences in translation efficiency; however, this strategy is, as a rule, supplemented with other methods. The choice of a particular method depends on the regulatory sequence or its context to be studied be it a full-sized 5'UTR or its regulatory elements, such as RNA G-quadruplex, IRES, uORFs, and so on, which is in part summarized above and is comprehensively described in the corresponding publications.

5. CONCISE SUMMARY

Translation is essential for the comprehensive execution of genetic information, and the recent insights into the complex, multilayered data encoded in the mRNA sequence are of utmost significance. Research on translation has uncovered numerous novel and intriguing insights regarding the structural and functional significance of mRNA regulatory sequences that facilitate differential translation. This accomplishment is attributed to the utilization of cutting-edge technology for evaluating the translational statuses of individual mRNA species on a genome-wide scale, in conjunction with computational algorithms and experimental verification approaches, as outlined here. (Figure 8).

Enhancing the understanding of regulatory settings in mRNA translation efficiency and their combinations necessitates advancements in both experimental methodologies and theoretical techniques. The researchers must have the opportunity to accurately ascertain the correlation between the observed fluctuations in transcript expression and the actual protein content in plants, as well as to precisely delineate and evaluate the contributions of individual regulatory contexts and their combinations within mRNAs. Thus, the necessity for the construction of a comprehensive information resource for this purpose appears highly justified. This resource will encompass information regarding (i) experimental methodologies for evaluating genome-wide translation alterations, including modifications and applicability to various plant species; (ii) tools for analyzing, interpreting, and visualizing polysome and ribosome profiling data; (iii) resources for constructing target sequences of plant transcripts and predicting their attributes; (iv) techniques for experimentally validating the regulatory codes of plant transcripts implicated in translation modulation; among other topics. This will establish the foundation for coordinating various investigations and provide insight into the intricate mechanisms governing the regulation of biological activities at the translation stage in plants. Furthermore, it will enhance the capabilities for future research and the potential applications of mRNA regulatory contexts, including their utilization in engineering novel plant genotypes with optimal allele combinations and the creation of new transgenes, incorporating genome editing technologies.

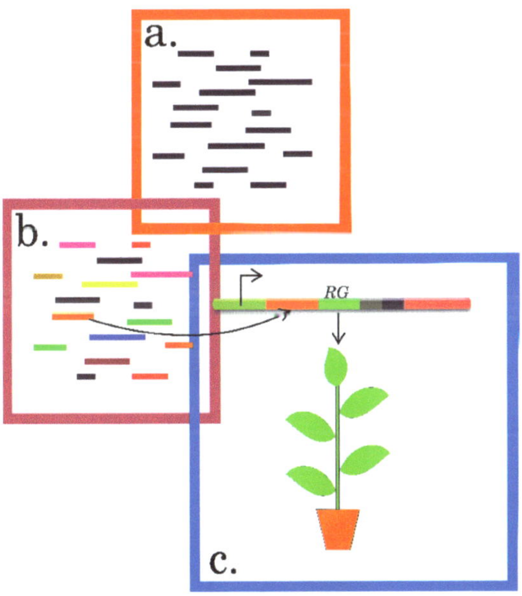

Figure 8. General strategy for identification, prediction, and experimental verification of the functional elements within transcripts that mediate their efficient translation. (a) Primary data on the transcripts differing in their translation efficiencies are obtained by sequencing the pools of the transcripts generated by polysome profiling, TRAP, or ribosome profiling. (b) In silico analysis of the transcript sequences identified the functional elements of transcripts. (c) Experimental verification of the predicted functional elements within transcripts (case study of reporter systems).

References

1. Baerenfaller, K.; Grossmann, J.; Grobei, M.A.; Hull, R.; Hirsch-Hoffmann, M.; Yalovsky, S.; Zimmermann, P.; Grossniklaus, U.; Gruissem, W.; Baginsky, S. Genome-scale proteomics reveals Arabidopsis thaliana gene models and proteome dynamics. Science 2008, 320, 938–941. [CrossRef] [PubMed]
2. Kawaguchi, R.; Bailey-Serres, J. mRNA sequence features that contribute to translational regulation in Arabidopsis. Nucleic Acids Res. 2005, 33, 955–965. [CrossRef] [PubMed]
3. Sablok, G.; Powell, J.J.; Kazan, K. Emerging Roles and Landscape of Translating mRNAs in Plants. Front. Plant Sci. 2017, 8, 1443. [CrossRef] [PubMed]
4. Mazzoni-Putman, S.M.; Stepanova, A.N. A Plant Biologist's Toolbox to Study Translation. Front. Plant Sci. 2018, 9, 873. [CrossRef] [PubMed]
5. Lecampion, C.; Floris, M.; Fantino, J.R.; Robaglia, C.; Laloi, C. An Easy Method for Plant Polysome Profiling. J. Vis. Exp. 2016. [CrossRef] [PubMed]
6. Sharma, V.; Salwan, R.; Sharma, P.N.; Gulati, A. Integrated Translatome and Proteome: Approach for Accurate Portraying of Widespread Multifunctional Aspects of Trichoderma. Front. Microbiol. 2017, 8, 1602. [CrossRef] [PubMed]
7. Yamasaki, S.; Matsuura, H.; Demura, T.; Kato, K. Changes in Polysome Association of mRNA Throughout Growth and Development in Arabidopsis thaliana. Plant Cell Physiol. 2015, 56, 2169–2180. [CrossRef]
8. Matsuura, H.; Takenami, S.; Kubo, Y.; Ueda, K.; Ueda, A.; Yamaguchi, M.; Hirata, K.; Demura, T.; Kanaya, S.; Kato, K. A computational and experimental approach reveals that the 5'-proximal region of the 5'-UTR has a Cis-regulatory signature responsible for heat stress-regulated mRNA translation in Arabidopsis. Plant Cell Physiol. 2013, 54, 474–483. [CrossRef]
9. Yamasaki, S.; Sanada, Y.; Imase, R.; Matsuura, H.; Ueno, D.; Demura, T.; Kato, K. Arabidopsis thaliana cold-regulated 47 gene 5t-untranslated region enables stable high-level expression of transgenes. J. Biosci. Bioeng. 2018, 125, 124–130. [CrossRef]
10. Matsuura, H.; Ishibashi, Y.; Shinmyo, A.; Kanaya, S.; Kato, K. Genome-wide analyses of early translational responses to elevated temperature and high salinity in Arabidopsis thaliana. Plant Cell Physiol. 2010, 51, 448–462. [CrossRef]
11. Juntawong, P.; Hummel, M.; Bazin, J.; Bailey-Serres, J. Ribosome profiling: A tool for quantitative evaluation of dynamics in mRNA translation. Methods Mol. Biol. 2015, 1284, 139–173. [CrossRef] [PubMed]
12. Zanetti, M.E.; Chang, I.F.; Gong, F.; Galbraith, D.W.; Bailey-Serres, J. Immunopurification of polyribosomal complexes of Arabidopsis for global

analysis of gene expression. Plant Physiol. 2005, 138, 624–635. [CrossRef] [PubMed]

13. Mustroph, A.; Juntawong, P.; Bailey-Serres, J. Isolation of plant polysomal mRNA by differential centrifugation and ribosome immunopurification methods. Methods Mol. Biol. 2009, 553, 109–126. [CrossRef]
14. Mustroph, A.; Zanetti, M.E.; Girke, T.; Bailey-Serres, J. Isolation and analysis of mRNAs from specific cell types of plants by ribosome immunopurification. Methods Mol. Biol. 2013, 959, 277–302. [CrossRef] [PubMed]
15. Lin, S.Y.; Chen, P.W.; Chuang, M.H.; Juntawong, P.; Bailey-Serres, J.; Jauh, G.Y. Profiling of translatomes of in vivo-grown pollen tubes reveals genes with roles in micropylar guidance during pollination in Arabidopsis. Plant Cell 2014, 26, 602–618. [CrossRef] [PubMed]
16. Jiao, Y.; Meyerowitz, E.M. Cell-type specific analysis of translating RNAs in developing flowers reveals new levels of control. Mol. Syst. Biol. 2010, 6, 419. [CrossRef] [PubMed]
17. Sorenson, R.; Bailey-Serres, J. Selective mRNA sequestration by OLIGOURIDYLATE-BINDING PROTEIN 1 contributes to translational control during hypoxia in Arabidopsis. Proc. Natl. Acad. Sci. USA 2014, 111, 2373–2378. [CrossRef] [PubMed]
18. Ingolia, N.T.; Ghaemmaghami, S.; Newman, J.R.; Weissman, J.S. Genome-wide analysis in vivo of translation with nucleotide resolution using ribosome profiling. Science 2009, 324, 218–223. [CrossRef]
19. McGlincy, N.J.; Ingolia, N.T. Transcriptome-wide measurement of translation by ribosome profiling. Methods 2017, 126, 112–129. [CrossRef]
20. Chotewutmontri, P.; Stiffler, N.; Watkins, K.P.; Barkan, A. Ribosome Profiling in Maize. Methods Mol. Biol. 2018, 1676, 165–183. [CrossRef]
21. Juntawong, P.; Girke, T.; Bazin, J.; Bailey-Serres, J. Translational dynamics revealed by genome-wide profiling of ribosome footprints in Arabidopsis. Proc. Natl. Acad. Sci. USA 2014, 111, E203–212. [CrossRef] [PubMed]
22. Planchard, N.; Bertin, P.; Quadrado, M.; Dargel-Graffin, C.; Hatin, I.; Namy, O.; Mireau, H. The translational landscape of Arabidopsis mitochondria. Nucleic Acids Res. 2018, 46, 6218–6228. [CrossRef] [PubMed]
23. Gerashchenko, M.V.; Gladyshev, V.N. Ribonuclease selection for ribosome profiling. Nucleic Acids Res. 2017, 45, e6. [CrossRef] [PubMed]
24. Eastman, G.; Smircich, P.; Sotelo-Silveira, J.R. Following Ribosome Footprints to Understand Translation at a Genome Wide Level. Comput. Struct. Biotechnol. J. 2018, 16, 167–176. [CrossRef] [PubMed]

25. Hsu, P.Y.; Benfey, P.N. Small but Mighty: Functional Peptides Encoded by Small ORFs in Plants. Proteomics 2018, 18, e1700038. [CrossRef] [PubMed]

26. Lei, L.; Shi, J.; Chen, J.; Zhang, M.; Sun, S.; Xie, S.; Li, X.; Zeng, B.; Peng, L.; Hauck, A.; et al. Ribosome profiling reveals dynamic translational landscape in maize seedlings under drought stress. Plant J. 2015, 84, 1206–1218. [CrossRef]

27. Hsu, P.Y.; Calviello, L.; Wu, H.L.; Li, F.W.; Rothfels, C.J.; Ohler, U.; Benfey, P.N. Super-resolution ribosome profiling reveals unannotated translation events in Arabidopsis. Proc. Natl. Acad. Sci. USA 2016, 113, E7126–E7135. [CrossRef]

28. Lukoszek, R.; Feist, P.; Ignatova, Z. Insights into the adaptive response of Arabidopsis thaliana to prolonged thermal stress by ribosomal profiling and RNA-Seq. BMC Plant Biol. 2016, 16, 221. [CrossRef]

29. Chotewutmontri, P.; Barkan, A. Multilevel effects of light on ribosome dynamics in chloroplasts program genome-wide and psbA-specific changes in translation. PLoS Genet. 2018, 14, e1007555. [CrossRef]

30. Andreev, D.E.; O'Connor, P.B.; Loughran, G.; Dmitriev, S.E.; Baranov, P.V.; Shatsky, I.N. Insights into the mechanisms of eukaryotic translation gained with ribosome profiling. Nucleic Acids Res. 2017, 45, 513–526. [CrossRef]

31. Lamesch, P.; Berardini, T.Z.; Li, D.; Swarbreck, D.; Wilks, C.; Sasidharan, R.; Muller, R.; Dreher, K.; Alexander, D.L.; Garcia-Hernandez, M.; et al. The Arabidopsis Information Resource (TAIR): Improved gene annotation and new tools. Nucleic Acids Res. 2012, 40, D1202–1210. [CrossRef]

32. Yanguez, E.; Castro-Sanz, A.B.; Fernandez-Bautista, N.; Oliveros, J.C.; Castellano, M.M. Analysis of genome-wide changes in the translatome of Arabidopsis seedlings subjected to heat stress. PLoS ONE 2013, 8, e71425. [CrossRef] [PubMed]

33. Liu, M.J.; Wu, S.H.; Chen, H.M.; Wu, S.H. Widespread translational control contributes to the regulation of Arabidopsis photomorphogenesis. Mol. Syst. Biol. 2012, 8, 566. [CrossRef] [PubMed]

34. Basbouss-Serhal, I.; Pateyron, S.; Cochet, F.; Leymarie, J.; Bailly, C. 5t to 3t mRNA Decay Contributes to the Regulation of Arabidopsis Seed Germination by Dormancy. Plant Physiol. 2017, 173, 1709–1723. [CrossRef] [PubMed]

35. Mustafaev, O.; Sadovskaya, N.S.; Tyurin, A.A.; Goldenkova-Pavlova, I.V. JetGene: An integrated database for analysis of regulatory regions or nucleotide contexts in plant differentially translated transcripts. manuscript in preparation.

36. Deyneko, I.V.; Kel, A.E.; Bloecker, H.; Kauer, G. Signal-theoretical DNA similarity measure revealing unexpected similarities of E. coli promoters. In Silico Biol. 2005, 5, 547–555. [PubMed]

37. Deyneko, I.V.; Kalybaeva, Y.M.; Kel, A.E.; Blöcker, H. Human-chimpanzee promoter comparisons: Property-conserved evolution? Genomics 2010, 96, 129–133. [CrossRef] [PubMed]

38. Bailey, T.L.; Johnson, J.; Grant, C.E.; Noble, W.S. The MEME Suite. Nucleic Acids Res. 2015, 43, W39–49. [CrossRef]

39. Williams, B.P.; Burgess, S.J.; Reyna-Llorens, I.; Knerova, J.; Aubry, S.; Stanley, S.; Hibberd, J.M. An Untranslated cis-Element Regulates the Accumulation of Multiple C4 Enzymes in Gynandropsis gynandra Mesophyll Cells. Plant Cell 2016, 28, 454–465. [CrossRef]

40. Merchante, C.; Stepanova, A.N.; Alonso, J.M. Translation regulation in plants: An interesting past, an exciting present and a promising future. Plant J. 2017, 90, 628–653. [CrossRef]

41. Thompson, S.R. So you want to know if your message has an IRES? Wiley Interdiscip. Rev. RNA 2012, 3, 697–705. [CrossRef]

42. Cobbold, L.C.; Spriggs, K.A.; Haines, S.J.; Dobbyn, H.C.; Hayes, C.; de Moor, C.H.; Lilley, K.S.; Bushell, M.; Willis, A.E. Identification of internal ribosome entry segment (IRES)-trans-acting factors for the Myc family of IRESs. Mol. Cell. Biol. 2008, 28, 40–49. [CrossRef] [PubMed]

43. Sheshukova, E.V.; Komarova, T.V.; Ershova, N.M.; Shindyapina, A.V.; Dorokhov, Y.L. An Alternative Nested Reading Frame May Participate in the Stress-Dependent Expression of a Plant Gene. Front. Plant Sci. 2017, 8, 2137. [CrossRef] [PubMed]

44. Andrews, S.J.; Rothnagel, J.A. Emerging evidence for functional peptides encoded by short open reading frames. Nat. Rev. Genet. 2014, 15, 193–204. [CrossRef] [PubMed]

45. Takahashi, H.; Takahashi, A.; Naito, S.; Onouchi, H. BAIUCAS: A novel BLAST-based algorithm for the identification of upstream open reading frames with conserved amino acid sequences and its application to the Arabidopsis thaliana genome. Bioinformatics 2012, 28, 2231–2241. [CrossRef] [PubMed]

46. Ebina, I.; Takemoto-Tsutsumi, M.; Watanabe, S.; Koyama, H.; Endo, Y.; Kimata, K.; Igarashi, T.; Murakami, K.; Kudo, R.; Ohsumi, A.; et al. Identification of novel Arabidopsis thaliana upstream open reading frames that control expression of the main coding sequences in a peptide sequence-dependent manner. Nucleic Acids Res. 2015, 43, 1562–1576. [CrossRef] [PubMed]

47. Tunney, R.; McGlincy, N.J.; Graham, M.E.; Naddaf, N.; Pachter, L.; Lareau, L.F. Accurate design of translational output by a neural network model of ribosome distribution. Nat. Struct. Mol. Biol. 2018, 25, 577–582. [CrossRef] [PubMed]

48. Hill, S.T.; Kuintzle, R.; Teegarden, A.; Merrill, E., 3rd; Danaee, P.; Hendrix, D.A. A deep recurrent neural network discovers complex biological rules to decipher RNA protein-coding potential. Nucleic Acids Res. 2018, 46, 8105–8113. [CrossRef] [PubMed]

49. Ching, T.; Zhu, X.; Garmire, L.X. Cox-nnet: An artificial neural network method for prognosis prediction of high-throughput omics data. PLoS Comput. Biol. 2018, 14, e1006076. [CrossRef]

50. Cao, Z.; Zhang, S. Simple tricks of convolutional neural network architectures improve DNA-protein binding prediction. Bioinformatics 2018. [CrossRef]

51. Tyurin, A.A.; Kabardaeva, K.V.; Gra, O.A.; Mustafaev, O.M.; Sadovskaya, N.S.; Pavlenko, O.S.; Goldenkova-Pavlov, I.V. Efficient expression of a heterologous gene in plants depends on the nucleotide composition of mRNA's 5'-region. Russ. J. Plant. Physiol. 2016, 4, 511–522. [CrossRef]

52. Anami, S.; Njuguna, E.; Coussens, G.; Aesaert, S.; Van Lijsebettens, M. Higher plant transformation: Principles and molecular tools. Int. J. Dev. Biol. 2013, 57, 483–494. [CrossRef] [PubMed]

53. Cho, H.; Cho, H.S.; Nam, H.; Jo, H.; Yoon, J.; Park, C.; Dang, T.V.T.; Kim, E.; Jeong, J.; Park, S.; et al. Translational control of phloem development by RNA G-quadruplex-JULGI determines plant sink strength. Nat. Plants 2018, 4, 376–390. [CrossRef] [PubMed]

54. Alvarez, D.; Voss, B.; Maass, D.; Wust, F.; Schaub, P.; Beyer, P.; Welsch, R. Carotenogenesis Is Regulated by 5'UTR-Mediated Translation of Phytoene Synthase Splice Variants. Plant Physiol. 2016, 172, 2314–2326. [CrossRef]

55. Dorokhov, Y.L.; Skulachev, M.V.; Ivanov, P.A.; Zvereva, S.D.; Tjulkina, L.G.; Merits, A.; Gleba, Y.Y.; Hohn, T.; Atabekov, J.G. Polypurine (A)-rich sequences promote cross-kingdom conservation of internal ribosome entry. Proc. Natl. Acad. Sci. USA 2002, 99, 5301–5306. [CrossRef] [PubMed]

56. Ali, Z.; Schumacher, H.M.; Heine-Dobbernack, E.; El-Banna, A.; Hafeez, F.Y.; Jacobsen, H.J.; Kiesecker, H. Dicistronic binary vector system-A versatile tool for gene expression studies in cell cultures and plants. J. Biotechnol. 2010, 145, 9–16. [CrossRef] [PubMed]

57. Jimenez-Gonzalez, A.S.; Fernandez, N.; Martinez-Salas, E.; Sanchez de Jimenez, E. Functional and structural analysis of maize hsp101 IRES. PLoS ONE 2014, 9, e107459. [CrossRef]

58. Cui, Y.; Rao, S.; Chang, B.; Wang, X.; Zhang, K.; Hou, X.; Zhu, X.; Wu, H.; Tian, Z.; Zhao, Z.; et al. AtLa1 protein initiates IRES-dependent translation of WUSCHEL mRNA and regulates the stem cell homeostasis of Arabidopsis in response to environmental hazards. Plant Cell Environ. 2015, 38, 2098–2114. [CrossRef]

59. Jorgensen, R.A.; Dorantes-Acosta, A.E. Conserved Peptide Upstream Open Reading Frames are Associated with Regulatory Genes in Angiosperms. Front. Plant Sci. 2012, 3, 191. [CrossRef]

60. Tanaka, M.; Sotta, N.; Yamazumi, Y.; Yamashita, Y.; Miwa, K.; Murota, K.; Chiba, Y.; Hirai, M.Y.; Akiyama, T.; Onouchi, H.; et al. The Minimum Open Reading Frame, AUG-Stop, Induces Boron-Dependent Ribosome Stalling and mRNA Degradation. Plant Cell 2016, 28, 2830–2849. [CrossRef]

61. Xu, G.; Yuan, M.; Ai, C.; Liu, L.; Zhuang, E.; Karapetyan, S.; Wang, S.; Dong, X. uORF-mediated translation allows engineered plant disease resistance without fitness costs. Nature 2017, 545, 491–494. [CrossRef]

62. Hayashi, N.; Sasaki, S.; Takahashi, H.; Yamashita, Y.; Naito, S.; Onouchi, H. Identification of Arabidopsis thaliana upstream open reading frames encoding peptide sequences that cause ribosomal arrest. Nucleic Acids Res. 2017, 45, 8844–8858. [CrossRef] [PubMed]

63. Li, M.Z.; Elledge, S.J. Harnessing homologous recombination in vitro to generate recombinant DNA via SLIC. Nat. Methods 2007, 4, 251–256. [CrossRef] [PubMed]

64. Smedley, M.A.; Harwood, W.A. Gateway(R)-compatible plant transformation vectors. Methods Mol. Biol. 2015, 1223, 3–16. [CrossRef] [PubMed]

65. Tsvetanova, B.; Peng, L.; Liang, X.; Li, K.; Yang, J.P.; Ho, T.; Shirley, J.; Xu, L.; Potter, J.; Kudlicki, W.; et al. Genetic assembly tools for synthetic biology. Methods Enzymol. 2011, 498, 327–348. [CrossRef] [PubMed]

66. Engler, C.; Gruetzner, R.; Kandzia, R.; Marillonnet, S. Golden gate shuffling: A one-pot DNA shuffling method based on type IIs restriction enzymes. PLoS ONE 2009, 4, e5553. [CrossRef] [PubMed]

67. Engler, C.; Kandzia, R.; Marillonnet, S. A one pot, one step, precision cloning method with high throughput capability. PLoS ONE 2008, 3, e3647. [CrossRef] [PubMed]

68. Quan, J.; Tian, J. Circular polymerase extension cloning. Methods Mol. Biol. 2014, 1116, 103–117. [CrossRef] [PubMed]

69. Engler, C.; Marillonnet, S. Golden Gate cloning. Methods Mol. Biol. 2014, 1116, 119–131. [CrossRef]

70. Sarrion-Perdigones, A.; Vazquez-Vilar, M.; Palaci, J.; Castelijns, B.; Forment, J.; Ziarsolo, P.; Blanca, J.; Granell, A.; Orzaez, D. GoldenBraid 2.0: A comprehensive DNA assembly framework for plant synthetic biology. Plant Physiol. 2013, 162, 1618–1631. [CrossRef]

71. Vyacheslavova, A.O.; Berdichevets, I.N.; Tyurin, A.A.; Shimshilashvili, K.R.; Mustafaev, O.N.; Goldenkova-Pavlova, I.V. Expression of heterologous genes in plant systems: New possibilities. Russ. J. Genet. 2012, 48, 1067–1079. [CrossRef]

72. Agarwal, P.; Garg, V.; Gautam, T.; Pillai, B.; Kanoria, S.; Burma, P.K. A study on the influence of different promoter and 5'UTR (URM) cassettes from Arabidopsis thaliana on the expression level of the reporter gene beta glucuronidase in tobacco and cotton. Transgenic Res. 2014, 23, 351–363. [CrossRef] [PubMed]

73. Tyurin, A.A.; Kabardaeva, K.V.; Berestovoy, M.A.; Sidorchuk, Y.V.; Fomenkov, A.A.; Nosov, A.V.; Goldenkova-Pavlova, I.V. Simple and reliable system for transient gene expression for the characteristic signal sequences and the estimation of the localization of target protein in plant cell. Russ. J. Plant. Physiol. 2017, 64, 672–679. [CrossRef]

5 Chapter

A SIGNIFICANT CHALLENGE IN SEQUENCING EXTENSIVE PLANT GENOMES IS DETERMINING THEIR ACTUAL SIZE

1. INTRODUCTION: AN OVERVIEW

As knowledge of the organization and function of plant and animal genomes expands, it becomes increasingly evident that a comprehensive understanding of genome function necessitates the acquisition of a whole sequence. The substantial throughput provided by contemporary short-read DNA sequencing technologies enables the sequencing of genomes of any magnitude with great sequencing depth. The assembly of single and low-copy sequences is facilitated; however, including repeating sequences is a significant problem. Additionally, sequence redundancy resulting from polyploidy constitutes a substantial barrier to obtaining gap-free long-range genome sequences.

A reference genome assembly seeks to accurately depict a complete genome sequence, preferably with each chromosome represented by a singular, contiguous pseudomolecule. Determining the completeness of an assembly is challenging, particularly with complex genomes characterized by repetitive DNA, segmental duplications, and, in polyploid genomes, homoeologs, all of which hinder the creation of an accurate assembly. Consequently, gaps, mis-assemblies, and collapsed tandem repeats are prevalent in most published genome sequences. A commonly employed computational technique for quantifying a nuclear genome is based on k-mer frequency analysis [1,2]. An alternate approach could involve quantifying the total number of complete LTR-retrotransposons. Given that their quantity escalates linearly with genome size, particularly in grass species, it may function as an indicator of assembly quality [3]. The genome size of an unidentified species may be inferred through extrapolation, utilizing data from species with established genome sizes. Nonetheless, as both methodologies depend on sequence data, the sole genuinely independent method to ascertain genome size is through the experimental measurement of DNA quantity in the nuclei.

2. ESTIMATION OF GENOME SIZE

Two experimental approaches have been developed to estimate nuclear DNA amounts: biochemical and cytometric. The former seeks to quantify the DNA harbored within a known mass of plant tissue [4]; its weakness lies in the errors inherent in the estimation of the number of nuclei present in the sample, in the unknown proportion of nuclei present at each of the various different cell cycle stages and the non-estimable proportion of endo-reduplicated nuclei present. As a result, cytometry-based estimations tend to be preferred, since these are designed to quantify the DNA present in a population of nuclei at a known cell cycle stage [5]. The attempt by [6] to derive relative nuclear DNA amounts present in several plant species using Feulgen micro-densitometry led to the development of the now universally understood C-value terminology, where un-replicated haploid nuclei contain a 1C DNA amount; the terminology has been refined in recent years [7]. Feulgen microdensitometry was phased out during the 1980s as a result of the throughput benefits offered by flow cytometry, which offers the possibility of analyzing large numbers of isolated nuclei in a short time [5].

It is important to note that flow cytometry does not quantify nuclear DNA directly, but rather achieves this by capturing the signal emitted from fluorochrome-stained nuclei. In order to determine a nuclear DNA amount in absolute units, the fluorescence of an unknown sample has to be compared with that of a reference standard of known genome size [8]. To avoid errors due to non-linearity, an ideal reference standard should not differ in size by more than two or three-fold from the test sample, implying that a set of reference standards is needed in order to cope with the large range of genome size encountered among higher organisms. The question then becomes how to calibrate these reference standards if none of the candidate species has itself been completely sequenced.

3. STANDARDIZATION

Not unexpectedly, the issue applies as much to animal to plant or fungal genomes. To enable a comparison of data obtained by different laboratories, Tiersch et al. [9] calibrated a set of animal reference standards, choosing human male leukocytes (7 pg DNA/2C) as the primary reference; the 7 pg figure was based on estimates derived from Feulgen micro-densitometry [10]. The experiments derived a 2C value of 2.5 pg DNA for domestic chicken, which was close to the value given by [11]. The domestic chicken genome has been adopted since this time as the most widely used reference standard for the sizing of animal genomes [12]. In an effort to enable comparisons between animal and plant genomes, Doležel et al. [8] recommended a set of plant reference standards (Table 1), also calibrated with respect to the human male leukocyte genome, assuming the 7 pg value assigned by Tiersch et al. [9]. Over the past three decades, hundreds of genome size estimates have been published, based mainly on the 7 pg value. The question is how close to reality these estimates really are, which relates in the main to how accurate the 7 pg figure is. According to the arguments made by Doležel and Greilhuber [13], the value most probably over-estimates the true value by 5–10%.

Table 1. Plant DNA reference standards calibrated for the estimation of nuclear DNA amounts in absolute units [8].

Plant Species and Cultivar *	2C DNA Content (pg DNA) **
Raphanus sativus L. 'Saxa'	1.11
Solanum lycopersicum L. 'Stupické polní rané'	1.96
Glycine max Merr. 'Polanka'	2.50
Zea mays L. 'CE-777'	5.43
Pisum sativum L. 'Ctirad'	9.09
Secale cereale L. 'Danˇkovské'	16.19
Vicia faba L. 'Inovec'	26.90
Allium cepa L. 'Alice'	34.89

* Seeds of the reference standards can be obtained from the corresponding author free of charge at dolezel@ueb.cas.cz. Since the year 2000, seed samples were provided to 615 research projects worldwide. ** Estimated after considering 7 pg DNA/2C for human [9].

4. THE HUMAN GENOME AS A UNIVERSAL REFERENCE STANDARD

Seventeen years have passed since the joint announcement of the human genome sequence [14,15]. This period has seen a number of attempts to complete the assembly, applying a variety of technologies [16,17]. All of these have reported a smaller genome size than what has, as of the end of 2017, been suggested in GRCh38.p12, the most recently released Genome Reference Consortium version, which comprises 3,257,319,537 bp. Assuming the Doležel et al. [18] conversion of 1 pg = 0.978 Gbp, 3.5 pg 1C DNA is equivalent to 3,423,000,000 bases. Thus, the 7 pg value represents an ~5.1% over-estimate of the GRCh38.p12 assembly prediction. This difference lies at the lower end of the error range predicted by Doležel and Greilhuber [13]. Given that the human reference genome is still incomplete, the expectation is that the gap between the 7 pg figure and the "real" human genome size will continue to diminish. Nevertheless, a 5% error is not dissimilar to the variation observed between estimates of nuclear DNA amounts of a given species produced by different laboratories [19,20]. Thus, the recommendation remains that the 7 pg figure continue to be used as the reference for measuring 2C values of both animal and plant genomes.

5. SIZING THE LARGE TRITICEAE GENOMES

Three species belonging to the tribe Triticeae—namely, bread wheat (Triticum aestivum), barley (Hordeum vulgare) and to a lesser extent, cereal rye (Secale cereale)—provide a major proportion of the calories used by humans and their livestock across the temperate world. The acquisition of their genome sequences will facilitate marker- and genomics-assisted breeding, gene editing and other novel breeding technologies currently under development. Reference genome sequences have been published for barley [21], wild emmer wheat (T. dicoccoides) [22] and hexaploid bread wheat (T. aestivum) [3], and one for cereal rye is currently being finalized (Nils Stein, pers. comm.). Here, flow cytometry was utilized to assess the nuclear DNA content of wild emmer, bread wheat, barley and cereal rye. To minimize errors due to copy number variants and intraspecific differences in genome size, the accessions of each species were those used for the acquisition of their genome sequences. The cereal rye cultivar Danˇkovské (16.19 pg/2C) and garden pea (Pisum sativum) cultivar Ctirad (9.09 pg/2C) were used as reference standards (Table 1). Rye was selected out of the calibrated reference standards (Table 1) as its 2C value was close to 2 C DNA amounts of tetraploid and hexaploid wheat and barley. However, this standard could not be used for another accession of rye and thus pea was employed as the second standard. The outcomes are summarized in Table 2.

Table 2. Estimation of nuclear DNA amounts in the four Triticeae species.

Species and Genotype	2C Nuclear DNA Content (pg) *		Reference Standard
	Mean	± SD	
Triticum aestivum cv.Chinese Spring	33.91	0.27	*Secale cereale* cv. Danˇkovské
Triticum dicoccoides cv. Zavitan	25.11	0.16	*Secale cereale* cv. Danˇkovské
Hordeum vulgare cv.Morex	10.31	0.09	*Secale cereale* cv. Danˇkovské
Secale cereale inbredlineLo7	15.95	0.11	*Pisum sativum* cv.Ctirad

* Considering 7 pg DNA/2C for human [9].

6.COMPLETENESSOFTHECURRENTTRITICEAEREFERENCEGENOME ASSEMBLIES

To estimate the completeness of reference genome assemblies of the four test-species, the sizes predicted by each of their assemblies were compared with their estimated genome sizes as derived by flow cytometry. Taking the [9] figure of 7 pg DNA/2C, the conclusion was that the Triticeae assemblies represent at least 85% of their full genome (Table 3). However, adopting the GRCh38.p12 with 1C genome size of 3,257,319,537 bases as the reference, increased the coverage to at least 88%. It should be noted, however, that the data on the size of genome assembly do not inform about its quality, i.e., the correct ordering and orientation DNA contigs. This parameter needs to be assessed using other methods than flow cytometry.

Table 3. The estimated level of completeness of the four Triticeae reference genome assemblies.

Flow Cytometric Estimation of 1C Genome Size ** GRCh38.12 [9]

Species	Reference Genome	Flow Cytometric Estimation of 1C Genome Size **			
		GRCh38.12		[9]	
	Assembly (Gbp) *	Genome Size (Gbp)	Assembly Coverage (%)	Genome size (Gbp)	Assembly Coverage (%)
H. vulgare	4.79	4.88	98	5.04	95
S. cereale	6.67	7.42	90	7.80	86
T. dicoccoides	10.50	11.87	88	12.28	85
T. aestivum	14.50	16.03	90***	16.58	87

* Reference genome assemblies: H. vulgare [21], T. dicoccoides [22], T. aestivum [3], S. cereale (Nils Stein, pers. comm.). ** Two different values were used for human 1C genome size as a primary reference standard: 3,257,319,537 bp (GRCh38.p12) and 3,423,000,000 bp [9]. *** Slightly higher value (92%) was estimated by the International Wheat Genome Sequencing Consortium [3] when considering human genome size of 3,253,848,404 bases (Human Genome Assembly GRCh38.p11).

7. MATERIALS AND METHODS

Grain of hexaploid bread wheat cultivar (cv.) Chinese Spring were obtained from P. Sourdille (INRA Clermont-Ferrand, Clermont-Ferrand, France), those of T. dicoccoides (accession Zavitan) from A. Distelfeld (Tel Aviv University, Tel Aviv, Israel), those of barley cv. Morex from Nils Stein (IPK, Gatersleben, Germany) and those of cereal rye inbred line Lo7 from Eva Bauer (Technische Universität Munich, Munich, Germany). Grains of cereal rye cv. Danˇkovské and seed of pea cv. Ctirad were obtained from, respectively, the Oseva Agro (Brno, Czech Republic) and Semo (Smržice, Czech Republic) breeding stations. Plants were raised in garden compost in pots and maintained in a greenhouse until they reached a height of 10–15 cm. Nuclei were extracted from leaves and suspended in preparation for flow cytometry following the methods given by [23]. Briefly, 10 mg of leaf tissue of each of the sample species and one of the two reference standards were chopped together in a 1 mL volume of LB01 solution [23] using a razor blade. The resulting homogenate was filtered through a 50-μm nylon mesh. The filtrate was made up to 50 μg/mL RNase and 50 μg/mL propidium iodide, and subjected to flow cytometry using a CyFlow Space flow cytometer (Sysmex Partec GmbH, Görlitz, Germany) equipped with a 532 nm green laser. The gain of the instrument was adjusted so that the peak representing G1 nuclei of the standard was positioned approximately on channel 100 on a histogram of relative fluorescence intensity when using a 512-channel scale. Five individual plants per each test species were sampled, and each sample was analyzed three times, each time on a different day. A minimum of 5000 nuclei per sample was analyzed and 2C DNA contents (in pg) were calculated from the means of the G1 peak positions by applying the formula (sample G1 peak mean) × (standard 2C DNA content)/(standard G1 peak mean). DNA contents in pg were converted to genome lengths in bp using the factor suggested by Doležel et al. [18], i.e., 1 pg DNA = 0.978 Gbp.

8. CONCISE SUMMARY

Cytometric techniques for estimating nuclear genome size, independent of DNA sequence data, necessitate a reference standard with a known genome size. The predominant animal and plant DNA reference standards have been calibrated using the human male leukocyte genome, with an assumed length of 3.42 Gbp/1C (and a 2C content of 7 pg DNA), despite this estimate being 5.1% greater than the latest assembly prediction; however, considering that the GRCh38.p12 assembly is likely still incomplete, the actual discrepancy may be less significant. Therefore, it becomes prudent to maintain this figure for the time being. The implementation of a standardized protocol will enable comparisons of results acquired across various laboratories. Upon achieving a greater precision in determining the size of the human genome, recalculating the sizes of previously estimated genomes will be uncomplicated.

REFERENCES

1. He, K.; Lin, K.; Wang, G.; Li, F. Genome sizes of nine insect species determined by flow cytometry and k-mer analysis. Front Physiol. 2016, 7, 569. [CrossRef] [PubMed]
2. Sun, H.; Ding, J.; Piednoël, M.; Schneeberger, K. findGSE: Estimating genome size variation within human and Arabidopsis using k-mer frequencies. Bioinformatics 2018, 34, 550–557. [CrossRef] [PubMed]
3. International Wheat Genome Sequencing Consortium. Shifting the limits in wheat research and breeding using a fully annotated reference genome. Science 2018, 361, eaar7191. [CrossRef] [PubMed]
4. Van't Hof, J. Cell population kinetics of excised roots of Pisum sativum. J. Cell Biol. 1965, 27, 179–189. [CrossRef] [PubMed]
5. Doležel, J.; Bartoš, J. Plant DNA flow cytometry and estimation of nuclear genome size. Ann. Bot. 2005, 95, 99–110. [CrossRef] [PubMed]

6. Swift, H. The constancy of desoxyribose nucleic acid in plant nuclei. Proc. Natl. Acad. Sci. USA 1950, 36, 643–654. [CrossRef] [PubMed]

7. Greilhuber, J.; Doležel, J.; Lysák, M.A.; Bennett, M.D. The origin, evolution and proposed stabilization of the terms 'genome size', and 'C-value' to describe nuclear DNA contents. Ann. Bot. 2005, 95, 255–260. [CrossRef] [PubMed]

8. Doležel, J.; Greilhuber, J.; Suda, J. Estimation of nuclear DNA content in plants using flow cytometry. Nat. Protoc. 2007, 2, 2233–2244. [CrossRef] [PubMed]

9. Tiersch, T.R.; Chandler, R.W.; Wachtel, S.S.; Elias, S. Reference standards for flow cytometry and application in comparative studies of nuclear DNA content. Cytometry 1989, 10, 706–710. [CrossRef] [PubMed]

10. Shapiro, H.S. Deoxyribonucleic acid content per cell of various organisms. In Handbook of Biochemistry and Molecular Biology; Fasman, G.D., Ed.; CRC Press: Cleveland, OH, USA, 1976; Volume 2, pp. 284–306.

11. Rasch, E.M.; Barr, H.J.; Rasch, R.W. The DNA content of sperm of Drosophila melanogaster. Chromosoma 1971, 33, 1–18. [CrossRef] [PubMed]

12. Gregory, T.R. Animal Genome Size Database. 2005. Available online: http://www.genomesize.com (accessed on 25 October 2018).

13. Doležel, J.; Greilhuber, J. Nuclear genome size: Are we getting closer? Cytometry 2010, 77, 635–642. [CrossRef] [PubMed]

14. International Human Genome Sequencing Consortium. Initial sequencing and analysis of the human genome. Nature 2001, 409, 860–921. [CrossRef] [PubMed]

15. Venter, J.C.; Adams, M.D.; Myers, W.W.; Li, P.W.; Mural, R.J.; Sutton, G.G. The sequence of the human genome. Science 2001, 291, 1304–1351. [CrossRef] [PubMed]

16. Seo, J.S.; Rhie, A.; Kim, J.; Lee, S.; Sohn, M.H.; Kim, C.U.; Hastie, A.; Cao, H.; Yun, J.Y.; Kim, J.; et al. De novo assembly and phasing of a Korean human genome. Nature 2016, 538, 243–247. [CrossRef] [PubMed]

17. Jain, M.; Koren, S.; Miga, K.H.; Quick, J.; Rand, A.C.; Sasani, T.A.; Tyson, J.R.; Beggs, A.D.; Dilthey, A.T.; Fiddes, I.T.; et al. Nanopore sequencing and assembly of a human genome with ultra-long reads. Nat. Biotechnol. 2018, 36, 338–345. [CrossRef] [PubMed]

18. Doležel, J.; Bartoš, J.; Voglmayr, H.; Greilhuber, J. Nuclear DNA content and genome size of trout and human. Cytometry 2003, 51, 127–128. [CrossRef] [PubMed]

19. Doležel, J.; Greilhuber, J.; Lucretti, S.; Meister, A.; Lysák, M.A.; Nardi, L.; Obermayer, R. Plant genome size estimation by flow cytometry: Inter-laboratory comparison. Ann. Bot. 1998, 82, 17–26. [CrossRef]

20. Praca-Fontes, M.M.; Carvalho, C.R.; Clarindo, W.R.; Cruz, C.D. Revisiting the DNA C-values of the genome size-standards used in plant flow cytometry to choose the "best primary standards". Plant Cell Rep. 2011, 30, 1183–1191. [CrossRef] [PubMed]

21. Mascher, M.; Gundlach, H.; Himmelbach, A.; Beier, S.; Twardziok, S.O.; Wicker, T.; Radchuk, V.; Dockter, C.; Hedley, P.E.; Russell, J.; et al. A chromosome conformation capture ordered sequence of the barley genome. Nature 2017, 544, 427–433. [CrossRef] [PubMed]

22. Avni, R.; Nave, M.; Barad, O.; Baruch, K.; Twardziok, S.O.; Gundlach, H.; Hale, I.; Mascher, M.; Spannagl, M.; Wiebe, K.; et al. Wild emmer genome architecture and diversity elucidate wheat evolution and domestication. Science 2017, 357, 93–97. [CrossRef] [PubMed]

23. Doležel, J.; Binarová, P.; Lucretti, S. Analysis of nuclear DNA content in plant cells by flow cytometry. Biol. Plant. 1989, 31, 113–120. [CrossRef]

Chapter 6

STRUCTURAL AND COMPARATIVE ANALYSIS OF THE COMPLETE CHLOROPLAST GENOME OF PYRUS HOPEIENSIS—"WILD PLANTS WITH A LIMITED POPULATION"—AND THREE ADDITIONAL PYRUS SPECIES

1. INTRODUCTION: AN OVERVIEW

Pyrus is classified within the Pyrus species of the Maloideae subfamily (Rosaceae), primarily encompassing temperate fruit trees. This genus has around 30 species, with 13 species found in China [1,2]. The pear has been cultivated in China for about 3000 years. It is the third most often cultivated fruit tree, following apple and citrus. China is the preeminent global producer of pears, representing 71.2% of the total pear cultivation area worldwide. The wild resources of Pyrus are exceedingly valuable. They are predominantly located in valleys, hillsides, and forest edges, exhibiting cold resistance, drought resistance, disease resistance, tolerance to barren conditions, saline-alkali tolerance, and robust adaptation. They provide material for screening quality rootstocks and for molecular breeding. The flowers, leaves, and fruit also have high ornamental value. The fruit of the pear is rich in fruit acids, vitamins, sugars, and many mineral elements essential for human life. It is sweet and refreshing and can be used to make dried and preserved pears, wine, and other products. Furthermore, it is regarded as having a high medicinal value and is used to reduce fevers, moisten the lungs, provide cough relief, and eliminate phlegm. Therefore, Pyrus is a valuable wild resource with a high exploitation value. P. hopeiensis is a rare wild resource of the genus Pyrus in the subfamily Rosaceae [3,4], which has been listed as one of the "wild plants with a tiny population" in China. It can be found at the edges of hillside jungles at 100–800 m above sea level. At present, only a few genotypes have been found in Changli, Hebei Province. So far, there have been few studies about P. hopeiensis. Successful sequencing of the chloroplast genome of P. hopeiensis in our study provides a foundation for further study of its chloroplast molecular biology and can effectively promote genetic breeding and help clarify the molecular evolution of P. hopeiensis. It also provides some basis for the evolutionary analysis and classification of the genus Pyrus. What is more, this study gives a systematic review of P. hopeiensis that is useful for the conservation and restoration of wild P. hopeiensis resources.

Chloroplasts serve as the primary locus for photosynthesis, facilitating the creation of fatty acids, carbohydrates, pigments, and various other substances [5]. They are autonomous from the plant nuclei and possess a highly preserved genomic architecture. Chloroplast DNA (cpDNA) possesses advantageous traits, including many copies, low molecular weight, and a straightforward structure. In contrast to the nuclear genome, characterized by a higher frequency of repeated sequences, and the mitochondrial genome, which often undergoes rearrangement, the chloroplast genome has a more conservative structure. The primary types of mutations are substitution and base insertion or deletion, with a modest mutation rate. Furthermore, the chloroplast genome is of moderate dimensions, facilitating its sequencing compared to intricate nuclear genomes. The chloroplast genome is inherited maternally in angiosperms, following a distinct evolutionary trajectory. Phylogenetic trees may be generated exclusively from cpDNA data, and the chloroplast genome exhibits significant collinearity among plant taxa. Sequencing data are very straightforward to evaluate, and the structural sequence information of the chloroplast genome can be utilized to investigate species origin, evolution, and interspecies interactions. Recently, advancements in high-throughput sequencing technology have led to the sequencing of more chloroplast genomes within the Rosaceae family. This study sequenced and compared the chloroplast genomes of two genotypes of P. hopeiensis and three more Pyrus species (P. ussuriensis Maxin. cv. Jingbaili, P. communis L. cv. Early Red Comice, and P. betulifolia) with other Rosaceae plants. The genomic architecture and phylogenetic relationships of Pyrus were clarified.

2. RESULTS AND ANALYSIS

2.1 Basic Characteristics of Chloroplast Genome of P. hopeiensis

The chloroplast genome of P. hopeiensis has a typical tetrad structure, including paired IRa and IRb sequences, encoding in opposite directions, and large and small single copy regions (Figure 1). The total chloroplast genome of P. hopeiensis HB-1 was 159,935 bp in length, 46 bp smaller than that of P. hopeiensis HB-2. The large single copy (LSC) region was 87,961 bp long, 46 bp smaller than that of P. hopeiensis HB-2. The length of the small single copy (SSC) region was 19,200 bp and the IR region was 26,387 bp, the same as those of P. hopeiensis HB-2. There was little difference in length between the two P. hopeiensis genotypes, and what difference existed was in the LSC region.

A total of 117 genes (Table 1) from the chloroplast genome of P. hopeiensis HB-1 were annotated, including 77 protein coding genes, 31 tRNAs, eight rRNAs, and two pseudogenes (clpP and atpF). P. hopeiensis HB-1 lacked only the MATK protein-coding gene that was associated with biosynthesis in P. hopeiensis HB-2. These 77 protein-coding genes can be divided into four categories. The first contains 28 self-replicating genes, including three subunits encoding the synthesis of chloroplast RNA polymerase. The second category contains 40 genes related to photosynthesis, including light systems I and II, a cytochrome b 6/f protein complex, and ATP synthase and other biosynthesis genes, including cytochrome-related genes (P. hopeiensis HB-1 contains three genes and P. hopeiensis HB-2 contains four genes). The fourth category contains five unknown genes, such as the ycf gene. The IR region of P. hopeiensis contains 32 genes, of which the ndhB gene is present only in the IRb region and is absent in the IRa region. In addition, rps12 is a mitotic gene with its 5 terminal located at the LSC region and 3 end with a copy is located in each of the two IR regions. This phenomenon is common in higher plants.

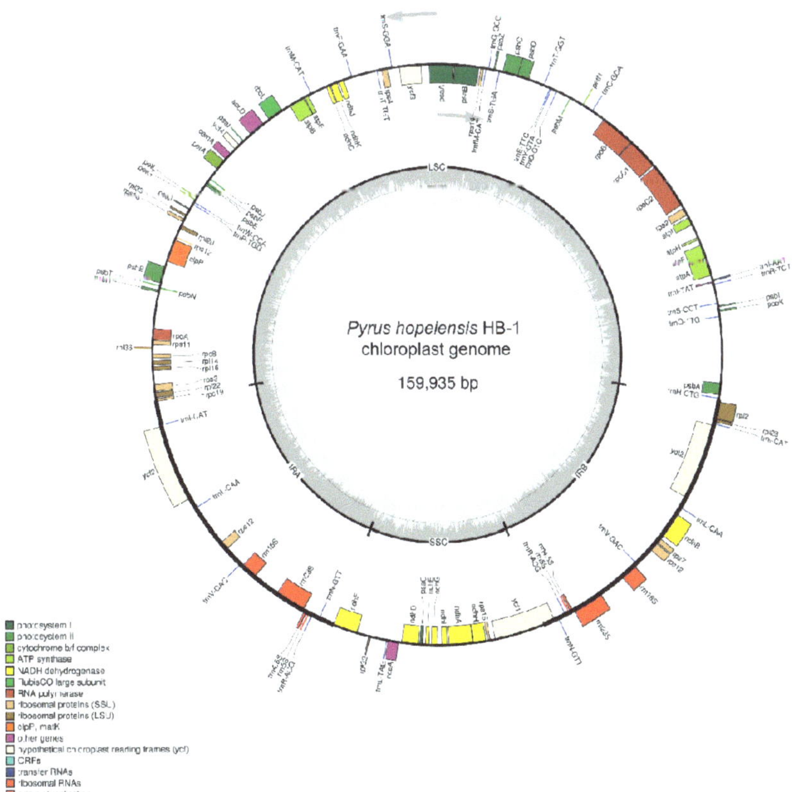

Figure 1. Gene maps of Pyrus hopeiensis HB-1chloroplast genomes.

The chloroplast genome of P. hopeiensis HB-1 contains 11 genes that harbor introns and one more trnI-TAT gene than P. hopeiensis HB-2. Of these 11 genes, two are tRNA genes (trnI-TAT and trnI-AAT) and nine are protein-coding genes (rpoC1, rpl22, rpl22, ndhA, rpl2, rpl2, rps12, rps12, rps12, ycf3), of which ycf3 contains two introns. Excepting the intron length of the trnI-AAT gene, P. hopeiensis HB-2 is larger than P. hopeiensis HB-1. The exon and intron lengths of the two P. hopeiensis genotypes are identical. Among the coding genes, ndhA was the longest at 1125 bp, and rpl22 was the shortest at 63 bp. Int. J. Mol. Sci. 2018, 19, 3262.

Table 1. Genes of the cp genome of P. *hopeiensis* HB-1.

Functions	Family Name	Code	List of Genes
Self-replication	Small subunit of ribosome	*rps*	*rps2, rps3, rps4, rps7* [a], *rps8, rps11, rps12* [a b e], *rps14, rps15, rps18, rps19*
	rRNA Genes	*rrn*	*rrn4.5S* [a], *rrn5S* [a], *rrn16S* [a], *rrn23S* [a]
	Large subunit of ribosome	*rpl*	*rpl2* [a b], *rpl14, rpl16, rpl20, rpl22* [b], *rpl23* [a], *rpl32, rpl33, rpl36*
	DNA dependent RNA polymerase	*rpo*	*rpoA, rpoB, rpoC1* [b], *rpoC2*
	tRNA Genes	*trn*	*trnC-GCA, trnD-GTC, trnE-TTC, trnF-GAA, trnfM-CAT, trnG-GCC, trnH-GTG, trnI-TAT* [b], *trnI-CAT* [a], *trnI-AAT* [b], *trnL-CAA* [a], *trnL-TAG, trnM-CAT, trnN-GTT* [a], *trnQ-TTG, trnP-TGG, trnR-TCT, trnR-ACG* [a], *trnS-GCT, trnS-TGA, trnS-GGA, trnT-GGT, trnT-TGT, trnV-GAC* [a], *trnW-CCA, trnY-GTA*
Genes for hotosynthesis	Subunits of ATP synthase	*atp*	*atpA, atpB, atpE, atpF* [d], *atpH, atpI*
	Subunits of protochlorophyllide reductase	*chl*	
	Subunits of NADH-dehydrogenase	*ndh*	*ndhA* [b], *ndhB* [b], *ndhC, ndhD, ndhE, ndhF, ndhG, ndhH, ndhI, ndhJ, ndhK*
	Subunits of cytochrome b/f complex	*pet*	*petA, petG, petL, petN*
	Subunits of photosystem I	*psa*	*psaA, psaB, psaC, psaJ, psaI*
	Subunits of photosystem II	*psb*	*psbA, psbB, psbC, psbD, psbE, psbF, psbH, psbI, psbJ, psbK, psbM, psbN, psbT, psbZ*
	Subunit of rubisco	*rbc*	*rbcL*
	Subunit of Acetyl-CoA-carboxylase	*acc*	*accD*
	Envelop membrane protein	*cem*	*cemA*
Other genes	c-type cytochrome synthesis gene	*ccs*	*ccsA*

	Protease	*clp*	*clpP* [d]
	Translational initiation factor	*inf*	
	Maturase	*mat*	*matK*
	Elongation factor	*tuf*	
Genes of unknown function	Conserved open reading frames	*ycf*	*ycf1,ycf2* [a]*,ycf3* [c]*,ycf4*
a—Two gene copies in IRs; b—Gene containing a single intron; c—Gene containing two introns; d—Pseudogene; e—Gene divided into two independent transcription units.			

2.2 Comparison of the Basic Characteristics of the Chloroplast Genome in Five Pyrus Species

The chloroplast genome of Pyrus is a typical ring structure 159,834–160,059 bp in length (Table 2). The chloroplast genome of P. communis L. cv. Early Red Comice is the shortest, and the longest is P. ussuriensis Maxin. cv. Jingbaili. The LSC length of Pyrus is 87,793–88,074 bp, the longest in P. ussuriensis Maxin. cv. Jingbaili and the shortest in P. communis L. cv. Early Red Comice. The SSC length is 19,201–19,261 bp, with the longest in P. betulifolia and the shortest in both genotypes of P. hopeiensis.

Table 2. Comparison of the basic characteristics of chloroplast Genome in five Pyrus species.

	Pyrus hopeiensis HB-1	Pyrus hopeiensis HB-2	Pyrus ussuriensis Maxin. cv. Jingbaili	Pyrus communis L. cv. Early Red Comice	Pyrus betulifolia
Length (bp)	**159,935**	**159,981**	**160,059**	**159,834**	**160,058**
GCcontent(%)	36.59	36.57	36.57	36.58	36.57
ATcontent(%)	63.41	63.43	63.43	63.42	63.43
LSClength(bp)	87,962	88,008	88,075	87,794	88,025
SSClength(bp)	19,201	19,201	19,212	19,260	19,261
IRlength(bp)	26,386	26,386	26,386	26,390	26,386
Genenumber	118	119	117	114	120
Pseudogene number	2	2	2	2	2
Genenumberin IRregions	32	32	31	31	32
Protein-coding gene number	77	78	75	74	77
Protein-coding gene (%)	64.25	65.55	64.10	64.91	64.17

rRNAgene number	8	8	8	8	8
rRNA(%)	6.78	6.72	6.84	7.02	6.67
tRNAgene number	31	31	32	30	33
tRNA(%)	26.27	26.05	27.35	26.32	26.50

The IR regions were of similar length in four of the five Pyrus species, and only differed by 4 bp, except in P. communis L. cv. Early Red Comice. The GC content was similar, 36.57–36.59%. The number of protein-coding genes ranged from 74 to 78. The chloroplast genome of the five Pyrus species contained 15 genes, including introns (nine protein-coding genes and six tRNAs, see Table 3), and the ycf3 gene contained two introns. The intron in P. betulifolia contained 13 genes, followed by 12 in P. ussuriensis Maxin. cv. Jingbaili, 11 in P. hopeiensis HB-1, 10 in P. hopeiensis HB-2, and 10 in P. communis L. cv. Early Red Comice. In the five Pyrus species, nine genes contained introns (rpoC1, ycf3, rpl22, rpl2, ndhA, ndhB, rpl2, rps12, and rps12), all of which were protein-coding genes. The intron in the ndhA gene was the longest, at 1125–1169 bp. Other than the short intron length of the trnI-TAT gene in P. ussuriensis Maxin. cv. Jingbaili, rpl22 harbored the smallest intron of the other four Pyrus species, at 63–83 bp. The results showed that the size, structure, sequence, and GC content of the chloroplast genome of P. hopeiensis were similar to those of the other three Pyrus species, which was characteristic of the slow evolution of the genus Pyrus [7] in comparison to the chloroplast genomes of P. hopeiensis, P. communis L. cv. Early Red Comice, P. ussuriensis Maxin. cv. Jingbaili, and P. betulifolia.

Table 3. Statistics of gene introns in the chloroplast genome of five Pyrus species.

Gene	Strand	Pyrus hopeiensis HB-1	Pyrus hopeiensis HB-2	Pyrus ussuriensis Maxin. cv. Jingbaili	Pyrus communis L. cv. Early Red Comice	Pyrus betulifolia
trnI-TAT	–	√	×	√	×	√
trnI-TAT	+	×	×	√	×	√
trnN-ATT	+	×	×	√	√	×
rpoC1	–	√	√	√	√	√
ycf3	–	√	√	√	√	√
rpl22	–	√	√	√	√	√
rpl2	–	√	√	√	√	√
ndhA	–	√	√	√	√	√
ndhB	+	√	√	√	√	√

rpl2	+	√	√	√	√	√
rps12	–	√	√	√	√	√
rps12	+	√	√	√	√	√
trnI-AAT	+	√	√	×	×	×
trnL-TAG	–	×	×	×	×	√
trnY-ATA	+	×	×	×	×	√
Total	15	11	10	12	10	13

2.3 Chloroplast Gene Gain–Loss Events

The chloroplast genome structure of most higher plants is stable, and the number, sequence, and composition of genes are conserved. However, the loss of chloroplast genome genes is common. For example, the chloroplast genome of the sweet orange has lost the infA gene [8]; the ycf1, ycf2, and accD genes have been lost in Gramineae [9], and the chloroplast genome of some legumes has been recombined several times, resulting in the deletion of a copy of the IR region [10]. The rpl22 gene was detected in the chestnut nuclear genome [11], presumably having derived from the chloroplast genome. In addition, the rpl32 gene was transferred to the nuclear genome in the poplar [12], and 17 similar chloroplast regions were found in the mitochondrial genome of papaya, suggesting that the chloroplast gene may have transferred to the mitochondria [13]. However, there have been few studies on how these genes are lost, transferred, and integrated into the nuclear and mitochondrial genomes. In this study, we compared the gain–loss events of eight Pyrus species (five Pyrus species that were sequenced in this study and three other Pyrus species, P. pyrifolia, P. spinose, and P. pashia, which were downloaded from NCBI) (Tables 4 and 5). The psbL, psbl, trnI-GAU, trnA, trnL, and trnY-AUA genes were most readily lost through evolution, followed by trnA-UGC, trnG-UCC, trnI-AAU, trnI-GAU, trnK-UUU, trnN-AUU, and trnV-UAC. P. hopeiensis HB-1 contained one less MATK gene than P. hopeiensis HB-2. Compared with the other three sequenced Pyrus species, trnI-AAU was only present in P. hopeiensis HB-1 and P. hopeiensis HB-2. The atpB gene was only lost in P. ussuriensis Maxin. cv. Jingbaili and P. communis L. cv. Early Red Comice, and petB, petD, rps16 and trnL-UAA were present in the five Pyrus species sequenced here, but missing in P. pyrifolia, P. spinosa, and P. pashia.

2.4 Synonymous (KS) and Nonsynonymous (KA) Substitution Rate Analysis

Nucleotide mutations that do not cause amino acid changes are known as synonymous mutations, whereas nonsynonymous mutations do cause changes to the amino acid sequence. The Ka/Ks ratio (or dN/ds) of nonsynonymous substitution (Ka) and synonymous substitution (Ks) is the selection pressure of an encoded protein, which can be used to determine whether the gene encoded by the protein is under selection pressure. If Ka/Ks > 1, the protein is considered to be positively selected; if Ka/Ks = l, the protein is neutral; and if Ka/Ks < 1, the protein is considered to have undergone purifying selection. It is generally believed that synonymous mutations are not subject to natural selection, whereas nonsynonymous mutations are Compared to P. hopeiensis HB-1, psaJ, rpl20, rps18, and ycf1 in P. hopeiensis HB-2 were subject to negative selection, and no positive selection gene (Figure 2 and Table S1–S4) was found. In P. betulifolia, atpE, ndhF, ndhI, rps18, and ycf2 were subject to positive selective pressure, whereas ndhD, ndhH, ndhK, rpl20, rpl22, rpoC2, rps11, and ycf1 were subject to negative selection. The psbC, psbK, rpoA, rps14, rps18, and ycf2 genes were subject to positive selective pressure in P. communis L. cv. Early Red Comice. Moreover, accD, atpA, atpE, cemA, matK, ndhA, ndhD, ndhF, ndhH, petA, psaA, psaB, rbcL, rpl22, rpoB, rpoC2, rps11, rps2, rps3, and ycf4 were subject to negative selection. In P. ussuriensis Maxin. cv. Jingbaili, atpE, atpI, cemA, ndhF, rps18, and ycf2 were subject to positive selective pressure, and ndhD, ndhH, psaA, psbC, rpl20, rpl22, rpoC2, rps11, and ycf4 were subject to negative selection. Compared with P. betulifolia, atpE was subject to positive selective pressure in P. betulifolia and P. ussuriensis Maxin. cv. Jingbaili, whereas atpE was subject to negative selection in P. communis L. cv. Early Red Comice. This shows that the chloroplast genome of Pyrus has been affected by different environmental pressures during evolution, which may account for the different gene numbers among the five Pyrus species.

Table 4. Genes from the chloroplast genomes of Pyrus.

Specie	atpB	matK	petB	petD	psaC	psbI	psbL	psbl	Rpl20	Rpl36	rps16	ycf1	trnI-GAU	trnA
Pyrus ussuriensis Maxin.cv.Jingbaili	0	1	0	0	1	1	0	0	1	1	0	0	0	0
Pyrus communis L.cv. EarlyRed Comice	0	1	0	0	1	1	0	0	0	1	0	0	0	0
Pyrus hopeiensis HB-1	1	0	0	0	1	1	0	0	1	1	0	1	0	0
Pyrus hopeiensis HB-2	1	1	0	0	1	1	0	0	1	1	0	1	0	0
Pyrus betulifolia	1	1	0	0	1	1	0	0	1	1	0	1	0	0
Pyrus pyrifolia	1	1	1	1	0	0	0	1	1	0	1	1	1	1
Pyrus spinosa	1	1	1	1	1	1	0	0	1	1	1	1	0	0
Pyrus pashia	1	1	1	1	1	1	1	0	1	0	1	2	0	0
Total number of missing genes	2	1	5	5	1	1	7	7	1	2	5	2	7	7

Table 5. Genes from the chloroplast genomes of Pyrus.

Specie	trnA-UGC	trnG-GCC	trnG-UCC	trnI-AAU	trnI-GAU	trnI-UAU	trnK-UUU	trnL	trnL-UAA	trnN-AUU	trnV-UAC	trnY-AUA
Pyrus ussuriensis Maxin.cv. Jingbaili	0	1	0	0	0	2	0	0	0	1	0	0
Pyrus communis L.cv.Early RedComice	0	1	0	0	0	0	0	0	0	1	0	0
Pyrus hopeiensis HB-1	0	1	0	1	0	1	0	0	0	0	0	0
Pyrus hopeiensis HB-2	0	1	0	1	0	1	0	0	0	0	0	0
Pyrus betulifolia	0	1	0	0	0	2	0	0	0	0	0	1
Pyrus pyrifolia	1	1	0	0	1	0	1	1	1	0	1	0
Pyrus spinosa	0	0	1	0	0	0	0	0	1	0	0	0
Pyrus pashia	2	1	1	0	2	0	1	0	1	0	1	0
Totalnumberofmissing genes	6	1	6	6	6	4	6	7	5	6	6	7

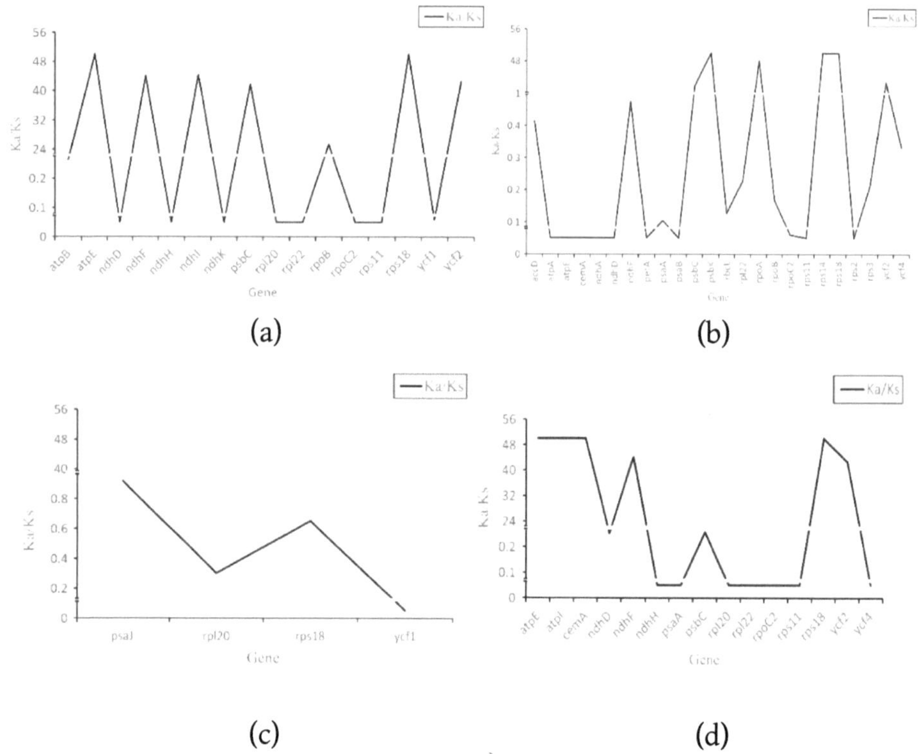

Figure 2. Ka/Ks value of five Pyrus species. (a)–(d) represent the Ka/Ks values of Pyrus betulifolia, Pyrus communis L. cv. Early Red Comice, Pyrus ussuriensis Maxin. cv. Jingbaili, and Pyrus hopeiensis HB-2, respectively, with respect to Pyrus hopeiensis HB-1.

2.5 Indel Identification and Relationship of the Five Pyrus cp Genomes

The nucleotide bases in coding and non-coding regions have different evolutionary mutation rates. DNA variations located in coding regions can lead to large phenotypic and functional variations; moreover, these often have a slower mutation rate, making them suitable for phylogenetic studies of higher order elements (families, orders, and higher). Mutations in non-coding regions have little effect on phenotype and fewer functional restrictions, and as they take no part in the transcription/translation process, they have a relatively high nucleotide replacement rate and hence rapid evolution, making them suitable for the phylogenetic study of lower order elements (species, genus) [14].

The chloroplast genome data of five Pyrus species were compared with those of P. hopeiensis HB-1 by multiple sequence alignment using MAFFT. All differentially expressed sites were extracted using a script from the comparison results, and differences in sites of indels ≥ 5 bp were screened out. The location of different chloroplast genome sites was determined and ggplot in R was used to create graphic plots that were then optimized using AI. The results indicated 15 mutation sites in

P. hopeiensis (Figure 3), which included 11 insertion and four deletion sites. All of these mutation sites were located in the LSC region; three were located in gene regions and 12 in intergenic regions. Among these, the longest was located in the ndh-trnM-CAT region, and as many as six mutations were located in the intergenic region rpl18–rps20. A total of 96 mutation sites were detected in the other four Pyrus species, 81 of which were located in the LSC region of the chloroplast genome and 11 in the SSC region, whereas only two mutation sites were found in the IRa and IRb regions in P. communis L. cv. Early Red Comice. There were more mutation sites in the SC region, and the IR region was more conserved. Indels were mainly located in the intergenic regions, and three indel loci were detected in the intron region (rpl22, trnN-ATT, ndhA). Because the protein-coding region is arranged by triplet codons, the tolerance of indels is poor. Therefore, only five indel loci were detected in the protein-coding region (trnL-TAT, trnN-ATT, rps18, rps19 and ycf1), but no indel loci were detected in the rRNA region. A comparison of the occurrence of these indel loci among the four Pyrus species revealed 15 indel loci in the chloroplast genome of P. hopeiensis, 32 in P. ussuriensis Maxin. cv. Jingbaili, 57 in P. communis L. cv. Early Red Comice, and 31 in P. betulifolia. The insertion or deletion frequency in the chloroplast genome of P. hopeiensis HB-2 was less than that in P. hopeiensis HB-1. The psbA-trnQ_TTG and rpl18-rps20 intergenic regions were the most variable regions with seven loci, followed by trnT-TGT_trnF_GAA (six) and the trnI-TAT gene-coding region (six). The largest indels were located in psbA_trnQ-TTG in the chloroplast genome of P. communis L. cv. Early Red Comice.

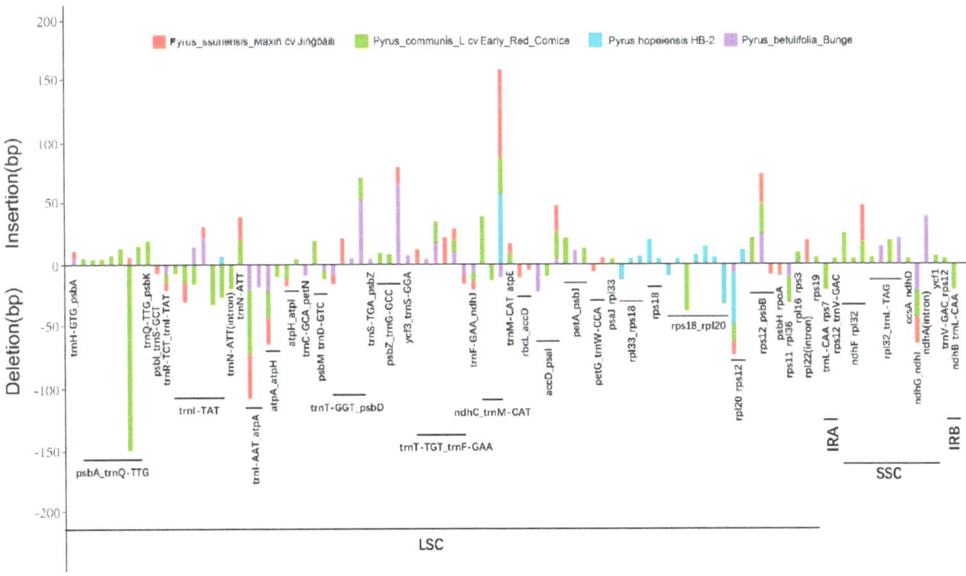

Figure 3. Indels (≥5 bp) identified based on multiple sequence alignment of five Pyrus cp genomes. Insertions are shown above and deletions below the horizontal axis. Indel distribution was positioned using Pyrus hopeiensis HB-1 as a reference.

2.6 Codon Preference Analysis

Codons have an important role in the transmission of genetic information. Codon use is not equal in many species, and the phenomenon of a specific codon use frequency being higher than that of its synonymous codon is known as codon preference [15]. Codon preference is formed during the long-term evolution of organisms, with different species having different codon preferences. Codon use is affected by natural selection, mutagenesis, tRNA abundance, the composition of base groups, hydrophilicity of codons, gene length, and expression levels [16]. Analysis of the codon use preferences of a species improves our understanding of the transmission of genetic information and the development of evolutionary and phylogenetic models.

The annotated files of plant genomes, including P. pashia, P. pyrifolia, Malus prunifolia, Prunus mume, and Chaenomeles japonica, were selected from the NCBI database, including the sequence files encoding CDS and proteins. According to the full-length CDS criterion, sequences with lengths <300 nt were deleted. The codon-use frequency of each genome was extracted from the annotated files of each genome and the corresponding frequency ratio was calculated. The final statistical results were clustered and mapped using the pheatmap package in R. The results showed obvious codon use preferences for both types of P. hopeiensis, among which ATT, AAA, GAA and AAT, and TTT were used most frequently (Figure 4). Statistical analysis of all the codons of P. hopeiensis, the three other Pyrus species, and the other Rosaceae showed a high A/T preference in the third chloroplast codon. This is common in the chloroplast genomes of higher plants [17–21].

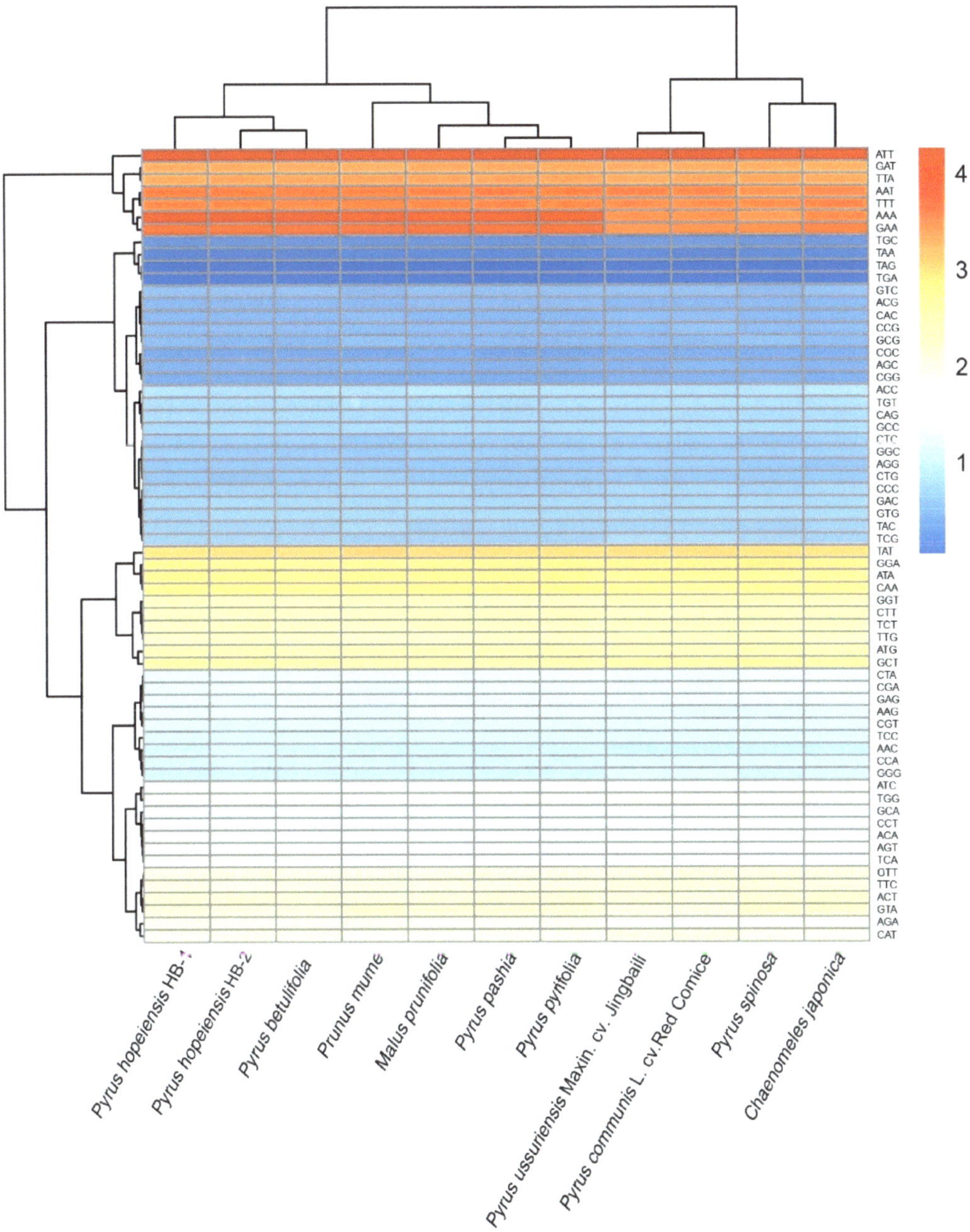

Figure 4. Codon distribution of all merged protein-coding genes. Red indicates a higher frequency and blue indicates a lower frequency.

2.7 Comparison of the Genome Structure in Rosaceae cp Genomes

The chloroplast genome structure of most higher plants is relatively stable and the number, sequence, and composition of their genes are conserved. However, because different plant groups have different evolutionary histories and genetic backgrounds, the chloroplast genome size, genome structure, and gene numbers vary. Insertion/deletion is the most frequent type of microstructural variation in the chloroplast genome, and it occurs frequently in some segments where the variation is high, such as trnH-psbA and trnS-G. In Rosaceae, an insertion/deletion of 277 bp in the intergenic region of the trnS-G gene was reported in peach plants [22], and an insertion/deletion of 198 bp in the intergenic region of trnL-F was identified in P. mume [23].

The collinear method was used to analyze and compare the chloroplast genomes of the two genotypes of P. hopeiensis, the other three sequenced Pyrus, and other related Rosaceae (P. pashia, P. pyrifolia, P. spinosa, M. prunifolia, P. mume, and C. japonica). The results showed optimal collinearity between P. hopeiensis HB-1 and P. hopeiensis HB-2, and only a few sites contained insertions and deletions (Figure 5). Compared with the other Rosaceae, the genome structure and gene sequences were highly conserved, with more linear relationships indicating high chloroplast genome homology among the different plants.

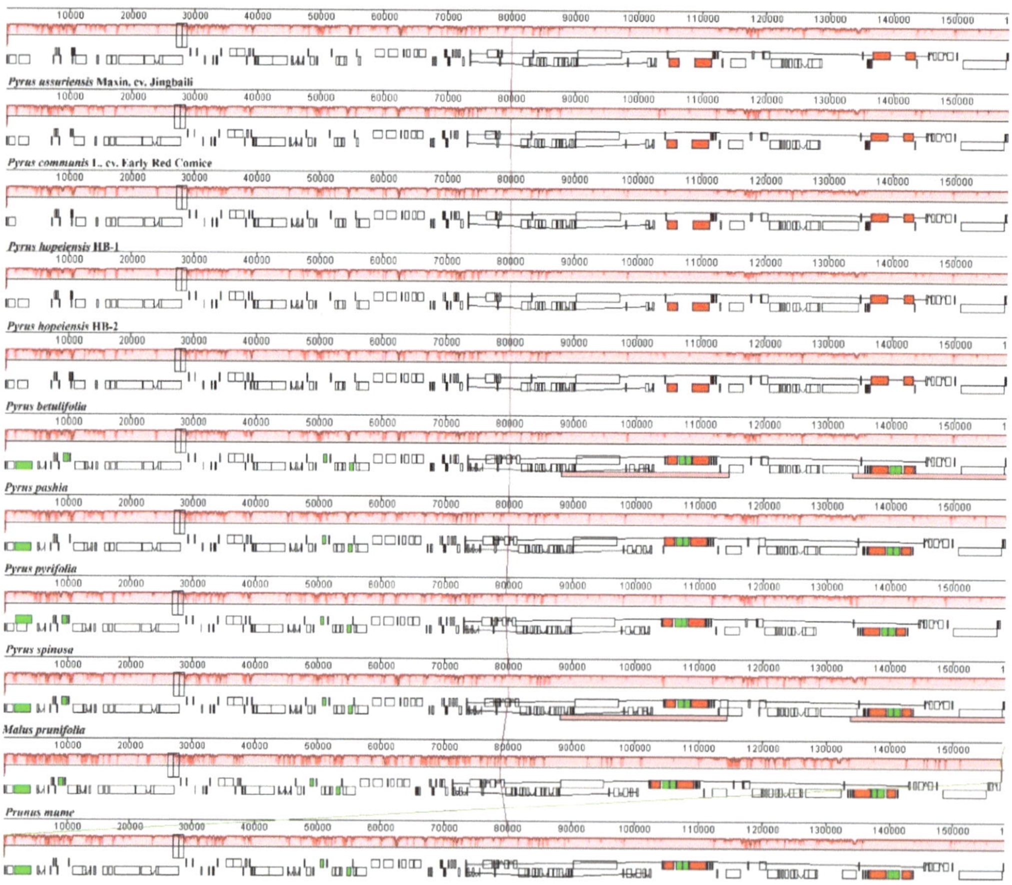

Figure 5. Co-linear analysis of various plant chloroplast genomes.

2.8 IR Contraction Analysis

The IR region is considered to be consistent and stable in the chloroplast genome. However, in the evolution of species, border region contraction and expansion are common. In this study, the IR boundaries of both genotypes of P. hopeiensis were compared. The IRa/LSC boundary extended into the rps19 gene, and 120 bp of rps19 extended into the IRa region. The IRa/SSC boundary extended into the ndhF gene, and 12 bp of ndhF extended into the IRa region. The IRb/SSC boundary extended into 1074 bp of ycf1 and the IRb/LSC border extended into the rpl2 gene, with the trnH-GTG gene located downstream.

The IR boundaries were compared among the Rosaceae, including the five Pyrus species sequenced, and P. pashia, P. pyrifolia, P. spinosa, M. prunifolia, P. mume, and C. japonica (Figure 6). The IRa/LSC boundary of these plants extended to the rps19 gene. The IRa/SSC boundaries were located upstream of the ndhF gene, except in M. prunifolia, whose IRa/SSC junction lost ndhF but extended to ycf1. The P. spinosa IRb/LSC boundary had no rpl2. All IRb/SSC boundaries expect those of P. ussuriensis Maxin. cv. Jingbaili, P. communis L. cv. Early Red Comice, and P. Spinosa extended to ycf1. The IRb/SSC boundary lost ycf1. These findings were similar to those in the Actinidiaceae, Theaceae, and Primulaceae, but differed markedly from those in Ericaceae. For the IRb/LSC boundary, all but that of P. spinosa extended into the rpl2 gene, and the IRb/LSC boundary of the trnH-GTG gene located downstream extended into the rpl23 gene in P. spinosa.

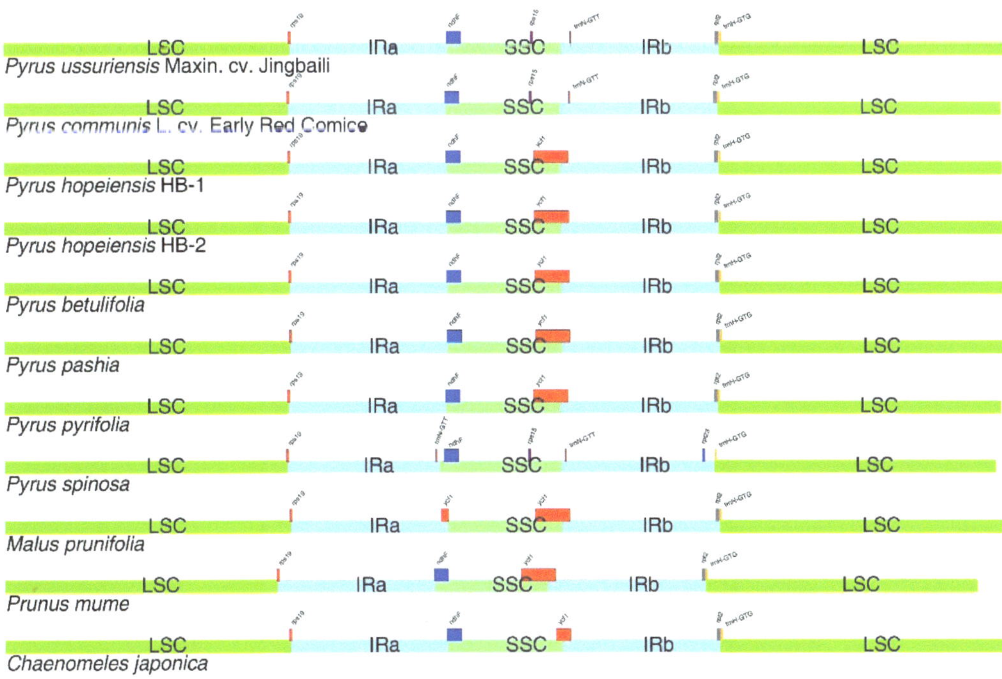

Figure 6. IR contraction analysis of Rosaceae.

2.9 Phylogenetic Analysis

To gain an insight into the position of Pyrus within the Rosaceae, a molecular phylogenetic tree was constructed using 57 protein-coding genes (accD, atpA, atpE, atpH, atpI, csA, cemA, ndhA, ndhB, ndhC, ndhD, ndhE, ndhG, ndhH, ndhJ, ndhK, petA, petG, petL, petN, psaA, psaB, psaI, psbA, psbB, psbC, psbD, psbE, psbF, psbH, psbJ, psbM, psbN, psbT, rbcL, rpl14, rpl16, rpl2, rpl22, rpl23, rpl33, rpoA, rpoC1, rps11, rps12, rps14, rps15, rps18, rps19, rps2, rps3, rps4, rps7, rps8, ycf2, ycf3, ycf4)from the chloroplast genomes of 36 Rosaceae, which were downloaded from GenBank, and using Arabidopsis thaliana as the outgroup. The resulting phylogenetic tree was consistent with the traditional plant morphological taxonomy (Figure 7), which can be divided into three sections: Maloideae, Prunoideae, and Rosoideae. The Maloideae include Pyrus, Malus, Sorbus L., and Eriobotrya. Prunus lies within the Prunoideae, and Fragaria is included in the Rosoideae. The phylogenetic relationship of the Prunoideae was closer than that of the Rosoideae to the Maloideae, and the relationship between Malus and Pyrus was the closest.

Within Pyrus, the relationship between P. hopeiensis HB-1 and P. hopeiensis HB-2 was the closest, and the relationship between P. betulifolia and P. ussuriensis Maxin. cv. Jingbaili was closer than that between other Pyrus and P. hopeiensis. In addition, it can be seen from the evolutionary tree that Rosoideae is a subfamily that split off from the evolutionary tree.

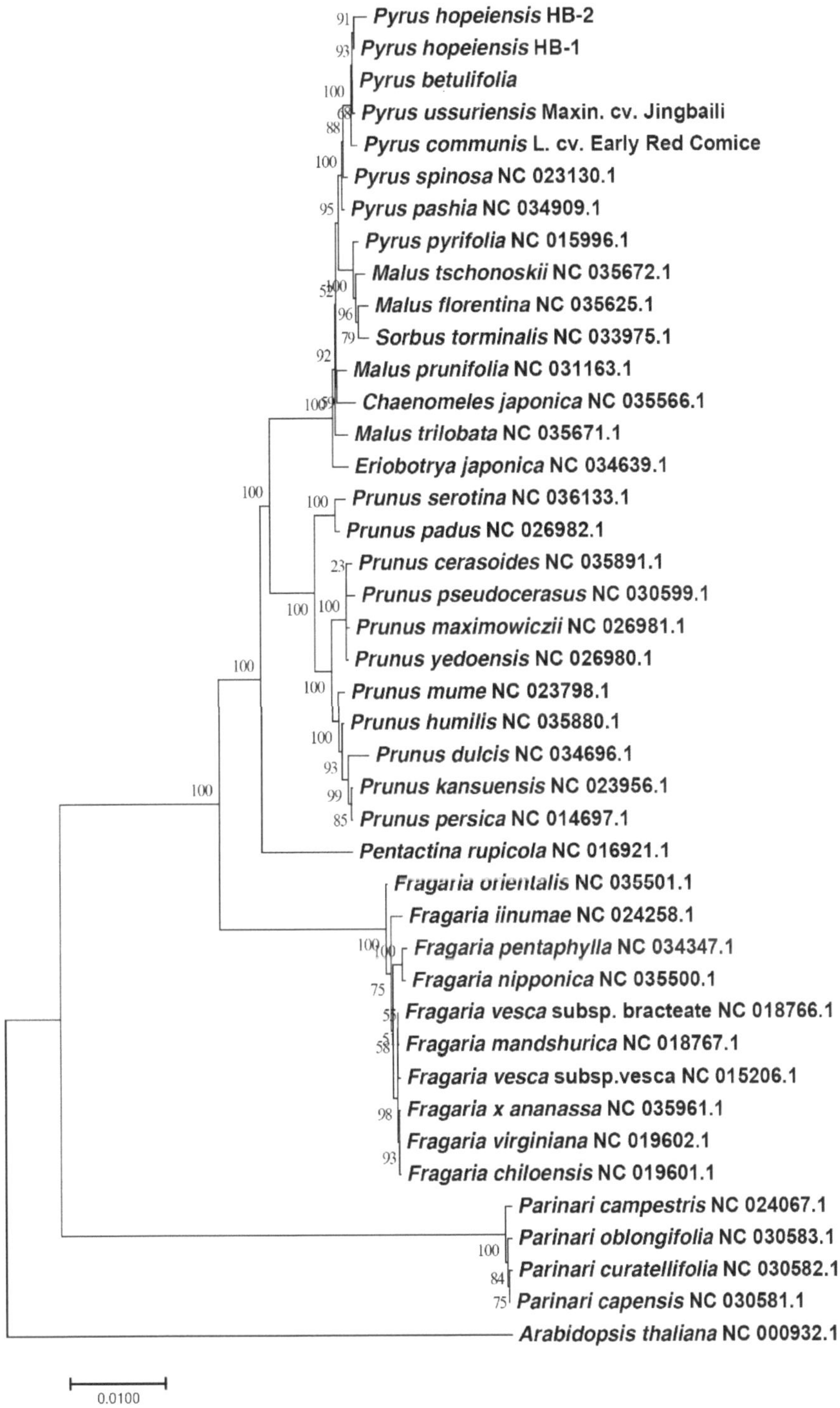

Figure 7. The ML phylogenetic tree of the Rosaceae clade based on same protein-coding genes. Numbers above or below the nodes are bootstrap support values.

3. DISCUSSION

Pyrus hopeiensis is a valuable wild resource of Pyrus, which belongs to the family Rosaceae. Because of its limited distribution and population decline, P. hopeiensis is listed among "the wild plants with tiny population" in China. It belongs to one of 13 species of Pyrus present in China. In this study, the chloroplast genomes of the two genotypes P. hopeiensis HB-1 and P. hopeiensis HB-2 and those of three other major pear plants, P. ussuriensis Maxin. cv. Jingbaili, P. communis L. cv. Early Red Comice, and P. betulifolia, were analyzed using high-throughput sequencing for comparative analysis. The chloroplast genome of Pyrus, like those of most higher plants, is a typical tetrad consisting of two reverse repeat IR regions and small and large single copy fragments [24]. There was a 46 bp difference in the chloroplast genome size between the two P. hopeiensis, which was located in the LSC region. Compared with the other three Pyrus species, the total genome length was <225 bp, and the gene number, gene type, gene sequence, LSC, IR, and SSC lengths and GC content were similar. This strongly suggests that chloroplast genomes are highly conserved [25]. The encoded genes of the chloroplast genome are divided into three categories based on their functions. The first is related to chloroplast gene expression, such as tRNAs, rRNAs, and three subunits encoding chloroplast RNA polymerase synthesis. The second is related to photosynthesis, and the third consists of other biosynthesis genes and some genes of unknown function, such as mat and ycf [26]. The chloroplast genes of Pyrus are similar in composition.

The genomic sequence of the P. hopeiensis HB-1 chloroplast was used as a reference sequence to detect single-nucleotide polymorphisms (SNPs) and indels in the other four Pyrus species. The results showed a significantly higher variation in the non-coding region than in the coding region and more mutation sites in the intergenic region of the psbA-trnQ_TTG, rpl18-rps20, and trnT-TGT_trnF_GAA genes, which could be used for evolutionary analysis of Pyrus. The chloroplast genomes of Pyrus show obvious codon preference and similar codon use frequency. Furthermore, the third chloroplast codon has a higher A/T preference. This phenomenon is common in the chloroplast genomes of other higher plants [27]. Although the IR region is highly conserved, the expansion and contraction of the IR region is a common characteristic of the chloroplast genome. The degree of expansion of the IR/SC boundary is similar among the five Pyrus species, the two P. hopeiensis genotypes contain few genes with different extension positions, and any differences are very small, which is useful in the classification of Pyrus, as it can be used as a basis to identify the evolution of the chloroplast genome. The classification and identification of pear species in this study can be utilized for the preservation of pear germplasm resources. The initial identification of species and varieties of Pyrus was mainly based on morphological features (leaves, petioles, floral organs, sepals, hairs, fruits, and ventricles) and geographical distribution. For example, based on an investigation of morphological characteristics and natural distribution, Chinese taxonomists believe that P. hopeiensis, P. phaeocarpa, P. sinkiangensis, and P. serrulata were all formed by natural crosses [28]. Yu [29] divided Pyrus from China into 13 species based on their serrated leaf margins, and these included P. hopeiensis, P. betulifolia, P. ussuriensis, P. phaeocarpa, P. bretschneideri, P. pyrifolia, P. pashia, P. armeniacaefolia, P. calleryana, P. pseudopashia, P. serrulata, P. sinkiangensis, and P. xerophila. Anatomical studies in Wang Yingzhong [30] revealed that the anatomical structures of P. betulifolia and P. ussuriensis, and P. bretschneideri, P. pyrifalia, and P. communis were similar. The results showed that the relationship between P. ussuriensis and P. betulifolia, P. bretschneideri, and P. pyrifalia was close. However, it is easy to cross Pyrus species and there are no obvious differences in the biological and morphological characteristics among species and varieties, which greatly increases the difficulty of establishing its phylogenetic evolution and classification.

Pollen morphological identification, cytological markers, isozymes, and other methods have also been studied with a view to classifying Pyrus. The pollen morphology of P. sinkiangensis is similar to that of the Western pear, indicating a close relationship [31]. The pollen morphology of the Western pear is obviously different from that of the Oriental pear. The pollen morphology of P. calleryana has many primitive characteristics, and it is a primitive species of Pyrus in China. The pollen morphology of P. bretschneideri, also present in China, has the characteristics of both P. pyrifalia and P. ussuriensis, and may be a natural hybrid of P. pyrifalia and P. ussuriensis. Cytological markers enabled the analysis of the number, banding, karyotype, and meiosis behavior of the chromosomes. P. phaeocarpa has a similar karyotype to that of P. betulifolia, and those of P. sinkiangensis, P. hopeiensis, and P. serrulata were also similar [32,33]. In 1983, Lin Bonian and Shen Dexu [34] proved, through the use of the peroxidase isozyme, that P. bretschneideri and P. pyrifolia were closely related. However, these methods have few characteristic sites, poor polymorphism, and low accuracy, and provide a limited amount of information. To date, the relationships among Pyrus species, their origin, evolution of cultivation systems, and the origin of some suspicious species and hybrids remain unclear.

In recent years, molecular markers based on DNA, such as restriction fragment length polymorphisms (RFLPs) and simple sequence repeats (SSRs) have been used to investigate the genetic relationships, genetic diversity, and germplasm of Pyrus. However, there remain some deficiencies in the study of the interspecific relationships and origins of hybrids. Results based on random amplification of polymorphic DNA (RAPD) showed that the origin of P. sinkiangensis involved the crossing of many Eastern and Western pear species and that the genetic relationship between P. bretschneideri and P. pyrifolia is very close [35]. RAPD, inter sequence simple repeats (ISSR), and other DNA markers showed that P. hopeiensis, P. betulifolia and P. phaeocarpa were closely related to each other. In the same way, P. phaeocarpa is considered to be a hybrid of P. betulifolia and P. ussuriensis, whereas P. hopeiensis is a hybrid of P. phaeocarpa and P. ussuriensis. In a study using RAPD, P. hopeiensis and P. phaeocarpa shared some spectral bands with P. betulifolia and P. ussuriensis [36–38]. Zheng et al. identified a close relationship between P. ussuriensis and P. hopeiensis using internal transcribed spacer (ITS) sequences, which is consistent with the results in our study [39].

Because the chloroplast genome is the second-largest genome after the nuclear genome, it is maternally inherited in most angiosperms; thus it reflects the maternal evolutionary history, and this helps us to understand the maternal ancestors of suspected hybrids. The coding and non-coding regions of the chloroplast genome evolve at different rates, making them suitable for systematic research at different levels. The coding region is highly conserved and is only suitable for phylogenetic studies of families, orders, and higher taxonomic levels, whereas the non-coding regions are less constrained by function and the rapid evolutionary rate is suitable for plant phylogenetic studies at interspecific and subspecies levels. At present, the successful design of a set of universal primers for the chloroplast gene non-coding regions (such as trnS-psbcc, trnL-trnF and accD-pasI) [40] has made the study of chloroplast non-coding regions a hot topic in studies of the systematic relationship of Pyrus. Phylogenetic trees based on combinations of the sequences of trnL-trnF and accD-psaI in the chloroplast non-coding regions have further confirmed the theory of an independent evolution of the Oriental pear and the Western pear from the background of matrilineal evolution, and have shown the close relationship between P. bretschneideri and P. pyrifolia [41]. A study of the trnL-trnF region of cpDNA showed that P. sinkiangensis is closely related to the Western pear and the Oriental pear; the relationship between P. betulifolia and P. ussuriensis is close; P. bretschneideri is a hybrid of P. ussuriensist, P. phaeocarpa, and P. pyrifolia; and that the Western pear and Oriental pear are related to each other [42]. The sequences of these regions are highly conserved, with only limited sites available to provide information to unravel the phylogeny of Pyrus. However, no comprehensive and systematic cpDNA sequence analysis of Pyrus exists in China. To further our understanding of the inter-species relationships of Pyrus and to reveal the origin of hybrids and explore the evolutionary model of Eastern and Western pears, a wider range of representative species and varieties of Eastern and Western pears must be selected, and the nuclear gene fragments inherited by their parents should be combined, especially the low copy nuclear gene introns.

4. MATERIALS AND METHODS

4.1 Plant Materials

In early May 2017, fresh leaves were collected from P. hopeiensis HB-1, P. hopeiensis HB-2, and three local Pyrus species, P. ussuriensis Maxin. cv. Jingbaili (which belongs to the P. ussuriensis family),\ P. communis L. cv. Early Red Comice (a high-quality variety of Western pear native to the United Kingdom) and P. betulifolia (most widely used in northern China as pear rootstocks) in Changli, Hebei Province and were stored before being transported to ORI-GENE Ltd., a science and technology company based in Beijing, China, for chloroplast genome sequencing.

4.2 DNA Sequencing, Genome Assembly, and Validation

The total DNA of fresh young leaves was extracted using a plant DNA extraction kit (Tiangen, Beijing, China). Agarose gel electrophoresis was used to detect DNA integrity, and purity and concentration were ascertained. The Illumina HiSeq platform was used to sequence the total DNA. After sequencing, the raw data was initially screened to remove low quality regions affecting the data quality and subsequent analysis needed to obtain the expected clean data. The SOAPdenovo2.01 [43] oligonucleotide analysis package was used to assemble the contig sequence. BLAT36 [44] was used to locate the assembled long sequence on the chloroplast reference genome of the relative species and to obtain the relative position of the contig sequence to enable splicing of the contig according to its relative position, and to correct assembly errors. A full-length frame map of the chloroplast genome was obtained. GapCloser software was used to fill gaps on the frame map sequence with high-quality short sequences. Any remaining gaps and suspected regions were supplemented and confirmed by generation sequencing, and the small single copy (SSC) and inverted repeat (IR) region junctions were verified. Finally, a complete ring chloroplast genome sequence was obtained.

4.3 Gene Annotation

CpGAVAS [45] was used to annotate the gene and the final annotation results were obtained by artificial correction. First, the results of Blastx, BLASTn, protein2genome, and est2genome [46] were integrated to predict the protein coding gene and the rRNA gene. Then, tRNA was identified using tRNAscan [47] and ARAGORN [48]. Finally, the reverse repeat region IR was identified using Vmatch [49]. Chloroplast genome mapping was performed using OrganellarGenomeDRAW [50] (http://ogdraw.mpimp-golm.mpg.de/index.shtml) based on the annotated results.

The protein and coding sequences (CDS) of each sample were extracted from the annotated files of each sample and the pairwise protein sequences aligned using MUSCLE software. The aligned protein sequences were converted to DNA sequences using PAL2NAL. KaKs_Calculator2.0 [51] software (https://sourceforge.net/projects/kakscalculator2) was then used to calculate Ka/Ks, which was used to analyze the selection pressure on different Pyrus species during the evolutionary process. Chloroplast genome sequences of 36 Rosaceae species were selected from NCBI and 57 common protein coding genes were used to explore the evolution of the Pyrus chloroplast genome, using Arabidopsis thaliana as the outgroup. The taxonomic status was confirmed. The annotated files of all of the genomes were downloaded from NCBI and the protein sequences of any genes shared among the chloroplast genomes of all of the species were extracted. Each gene was placed in a file in which each genome contained only one protein sequence. MUSCLE was used to make multiple sequence alignments for each file. The first and last sequences were aligned according to the genome source to obtain a growing alignment sequence: final.fa. MEGA7.0 software was then used to construct a neighbor-joining tree and the CGView Server was used to analyze the genetic variation among the chloroplast genomes of the five Pyrus species.

5. CONCISE SUMMARY

This paper presents the de novo sequencing results of the chloroplast genomes of P. hopeiensis. The chloroplast genome of P. hopeiensis HB-1 measures 159,935 bp, which exceeds the length of P. hopeiensis HB-2 by 46 bp. The SSC and IR areas of the two Pyrus genotypes had identical lengths, with the sole distinction being in the LSC region. A total of 118 genes were found in P. hopeiensis HB-1, which was deficient in the MATK protein-coding gene linked to biosynthesis in P. hopeiensis HB-2. The GC content of P. hopeiensis HB-1 exceeded that of P. hopeiensis HB-2 by merely 0.02%. In the chloroplast genome of P. hopeiensis HB-1, 11 genes included introns, along with an extra trnI-TAT gene absent in P. hopeiensis HB-2. ycf3 is the sole gene that possesses two introns. The chloroplast genome structure and size, gene species, gene count, and GC content of P. hopeiensis were analogous to those of the other three examined Pyrus species. Nearly all protein-coding sequences and amino acid codons exhibited a distinct codon preference. Analysis of selection pressure indicated that the chloroplast genomes of various pear species were influenced by distinct environmental forces throughout evolution, potentially explaining the variation in gene numbers across the five Pyrus species. Phylogenetic study robustly corroborates the classification of Pyrus within the Rosaceae family. This study enhances our understanding of the molecular evolution of Pyrus and will be beneficial for genetic breeding and chloroplast engineering of the species.

REFERENCES

1. Chai, M.L.; Shen, D.X. Status and Prospects of Pear Breeding in China. J. Fruit Sci. 2003, 5, 379–383.

2. Yu, D.J. Taxonomy of the Fruit Tree in China; China Agriculture Press: Beijing, China, 1979; p. 122.
3. Jiang, X.F.; Chu, Q.G.; Zhang, C.S. Studies on the classification and evolution of the genus pyrus in China. J. Laiyang Agric. Coll. 1992, 9, 18–21.
4. Pu, F.S.; Huang, L.S.; Sun, B.J.; Li, S.L. Study on the chromosome number of wild and cultivated pears (Pyrus sp.) in China. Acta Hortic. Sin. 1985, 12, 155–158.
5. Neuhaus, H.E.; Ernes, M.J. Nonphotosyntiietic Metabolism in Plastids. Annu. Rev. Plant Physiol. Plant Mol. Biol. 2000, 51, 111–140. [CrossRef] [PubMed]
6. Dumolin, S.; Demesure, B.; Etit, R.J. Inheritance of chloroplast and mitochondrial genomes in pedunculate oak investigated with an efficient PCR method. Theor. Appl. Genet. 1995, 91, 1253–1256. [CrossRef] [PubMed]
7. Xing, S.C. Progress in Chloroplast Genome Analysis. Prog. Biochem. Biophys. 2008, 35, 21–28.
8. Bausher, M.G.; Singh, N.D.; Lee, S.B.; Jansen, R.K.; Daniell, H. The complete chloroplast genome sequence of Citrus sinensis (L.) Osbeck var 'Ridge Pineapple': Organization and phylogenetic relationships to other angiosperms. BMC Plant Biol. 2006, 6, 21. [CrossRef] [PubMed]
9. Guisinger, M.M.; Chumley, T.W.; Kuehl, J.V.; Boore, J.L.; Jansen, R.K. Implications of the plastid genome sequence of Typha (Typhaceae, Poales) for understanding genome evolution in Poaceae. J. Mol. Evol. 2010, 70, 149–166. [CrossRef] [PubMed]
10. Tao, X.L.; Ma, L.C.; Nie, B.; Wang, Y.R.; Liu, Z.P. The draft and characterization of the complete chloroplast genome of Vicia sativa cv. Lanjian No. 3. Pratacult. Sci. 2017, 34, 321–330.
11. Jansen, R.K.; Saski, C.; Lee, S.B.; Hansen, A.K.; Daniell, H. Complete plastid genome sequences of three Rosids (Castanea, Prunus, Theobroma): Evidence for at least two independent transfers of rpl22 to the nucleus. Mol. Biol. Evol. 2011, 28, 835–847. [CrossRef] [PubMed]
12. Ueda, M.; Fujimoto, M.; Arimura, S.I.; Murata, J.; Tsutsumi, N.; Kadowaki, K.I. Loss of the rp132 gene from the chloroplast genome and subsequent acquisition of a preexisting transit peptide within the nuclear gene in populus. Gene 2007, 402, 51–56. [CrossRef] [PubMed]
13. Cheng, H.; Ge, C.F.; Zhang, H.; Qiao, Y. Advances on Chloroplast Genome Sequencing and Phylogenetic Analysis in Fruit Trees. J. Nucl. Agric. Sci. 2018, 32, 58–69.
14. Clegg, M.T.; Gaut, B.S.; Learn, G.H.; Morton, B.R. Rates and patterns of chloroplast DNA evolution. Proc. Natl. Acad. Sci. USA 1994, 91, 6795–6801. [CrossRef] [PubMed]

15. Jiang, W.; Lyu, B.B.; He, J.H.; Wang, J.B.; Wu, X.; Wu, G.G.; Bao, D.P.; Chen, M.J.; Zhang, J.S.; Tan, Q.; et al. Codon usage bias in the straw mushroom Volvariella volvacea. Chin. J. Biotechnol. 2014, 30, 1424–1435. (In Chinese)
16. Li, Y.; Kuang, X.J.; Zhu, X.X.; Zhu, Y.J.; Sun, C. Codon usage bias of Catharanthus roseus. China J. Chin. Mater. Med. 2016, 41, 4165–4168.
17. Zuo, L.H.; Shang, A.Q.; Zhang, S.; Yu, X.Y.; Ren, Y.C.; Yang, M.S.; Wang, J.M. The first complete chloroplast genome sequences of Ulmus species by de novo sequencing: Genome comparative and taxonomic position analysis. PLoS ONE 2017, 12, e0171264. [CrossRef] [PubMed]
18. Yang, M.; Zhang, X.; Liu, G.; Yin, Y.; Chen, K.; Yun, Q.; Zhao, D.; Al-Mssallem, I.S.; Yu, J. The complete chloroplast genome sequence of date palm (Phoenix dactylifera L.). PLoS ONE 2010, 5, el2762. [CrossRef] [PubMed]
19. Nie, X.; Lv, S.; Zhang, Y.; Du, X.; Wang, L.; Biradar, S.S.; Tan, X.; Wan, F.; Weining, S. Complete chloroplast genome sequence of a major invasive species, crofton weed (Ageratina adenophord). PLoS ONE 2012, 7, e36869. [CrossRef] [PubMed]
20. Tangphatsornruang, S.; Sangsrakru, D.; Chanprasert, J.; Uthaipaisanwong, P.; Yoocha, T.; Jomchai, N.; Tragoonrung, S. The chloroplast genome sequence of mung bean (Vigna radiatd) determined by high- throughput pyrosequencing: Structural organization and phylogenetic relationships. DNA Res. 2010, 17, 11–22. [CrossRef] [PubMed]
21. Yi, D.K.; Kim, K.J. Complete chloroplast genome sequences of important oil seed crop Sesamum indicum L. PLoS ONE 2012, 7, e35872. [CrossRef] [PubMed]
22. Quan, X.; Zhou, S.L. Molecular identification of species in Prunus sect. Persica (Rosaceae), with emphasis barcodes for plants. J. Syst. Evol. 2011, 49, 138–145. [CrossRef]
23. Wand, L.; Dong, W.P.; Zhou, S.L. Structural Mutations and Reorganizations in Chloroplast Genomes of Flowering Plants. Acta Bot. Boreal. Occident. Sin. 2012, 32, 1282–1288.
24. Yu, D.J.; Gu, C.Z. Flora Reipublicae Popularis Sinicae; Bejing Science Press: Beijing, China, 1974; Volume 36. (In Chinese)
25. Jansen, R.K.; Raubeson, L.A.; Boore, J.L.; Pamphilis, C.W.; Chumley, T.W.; Haberle, R.C.; Wyman, S.K.; Alverson, A.J.; Peery, R.; Herman, S.J.; et al. Methods for obtaining and analyzing whole chloroplast genome sequences. Methods Enzymol. 2005, 395, 348–384. [PubMed]
26. Wu, Y.; Zhou, H. Research progress of sugarcane chloroplast genome. J. South. Agric. 2013, 44, 17–22.

27. Luo, R.; Liu, B.; Xie, Y.; Li, Z.; Huang, W.; Yuan, J.; He, G.; Chen, Y.; Pan, Q.; Liu, Y.; et al. SOAPdenovo2: An empirically improved memory-ecient short-read de novo assembler. Gigascience 2012, 1, 18. [CrossRef] [PubMed]

28. Mardis, E.R. The impact of next-generation sequencing technology on genetics. Trends Genet. 2008, 24, 133–141. [CrossRef] [PubMed]

29. Szmidt, A.E.; Aldén, T.; Hällgren, J.E. Paternal inheritance of chloroplast DNA in larix. Plant Mol. Biol. 1987, 9, 59–64. [CrossRef] [PubMed]

30. Palmer, J.D. Comparative organization of chloroplast genomes. Annu. Rev. Genet. 1985, 19, 325–354. [CrossRef] [PubMed]

31. Nock, C.J.; Waters, D.L.; Edwards, M.A.; Bowen, S.G.; Rice, N.; Cordeiro, G.M.; Henry, R.J. Chloroplast genome sequences from total DNA for plant identification. Plant Biotechnol. J. 2010, 9, 328–333. [CrossRef] [PubMed]

32. Pu, F.S.; Lin, S.H.; Chen, R.Y.; Song, W.Q.; Li, X.L. Studies on karyotype of pyrus in China(II). Acta Horticult. Sin. 1986, 13, 87–90.

33. Pu, F.S.; Lin, S.H.; Song, W.Q.; Chen, R.Y.; Li, X.L. Studies on karyotype of pyrus in China(I). J. Wuhan Bot. Res. 1985, 3, 381–387.

34. Lin, B.N.; Shen, D.X. Studies on the germplasmic characteristics of Pyrus by use of isozymic patterns. Acta Agric. Univ. Zhejiang Gensis 1983, 9, 235–242.

35. Teng, Y.W.; Tanabe, K.; Tamura, F.; Itai, A. Genetic relationships of pear cultivars in Xin Jiang, China, as measured by RAPD markers. J. Horticult. Sci. Biotechnol. 2001, 76, 771–779. [CrossRef]

36. Teng, Y.; Tanabe, K.; Tamura, F.; Itai, A. Genetic relationships of Pyrus species and cultivars native to East Asia revealed by randomly amplified polymorphic DNA markers. J. Am. Soc. Hortic. Sci. Biotech. 2002, 127, 262–270.

37. Teng, Y.W.; Tanabe, K. Reconsideration on the origin of cultivated pears native to East Asia. 4thInternational Symposium of Taxonomy and Nomenclature of Cultivated Plants. Acta Hortic. 2004, 634, 175–182. [CrossRef]

38. Teng, Y.W. Advances in the research on phylogeny of the genus Pyrus and the origin of pear cultivars native to East Asia. J. Fruit Sci. 2017, 34, 370–378.

39. Zheng, X.Y.; Cai, D.Y.; Yao, L.H.; Teng, Y.W. Non-concerted ITS evolution, early origin and phylogenetic utility of ITS pseudogenes in Pyrus. Mol. Phylogenet. Evol. 2008, 48, 892–903. [CrossRef] [PubMed]

40. Taberlet, P.; Gielly, L.; Pautou, G.; Bouvet, J. Universal primers for amplification of three non-coding regions of chloroplast DNA. Plant Mol. Biol. 1991, 17, 1105–1109. [CrossRef] [PubMed]

41. Hu, C.Y.; Zheng, X.Y.; Teng, Y.W. Characterization and Phylogenetic Utility of Non-coding Chloroplast Regions trnL-trnF and accD-psaI in Pyrus. Acta Hortic. Sin. 2011, 38, 2261–2272.

42. Liu, Y. Studies on Chloroplast DNA Diversity of Chinese Pear; Capital Normal University: Beijing, China, 2006.

43. Raghvendra, A.S. Photosynhcsis: A Comprehensive Treatise; Cambridge University Press: Cambridge, UK, 1998; pp. 72–86.

44. Kent, W.J. BLAT-the BLAST-like alignment tool. Genome Res. 2002, 12, 656–664. [CrossRef] [PubMed]

45. Liu, C.; Shi, L.; Zhu, Y.; Chen, H.; Zhang, J.; Lin, X.; Guan, X. CpGAVAS, an integrated web server for the annotation, visualization, analysis, and GenBank submission of completely sequenced chloroplast genome sequences. BMC Genom. 2012, 13, 715. [CrossRef] [PubMed]

46. Mott, R. EST_GENOME: A program to align spliced DNA sequences to unspliced genomic DNA. Comput. Appl. Biosci. 1997, 13, 477–478. [CrossRef] [PubMed]

47. Lowe, T.M.; Eddy, S.R. tRNAscan-SE: A program for improved detection of transfer RNA genes in genomic sequence. Nucleic Acids Res. 1997, 25, 955–964. [CrossRef] [PubMed]

48. Laslett, D.; Canback, B. ARAGORN, a program to detect tRNA genes and tmRNA genes in nucleotide sequences. Nucleic Acids Res. 2004, 32, 11–16. [CrossRef] [PubMed]

49. Abouelhoda, M.I.; Kurtz, S.; Ohlebusch, E. Replacing suffix trees with enhanced suffix arrays. J. Disc. Algo. 2004, 2, 53–86. [CrossRef]

50. Lohse, M.; Drechsel, O.; Kahlau, S.; Bock, R. OrganellarGenomeDRAW—A suite of tools for generating physical maps of plastid and mitochondrial genomes visualizing expression data sets. Nucleic Acids Res. 2013, 41, W575–W581. [CrossRef] [PubMed]

51. Wang, D.P.; Zhang, Y.B.; Zhang, Z.; Zhu, J.; Yu, J. KaKs_Calculator 2.0: A toolkit incorporating gamma-series methods and sliding window strategies. Genom. Proteom. Bioinform. 2010, 8, 77–80. [CrossRef]

7 Chapter

NATURAL HISTORY OF A SATELLITE DNA FAMILY: FROM ANCESTRAL GENOME COMPONENTS TO SPECIES-SPECIFIC SEQUENCES, INCLUDING CONCERTED AND NON-CONCERTED EVOLUTION

1. INTRODUCTION: AN OVERVIEW

Genome evolution is the multifaceted process of variation in nuclear genome components over time. The process is diverse, with various genomic fractions evolving at distinct speeds. The most fast alterations were observed in repeatomes, which constitute the foundation of most eukaryotic genomes and comprise repeated and repeat-derived sequences [1,2]. The repeatomes of diverging species predominantly evolve in a concerted manner, leading to greater sequence similarity of repeating units within species than among them [3]. Repetitive DNA complexes significantly contribute to evolutionary genome transformation, and elucidating their origin, composition, and dynamics is essential for comprehending genomic diversity [4]. The repeatome has many major types, predominantly transposable elements (TEs) and satellite DNA (satDNA) [5,6]. The latter comprises extensive, late-replicating, non-coding arrays of tandemly organized monomers [5,7]. These sequences are frequently distinct to species or genus and represent the most changeable component of the eukaryotic genome, hence indicating patterns of short-term evolutionary change [8–11]. Recent investigations indicate that satDNA, primarily located in the heterochromatic sections of chromosomes, plays a role in diverse processes, including chromosomal organization, pairing, cellular metabolism, and modulation of gene activity [12–16]. Despite their critical role in elucidating genome function and reorganization during micro- and macroevolutionary processes, as well as the increasing recognition of their structural and functional significance, knowledge regarding the origin and dynamics of satellite DNA remains fragmented, particularly in non-model species.

It is generally accepted that an intraspecific monomer change in various satDNA families is permanent [17]. Related species share a common satDNA library that was present in the common ancestor. Differential amplification of satellites from this library and acquisition of mutations in diverse lineages results in interspecific differences in that fraction [18]. Spreading of a new variant processed by non-Mendelian molecular mechanisms is followed by the fixation of the new variant within a population by sexual reproduction [19–21]. Thus, intraspecific homogenization of the satDNA family and fixation of species-specific polymorphisms occur simultaneously [22], and the main trend of satDNA conversion can be considered as a transformation from the common ancestral to the species-specific tandem repeats. The process appears to be a significant part of speciation at the molecular level [4]. Recently, the possibility of unraveling details of this ubiquitous phenomenon by next-generation sequencing (NGS) technology appeared through comparative analysis of the entire species repeatome. Importantly, this method is applicable not only for model organisms but also for a wide range of wild species, which allows the construction of a generalized model.

This study aimed to analyze NGS data utilizing the RepeatExplorer (RE) pipeline [23] to deduce the evolutionary dynamics of satDNA in the genomes of Chenopodium s. str. (commonly known as the Chenopodium album aggregation). Species within the C. album aggregate are globally widespread, exhibiting the greatest species diversity in temperate regions [24]. The majority of these diploid-polyploid species exhibit remarkable phenotypic plasticity, with some being extensively spread and capable of thriving under diverse environments. We concentrated on diploid species (2n = 2x = 18) in the aggregate that signify distinct lineages. Specifically: (i) “clade A” includes species indigenous to America and East Asia, represented by C. bryoniifolium Bunge; (ii) “clade B” encompasses the Eurasian temperate species C. ficifolium Sm. and the boreal species C. suecicum Murr.; (iii) “clade D” consists solely of the East and Central Asian species C. acuminatum Willd; (iv) “clade E” is represented by the Central Asian species C. pamiricum Iljin and C. iljinii Golosk.; (v) “clade H” presumably includes European and southwest Asian species C. vulvaria L; and clades C, F, and G comprise polyploid species. The presence of fundamental diploid lineages elucidates the genesis of most Eurasian polyploid species as a result of hybridization among these diploid lineages, which produced subgenomic combinations of distinct polyploid taxa (see to [27] for further details) (Figure 1). This group was chosen according to the subsequent two criteria: (i) Analyzed species of the genus Chenopodium exemplify a diploid/polyploid complex [26,27], which is characteristic of angiosperms and can be considered a standard model for the divergent evolution of higher plants; and (ii) a fundamental repeat unit with pan-chromosomal distribution, also associated with the satellite monomer of Beta corolliflora, was previously identified in the genome of a Chenopodium species [28,29]. This confluence of advantageous circumstances renders the study potential for elucidating satDNA family evolution within a representative collection of flowering plants. Considering the global distribution of the C. album aggregate and its extensive evolutionary history spanning tens of millions of years [27], we propose the existence of several modifications of satDNA families in divergent lineages.

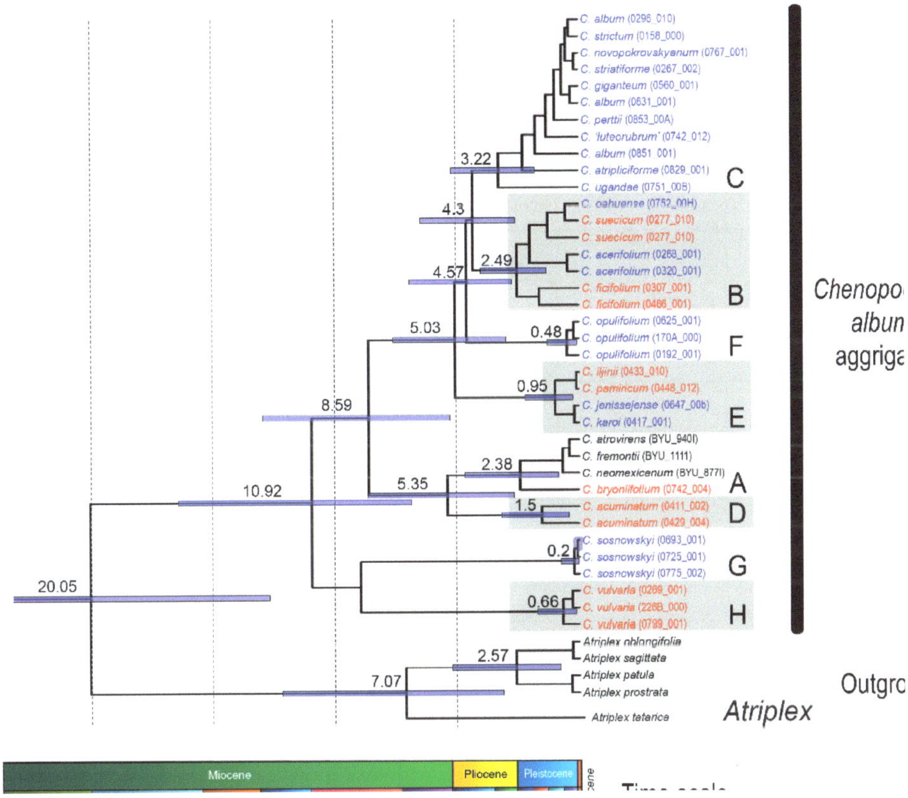

Figure 1. Phylogenetic tree calculated using Bayesian inference within the C. album aggregate estimated based on the concatenated dataset of three chloroplast DNA spacers (adapted from [27]). Major evolutionary lineages (A–H) are marked by grey rectangles. The numbers above branches correspond to the ages of the particular clades (in millions of years) as inferred by the analysis in BEAST2. Positions of explored diploid species are shown in red. Polyploid species are shown in blue. The schematic stratigraphic time scale (Miocene–Holocene) is shown at the bottom of the figure.

2. RESULTS

2.1 Clustering Results and Identification of satDNA Clusters

Application of the RE pipeline clustering tool for Illumina reads of seven diploid Chenopodium species (Table 1) (genome coverage 41.3–58.2%) resulted in the identification of clusters that represent different families of TEs, their derivatives and satDNAs. The latter was the main aim of our research. Several valuable outcomes from the present study are shown in Table 2. C. vulvaria and C. acuminatum possess the smallest genomes in the group, while those of C. ficifolium and C. suecicum are the largest. C. vulvaria exceeds all investigated species in the number of RE clusters and RE singlets, which emphasizes for its genome diversity.

Table 1. The accessions and geographic origin of Chenopodium diploid species used for satDNA cluster analysis (NGS), probe preparation (cloning) and fluorescent in situ hybridization (FISH).

Species	Clade	ID Number	Locality
C. acuminatum	D	429/3	China, Burquin
C. bryoniifolium	A	742/4	Russian Federation, Nakhodka
C. ficifolium	B	330/2	Czech Republic, Slatina
C. iljinii	E	433/9	China, Hoboksar
C. pamiricum	E	830/3C	Tajikistan, Gorno-Badakhshan
C. suecicum	B	328/10	Czech republic, Švermov
C. vulvaria	H	771/1	Iran, Shahr

Table 2. Summary of chromosome parameters, genome size, RE clusters and percentage of CficCl-61-40 satDNA family in the genomes of C. album aggregate diploid species.

Species	Chr. Numb. 2n	Chr. Size µm	Genome Size 2C Values Mb [26]	RE Clusters #	RE Singlets #	CficCl-61-40 % in Genome
C. acuminatum	18	0.8–1.5	960	393251	34269	3.80
C. bryoniifolium	18	0.7–0.9	1200	307778	38905	2.25
C. ficifolium	18	1.5–4.5	1785	369861	20661	0.31
C. iljinii	18	1.7–3.3	1144	327760	82679	0.42
C. pamiricum	18	1.2–2.5	1154	249599	42427	0.25
C. suecicum	18	2.5–5.0	1775	369583	72167	0.27
C. vulvaria	18	1.5–2.0	924	542674	93278	0.79

The satellite monomer of ~40 bp was found during the RE analysis of satDNA in each genome of analyzed species. According to BLAST results, these monomers were related to each other and to the tribe-specific repetitive sequence (GenBank ID HM641822.1), found in Chenopodium quinoa by Kolano et al. [29], and to the satellite sequence with the GenBank ID AJ288880.1, which was found in Beta corolliflora by Gao et al. [28] (S1). It was thus assumed that in the genomes of the several Chenopodium diploids under study, the most abundant and the evolutionarily oldest component (Chenopodium and Beta diverged approximately in the Paleogene) is present. In the remainder of this paper, tandem arrays from the genomes of Chenopodium diploid species related to GenBank accession HM641822.1 will be termed the "CficCl-61-40 satDNA family". This refers to the analysis of NGS data from the C. ficifolium genome, RE Cluster #61 (the single cluster in genome of C. ficifolium that contains the basic repeat unit), with a length of 40 bp. A further thorough analysis of the interspecies divergence of the sequences of this family was also conducted to identify the main phases of transformation over time.

2.2 Sequence Analysis in the CficCl-61-40 satDNA Family

Among the multitude of clusters produced by RE pipeline, a BLAST search determined a single cluster that belongs to the CficCl-61-40 satDNA family in the genomes of C. acuminatum, C. bryoniifolium, C. ficifolium, C. iljinii, C. pamiricum, and C. suecicum and seven clusters in genome of C. vulvaria (supplementary data 1, Figure 2). The highest percentages of the CficCl-61-40 satDNA family were observed in the C. acuminatum and C. bryoniifolium genomes (Table 2). Subsequent tandem repeat finder (TRF) analysis allows determination of consensus monomer(s) (supplementary data 1). The algorithm of TRF looks for tandem repeats that are often hidden in larger homologous regions or which may fall well below the level of significance required for other programs to report a match. The detection criteria are based on a stochastic model of tandem repeats specified by percent identity and frequency of insertions and deletions rather than some minimal alignment score and align repeat copies against a consensus sequence, revealing patterns of common mutations [30]. Nucleotide sequence divergence among monomers within satDNA arrays is usually quite low, generally, not exceeding a few percent, and for the purpose of sequence analysis, it is acceptable to manipulate with the satDNA consensus sequence [17]. For C. ficifolium, C. pamiricum and C. suecicum a single monomer of ~40 bp was detected. However, for C. acuminatum, C. bryoniifolium, C. iljinii, and C. vulvaria, several derivatives from CficCl-61-40 satDNA family monomers were found inside the single cluster. The following two levels of CficCl-61-40 satDNA family variability in the genomes of C. album aggregate diploid species were thus observed: (i) at the inter-cluster level, namely single or multiple RE clusters, and (ii) at the intra-cluster level, namely single monomer or a set of related monomers of different lengths detected by TRF.

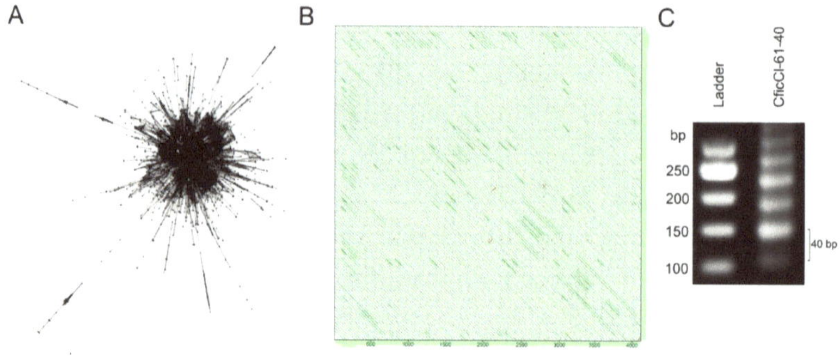

Figure 2. RepeatExplorer (RE) analysis of next-generation sequencing (NGS) data in Chenopodium diploids. (A) Cluster 61 of C. ficifolium demonstrate layouts that are typical for tandem repeats where nodes represent the sequence reads and edges between the nodes correspond to similarity hits; (B) Self-to-self comparisons of the contig 25 cluster 61 displayed as dot plots (genomic similarity search tool YASS program output) where parallel lines indicate tandem repeats (the distance between the diagonals equals the lengths of the motifs ~40 bp); (C) Agarose gel electrophoresis of PCR products obtained with primers designed from consensus monomer sequence of C. ficifolium (Cluster 61) showing typical ladder structure of tandem array.

2.3 Reconstruction of the Ancestral Monomer

BLAST-detected relatedness between satellite monomers of the CficCl-61-40 satDNA family allowed determination of the major part of the ancestral monomer. For this reconstruction, satellites of C. bryoniifolium (consensus monomer from one RE cluster) and C. vulvaria (consensus monomers from seven RE clusters) that show relatedness to both C. quinoa and B. corolliflora satellites were aligned. DNA fragments with 100% BLAST matches in combination formed the most conservative fragment of the basic monomer (supplementary data 2). This approach is quite similar to the method of ancient paralogs (LUCA) [31,32]. The sequence of 37 bp was as follows: TCAAACAAAGCTAATTGAATCAAATGAAAGTCAAATG. This sequence was used as a basis for the subsequent comparison of the monomer divergence in Chenopodium lineages. Analysis of basic satellite alterations revealed point mutations, indels, and shifts that were present with different frequencies in the genomes of the studied diploid Chenopodium species (supplementary data 1). K-mer based distance estimation revealed a phylogenetically reliable tree with the ancestral monomer as a base, B. corolliflora is located separately and rather close to the root, the analyzed diploids that form fairly natural groups with species of clades B and E located nearby, C. bryoniifolium, C. acuminatum aside, and polyploid C. quinoa at the maximum distance from the ancestral monomer (Figure 3).

Clade H (C. vulvaria) deserves separate attention. The RE pipeline divided the variety of CficCl-61-40 satDNA family sequences in the genome of C. vulvaria into seven clusters (supplementary data 1), indicating valuable heterogeneity. On one hand, all the basic monomers of the clusters contain BLAST-recognizable fragments of the ancestral monomer. On the other hand, the observed variability exceeds that for all clades taken together (Figure 3). An important question is whether all these clusters from the C. vulvaria genome belong to the same CficCl-61-40 satDNA family. RE output includes not only the row of clusters but also detailed cluster characteristics, including the cluster neighborhoods of connected components. The analysis showed that all clusters that we classified as belonging to the CficCl-61-40 satDNA family are related to each other and to the repetitive sequences with the GenBank IDs HM641822.1 and AJ288880.1. Additionally, these satDNA clusters possess a limited number of similarity hits with TEs clusters (mainly with Ty3-gypsy retrotransposons) which may indicate for splitting of satDNA arrays by the insertion of TEs.

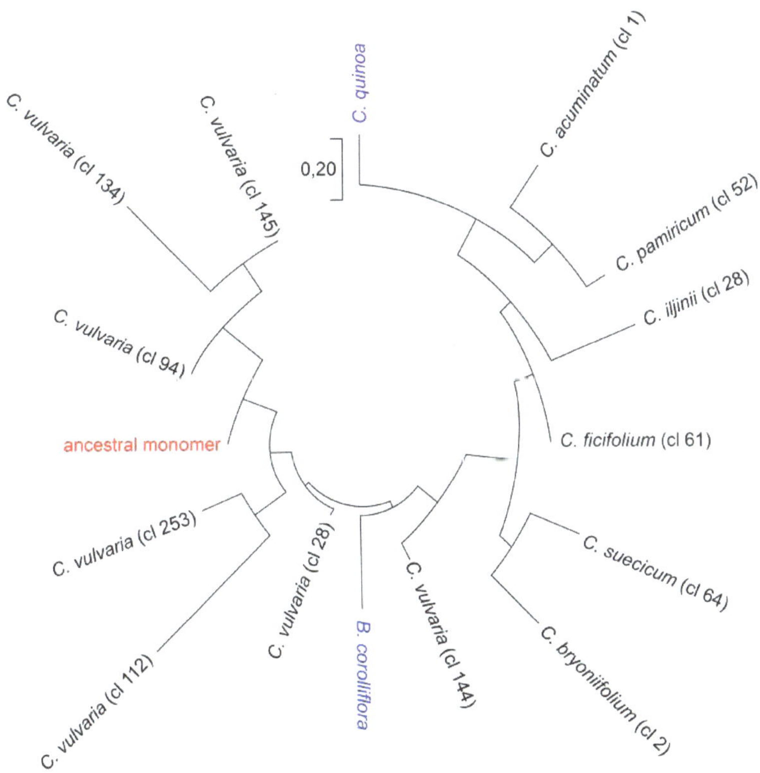

Figure 3. Phylogenetic relationships of the CficCl-61-40 satDNA family sequences. Phylogenetic tree based on the k-mer analysis.

2.4 High Order Repeat (HOR) Detection in the CficCl-61-40 satDNA Family and Determination of Its Physical Counterpart

TRF analysis of the CficCl-61-40 satDNA family in seven diploid species of Chenopodium revealed different structures of the arrays. In C. ficifolium, C.

pamiricum, and C. suecicum, uniform tandem arrays with basic satellite motifs of ~40 bp (87–96% matches between monomers and copy numbers of 79.2–153.4) were identified by TRF. In C. acuminatum, C. bryoniifolium, C. iljinii and C. vulvaria, derivatives from CficCl-61-40 satDNA family repeats ranging up to 332 bp and of different repeatability were found. It was proposed that in the latter species, HORs could be formed by concurrent amplification and homogenization of modified monomers.

Here, it is necessary to elucidate the TRF algorithm using an example of the detection of a 117 bp monomer in the genome of C. acuminatum (later used as a probe in fluorescent in situ hybridization (FISH) experiments). Analysis of the RE Cluster-1 sequence by TRF produced a table of monomers with the most frequent of 117 bp (consensus size) (supplementary data 1). However, when the consensus sequence was manually analyzed, it decomposed into three 39 bp long subrepeats. Nevertheless, it can be argued that the 117 bp fragment is the basic monomer and that the formation of a HOR unit is based on an ~40 bp monomer. The program finds likely patterns (monomers) and then refines them into a consensus sequence. Patterns are detected by a high percentage of matches at the candidate pattern length. For 39 bp not enough matches were found, but a very high number for 117 bp. This indicates that the unit of duplication was 117 bp and not 39 bp. Furthermore, the mismatches and indels are more consistent with a 117 bp monomer than with a 39 bp monomer (Gary Benson, personal communication). Following sequencing of physical counterparts of CacuCl-1-117 consensus sequence (see below) revealed that the physical components of the CacuCl-1-117 HOR unit did not coincide completely (as in consensus) but varied within the interval of 82% to 86% similarity, which confirmed the accuracy of the TRF algorithm. Additionally, it can be considered that the TRF analysis of all RE clusters belonging to the CficCl-61-40 satDNA family was performed with the same parameters, and in genomes of tree species, only homogeneous arrays were identified while the four other arrays were heterogeneous, which reflects the real structure of satDNA.

A total of three to four different proposed HOR units were detected in the genomes of C. acuminatum, C. bryoniifolium and C. iljinii. However, approximately 23 such units were found in genome of C. vulvaria (supplementary data 1). The genome of C. vulvaria is thus again the most variable according to this parameter.

While there are multiple studies that demonstrate that RE is efficient in repeat identification using NGS, there are some limitations regarding sequence analysis of satellite repeats. The most important one was that generation of consensus sequences by assembling reads to contigs. While this works well for most dispersed repeats like TEs, this is problematic for satellites due to their tandem structure. Consequently, contigs vary in their coverage by reads and their sequences could be partially chimeric (producing sequence variant combinations that in fact do not exist in the genome). To confirm the existence in genomes the physical counterparts of computer-generated consensus monomers we analyzed the sequence variation of CficCl-61-40 and proposed HOR units CacuCl-1-117, CvulCl-28-118, CvulCl-28-397, CvulCl-112-117, CvulCl-134-117 and Cvul-145-129 by cloning. We then compared the obtained sequences with the consensus sequence from the TRF output (supplementary data 3, Figure 4). For all monomers, we obtained several clones that differed from each other as well as from the consensus sequence (supplementary data 4). The CficCl-61-40 monomer is rather uniform with a few point mutations and sequence similarity between clones. The consensus sequence ranged from 90.2% to 95%. For the four obtained clones of the CacuCl-1-117 monomer, two sequence types were found with generally higher similarity to the consensus sequences as well as to each other (similarity value ranges 89.8–91.5 and 90.7–99.2, respectively). This once again confirmed the correctness of the TRF algorithm.

More variability was detected for the proposed HOR units in the C. vulvaria genome, which once again highlights the complexity of the satDNA fraction in this species. Thus, among tree clones obtained for the CvulCl-28-118-proposed HOR unit, two sequence types were found with generally high similarity to each other than to the consensus monomer (similarity value ranges 88.2–98.3 and 76.4–79.1, respectively). For CvulCl-28-397-proposed HOR unit sequences amplified by primers (supplementary data 3) also shows more relatedness to each other than to consensus sequence (supplementary data 4). For the CvulCl-112-117- and CvulCl-134-117-proposed HOR units, two types were found among cloned sequences. One showed high relatedness to the consensus monomer (83.3%–90.7%) and the other clones were 100% related to each other and less to the consensus monomer and to the first variant (supplementary data 4). This most likely suggests that several related HOR units could be formed simultaneously. For Cvul-145-129-proposed HOR unit clones possess high similarity to the consensus sequences as well as to each other (82.9%–100.0%). Part of the cloned sequences was submitted to GenBank (accession numbers MH257681–MH257687). However, it should also be noted that we were not able to amplify part of the proposed HOR units generated by TRF analysis (for example CvulCl-28-355 and Cvul-134-148) (supplementary data 1, far right line on Figure 4). These sequences could be attributed most likely to computer-generated chimeric sequences (i.e., method error). However, for the majority of the proposed HOR units its physical counterparts were discovered in the genomes.

2.5 Comparison of CficCl-61-40 and Proposed HOR Unit CacuCl-1-117 Chromosomal Distribution

For further confirmation of the existence of HOR units' in the genomes of Chenopodium species an alternative method of FISH was used. Two distant clusters according to phylogenetic analysis, cluster 61 of C. ficifolium (CficCl-61-40) and cluster 1 of C. acuminatum (CacuCl-1-117) from the RE output, were selected as sources for in situ probes for comparative molecular cytogenetic analysis (Figure 5, supplementary data 1 and 5). FISH experiments were performed to verify if (i) C. acuminatum-specific tandem repeats that were proposed to be HOR units (CacuCl-1-117) are species-specific and do not hybridize to chromosomes of the other six species and (ii) if the chromosomal positions of the C. acuminatum-specific tandem repeat (CacuCl-1-117) are similar to or different from the positions of the tribe-specific repeat (CficCl-61-40) on the chromosomes of C. acuminatum. It should be noted, however, that accurate FISH-based karyotyping and chromosome mapping of CficCl-61-40 satDNA family tandem repeats is challenging in Chenopodium due to the small chromosome sizes and to the large number of clusters (Table 2, Figure 5).

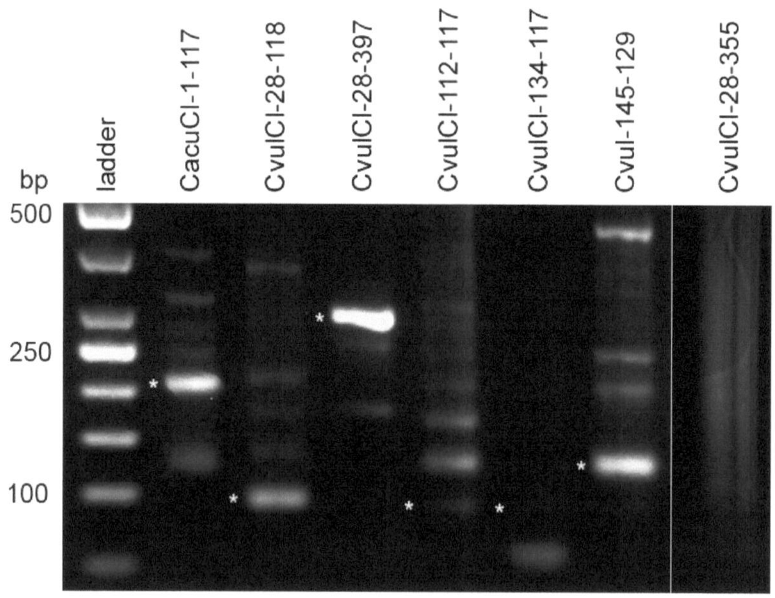

Figure 4. Agarose gel electrophoresis of PCR products obtained with primers designed from consensus monomer sequence of proposed high order repeat (HOR) units for determination of their physical counterparts. Cloned DNA fragments are shown by asterisks. The far-right line is an example of negative amplification of a computer-generated proposed HOR unit.

FISH experiments confirmed species specificity and sometimes separate chromosomal positions of newly formed HOR units. Probe CficCl-61-40 hybridized to the chromosomes of all analyzed species except C. iljinii (similarity, copy number, or both of the particular FISH probe in C. iljinii genome is likely much less in comparison with other species), which demonstrates the presence of a tribe-specific satellite, while CacuCl-1-117 hybridized only to chromosomes of C.

acuminatum with no signal on the chromosomes of the other six species (Figure 5, supplementary data 5, the minor green signal in C. pamiricum, C. suecicum and C. vulvaria in supplementary data 5 is epifluorescence). In addition, the simultaneous hybridization of CacuCl-1-117 and CficCl-61-40 on the chromosomes of C. acuminatum shows that in many cases these tandem arrays form separate clusters that create a species-specific chromosomal pattern (Figure 5).

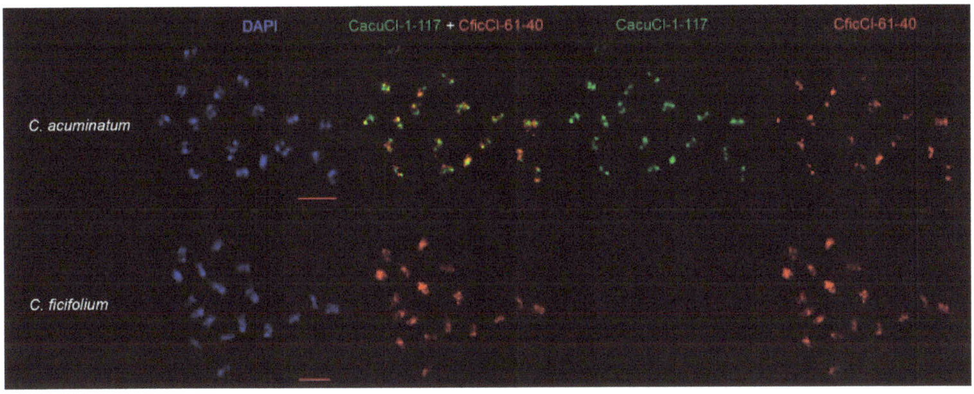

Figure 5. Chromosomal distribution CficCl-61-40 satDNA family sequences. CficCl-61-40 is labelled red; C. acuminatum-specific HOR unit CacuCl-1-117 is labelled green. Bar represent 5 μm.

3. DISCUSSION

Application of the RE pipeline for analysis of whole genome shotgun Illumina reads from the genomes of seven diploid Chenopodium species from divergent lineages revealed that the investigated CficCl-61-40 satDNA family is the most abundant and oldest component of the Chenopodium genome, given that related sequences were found in both Chenopodium and Beta species. Regarding these two genera, it is essential to note that the genome of Beta should be recognized as more static, at least because it contains many fewer species (approximately 7–8 species in total, [33]) in comparison with Chenopodium (approximately 150 species [24]). Alignment of the satellite monomers allowed identification of the ancestral DNA fragment of 37 bp that showed 100% identity between B. corolliflora from one side and C. bryoniifolium and C. vulvaria from the other (supplementary data 2). The latter two are species that split off early and possess a modified sequence that is still recognizable by BLAST as a ~40 bp variant of the ancestral monomer. The identified DNA fragment served as a benchmark for our subsequent analyses, in which we intended to characterize intra-unit evolutionary transformations in the diverse Chenopodium lineages.

Remarkably, the evolutionary history of the C. album aggregate revealed by cpDNA spacers and two low-copy genes [27] correlates fairly well with significant paleoclimatic events. Thus, the early differentiation coincides with the beginning of the Miocene Climatic Optimum in the Burdigalian Age (approximately 20 Mya) (Figure 1). Clade H (C. vulvaria) separated upon transition between the Serravallian and Tortonian Ages, ~11 Mya. However, the main lineages were formed in the Pliocene, when due to a cooler and dry, seasonal climate, grasslands spread on all continents, and savannahs and deserts appeared in Asia and Africa. Subsequent speciation within the lineages and the appearance of the majority of polyploids occurred in the Quaternary Period, when the glacial and interglacial epochs succeeded each other. During this time, since there were no places on Earth with identical climate history and since the species of aggregate were spread widely, the CficCl-61-40 satDNA arrays evolved divergently. Excluding clade H, which split off early and is now very different, k-mer-based distance estimation of basic monomer show the most significant differences in genomes of species from clades A and D. It is most likely that both lineages separated early from the ancestral group and evolved independently. This is consistent with the present species distribution ranges and with molecular phylogenetic data [26,27]. However, the pace of evolution of these clades was probably different and is most likely connected with the climatic history of the species distribution areas. In clades B and E, the species are much more similar in the CficCl-61-40 satDNA family structure (Figure 2).

The concept of "molecular drive" [19] postulates that mutations can gradually spread throughout a satDNA family by several of ubiquitous mechanisms of DNA turnover (homogenization) and become fixed in a population. SatDNA families can show a rapid rate of inter-specific evolutionary changes concerning DNA sequence and high levels of conservation between species separated for long evolutionary times [22,34,35]. Although these trends are also true for the CficCl-61-40 satDNA family when monomers are homogenized on the species level in the genomes of different Chenopodium lineages, each of them has its own mode and tempo. Although the genome of C. vulvaria presents an exception, it seems that concerted evolution does not operate there. This example of non-concerted evolution will be discussed below.

In addition to mutations in basic satellite monomers, a distinct trend toward increased complexity and length of the monomer (HOR unit formation) was recorded in the species of Clades A, D, E and H of the C. album aggregate. HORs occur by concurrent amplification and homogenization of different monomers in the original satDNA when a complex monomer is first formed, after which it merges into a more complex HOR unit [17].

The origin of such structures has been described for the alpha satellite of primates [36], for the satellite families in bovids [37,38] and for the plant species Vicia grandiflora [39]. A detailed analysis of the CficCl-61-40 satDNA family tandem arrays in the genomes of C. acuminatum, C. bryoniifolium, C. iljinii and C. vulvaria along with the basic ~40 bp monomer revealed related but longer monomers of up to 332 bp, suggesting the generation of new species-specific HOR units. Cloning of PCR-amplified DNA fragments in most cases confirmed the accuracy of the monomer/array compilation produced by the RE pipeline, and the physical counterparts were mostly in agreement with the consensus sequences. However, the exact satDNA array structure of the species could be determined by complete genome sequencing, assembly and annotation [40].

FISH experiments further prove the genesis of species-specific HOR units and their separate locations on the chromosomes. CficCl-61-40 arrays were thus found in all species. On the other hand, related CacuCl-1-117 arrays were found exclusively in C. acuminatum, where they form multiple, sometimes separate chromosome clusters, thus creating a species-specific chromosomal pattern (Figure 5). Formation of HOR units based on two or more monomers has been reported in primates and bovids (for a review, see [17]). We observed a similar process but based on the single tribe-specific monomer when unequal changes in the initial sequence in diverging satDNA sets led to monomer alterations with the subsequent merging of the modified monomers in a complex HOR unit. A similar process (i.e., HOR formation based on one initial repeated unit in Vicia sp.) was reported by Macas et al. [39]. Presumably, the process of HOR formation on the basis of a single monomer can take more time (in our research, it appears predominantly in ancient species) than that involving two or several monomers, although it apparently contributes to satDNA divergence.

We might next ask whether the formation of HORs is common for plant satDNA evolution. As another example of supposed HOR formation in plants, we can provide a complex structure of the Hieracium species centromeric tandem array [41]. Analysis of both RE clusters and the sequenced physical counterparts revealed a complex structure with 21 repetitive elements identified by TRF (ranging from 21 bp to 348 bp) and with two abandoned motifs of 21 and 23 bp. Eventually, we can also observe the stages of HOR formation based on the two short monomers in centromeric regions. It is essential to note that although chromosome segregation machinery is highly conserved across all eukaryotes, centromeric DNA evolves rapidly, and discovered tandem repeats are absent in related Pilosella species. Incompatibilities between rapidly evolving centromeric components may be responsible for both the organization of centromeric regions and the reproductive isolation of emerging species [42].

The above examples and the fact that the presented species refer to different large clades of flowering plants suggests that the HOR formation process may not only occur in the Chenopodium, Hieracium, and Vicia genomes but that this mechanism is also ubiquitous for at least angiosperms and could underlie satDNA divergence in related plant species, as it does in animal genomes. It should also be noted that HOR formation is presumably a species-specific event; in clade B (C. ficifolium and C. suecicum), neither species showed any sign of HORs. In contrast, in clade E, CficCl-61-40 satDNA family arrays of C. pamiricum are uniform, while HORs were detected in the C. iljinii genome. However, it is still not clear what triggers the HOR formation in a particular genome [17].

In generalizing the life history of the CficCl-61-40 satDNA family stretching from the ancestral basic repeat unit to species-specific sequences, it is worth noting that the family consists of an extensive group of related, divergent repeats. It is a dominant and old component of Chenopodium species genomes and can be characterized by a high complexity of evolution. Independently amplified in each genome, it ultimately acquires lineage-specific profiles due to differential stochastic amplifications, contractions or both. Additionally, in several lineages, a clear trend toward increased complexity and satellite monomer length was observed. Long tandem arrays are characterized by HOR units whose organization and nucleotide sequence are specific for a particular species. Analysis of the sequence organization of these diverged subsets provides a framework for considering mechanisms of sequence diversity generation and for understanding the evolutionary processes of satDNA family homogenization and polymorphism [37]. Homogenization of satellite repeats driven by molecular mechanisms of nonreciprocal sequence transfer occurs simultaneously, which makes satDNA evolve mostly in a concerted manner [3]. Nevertheless, as mentioned above, the small genome of C. vulvaria (2C value 0.945 pg) is an exception to this rule. The observed variability indicates a low level of CficCl-61-40 satDNA family homogenization, with multidirectional trends in the C. vulvaria genome (non-concerted evolution). Although the data are unusual, our unpublished results on the NGS-based qualitative analysis of TEs in genomes of the same Chenopodium diploid species (where we observed that C. vulvaria possesses a unique pool of different and diverse retrotransposons [43]) make it possible to hypothesize a link between the TE dynamics and abnormalities in the homogenization of satDNA families, given that satDNA could be a target for TE insertions [44] and evolve further to species-specific tandem repeats [45]. Suppression of concerted evolution resembles those described for termites by Luchetti et al. [46]. This was proposed to be evoked by the limited number of reproducers, especially considering that C. vulvaria is an ancient species, restricted to nutrient-rich bare soil largely of anthropogenic impact and not tolerant of competition [47]. Specific habitats may presumably cause abnormal repeatome composition that, in turn, may support the models assuming that genotypes from marginal populations are evolutionarily significant [48–51]. Despite the causes, discovered suppression of homogenization itself may result in alteration of satDNA libraries, ultimately leading to spontaneous transformation of the entire repeatome, thus producing a novel set of satDNA families for the next round of the conversion cycle, and genomes undergoing non-concerted evolution can be proposed as a significant source of genomic diversity.

4. MATERIAL AND METHODS

4.1 Plant Material, DNA Extraction, Library Preparation and Illumina Sequencing

For both preparation of the DNA libraries and cytogenetic experiments, plants of the diploid species C. acuminatum, C. bryoniifolium, C. ficifolium, C. iljinii, C. pamiricum, C. suecicum and C. vulvaria, which represent the main lineages of the C. album aggregate as described in Mandák et al. [27], were used (Table 1). For our research, we sampled genotypes that, according to our previous data, have average parameters for the lineage [27]. All plants were cultivated at the experimental garden of the Institute of Botany, Czech Academy of Sciences, Pru°honice, Czech Republic (49.9917° N, 14.5667° E, ca. 320 m above sea level). Leaves were collected, and DNA was extracted using the DNeasy Plant Mini Kit (Qiagen, Venlo, The Netherlands) according to the manufacturer's instructions. For in situ hybridization experiments, root tips of young, fine roots were collected and fixed as described in Mandák et al. [26] and stored until use. For all analyzed accessions, the ploidy level was verified by flow cytometry as described in Vít et al. [52].

One individual per species was used for library preparation and NGS. One microgram of extracted DNA was sheared to fragments of approximately 500 to 600 bp using a Bioruptor Pico sonication device (Diagenode, Liège, Belgium). The NEBNext adaptors for Illumina were ligated to the resulting fragments using the NEBNext Ultra DNA Library Prep Kit for Illumina (New England BioLabs, Ipswich, MA, USA), following the manufacturer's instructions. The QIAquick PCR Purification Kit (Qiagen) was used to clean the samples from unbound adapters and to concentrate the samples to a total volume of 30 μL. Afterwards, the samples were loaded onto a 1% agarose gel in low EDTA/TAE buffer. Fragments with sizes ranging from 500 to 750 bp were excised and purified using the Zymoclean Gel DNA Recovery Kit (Zymo Research, Irvine, CA, USA) and eluted into 20 μL of ddH2O Concentration was estimated with a Qubit fluorometer using the Qubit HS Assay kit (Thermo Scientific, Waltham, MA, USA). The individual libraries (corresponding to individual species) were enriched and indexed by unique barcodes using PCR with NEBNext Q5 HotStart HiFi PCR Master Mix and NEBNext Multiplex Oligos for Illumina (New England BioLabs) according to the manufacturer's instructions. The enriched libraries were purified twice using AMPure magnetic beads (Beckman Coulter, Pasadena, CA, USA). The bead:library ratio was 0.7:1 in the first purification and 1:1 in the second purification. The libraries were verified on 1% agarose gels after each purification step. Concentration was measured using the Qubit HS

Assay kit (Thermo Scientific) after the final purification step. Libraries of all seven species were pooled and sequenced on an Illumina MiSeq system at Macrogen Inc., to obtain 2x 300 bp paired-end reads.

4.2 Clustering of Repeatome Elements

To process Illumina NGS data and to compare the repetitive DNA fraction of the studied species, a public web server running RE version 1 (http://www.repeatexplorer.org) (Cˇ eské Budeˇjovice, Czech Republic) was used [53]. The discovery and characterization of repetitive elements in the genome was performed using "clustering" tools. An all-to-all sequence comparison of sequencing reads was performed using the mgblast tool. All hits with similarities above 90% over at least 55% of the sequence length were recorded, thus identifying a set of related DNA fragments. The information on similarity hits was used for construction of a graph in which nodes represent sequence reads and the edges between nodes correspond to similarity hits (Figure 2A). This algorithm was first applied to each species separately and subsequently for the seven species in conjunction for the comparative analysis of repeatome quantitative values. For comparative analysis, sampling was performed proportionally to the genome size of the species (Table 2) [26].

4.3 Satellite DNA Clusters Screening for Tandem Repeats

All genomically abundant clusters containing at least 0.01% of the input reads were examined manually to select those that potentially possess tandemly organized DNA. Primary selection of the clusters was performed based on their form (Figure 2A). Contigs of each selected cluster were analyzed using the following publicly available online tools: (i) the YASS genomic similarity tool, which enables searches of more fuzzy repeats for potential tandem organization (http://bioinfo.lifl.fr/yass/yass. php) [54], with each contig compared against itself and visualized by dot plots (Figure 2B); (ii) BLAST was used to confirm that the cluster belongs to the CficCl-61-40 satDNA family (supplementary data 1); and (iii) primers were designed from the consensus sequence for PCR conformation of the typical tandem array structure (Figure 2C). Each search was performed for each of the analyzed species.

4.4 Sequence Analysis

NGS clusters of C. acuminatum, C. bryoniifolium, C. ficifolium, C. iljinii, C. pamiricum, C. suecicum and C. vulvaria falling into the CficCl-61-40 satDNA family were investigated on the intra-unit (analysis of changes in single monomer) and inter-unit (analysis of changes in array components) levels with TRF software (https://tandem.bu.edu/trf/trf.html). As a result, performance tables with data on monomer sizes, copy numbers, percent matches, percent indels and consensus patterns were obtained (supplementary data 1).

For the reconstruction of phylogenetic relationships among the analyzed monomers k-mer based distance estimation was performed [55]. We have chosen the k-mer value equal to 9, as the most optimal for the analyzed sequences. For calculation of distances method based on fractional common k-mer count was used [56]. The phylogenetic relationships among the sequences are then reconstructed from the pairwise distance matrix [57]. The distance matrix thus obtained can be used to construct a phylogenetic tree using the Minimum Evolution method. The construction of the phylogenetic tree was performed in the MEGA program (Figure 3) [58]. The ancestral monomer (root) was reconstructed as follows: nucleotide-BLAST was used to align contigs of each cluster that, according to BLAST searches, show relatedness between satellite monomers of Chenopodium and Beta species. DNA fragments with 100% similarity were selected and aligned with each other (supplementary data 2). As a result, a fragment of the ancestral monomer was reconstructed.

4.5 Detection Physical Counterparts of Basic Monomer and Proposed HOR Units

RE identifies consensus sequences of the most abundant repetitive elements in the genome. However, these consensuses are only virtual assemblies of short reads originating from many different interspersed loci. To reveal the sequences' physical counterparts and sequence variation within the selected repetitive elements that are proposed to be HOR units, primers were designed based on the consensus sequences (supplementary data 3). PCRs were performed in 25 μL reactions and contained 1× TopBio Plain PP Master Mix (TopBio, Vestec, Czech Republic), each primer at 0.2 mM and 10 to 50 ng of genomic DNA. The cycling conditions were as follows: 4 min at 95 ◦C followed by 35 cycles of 95 ◦C for 30 s, sequence-specific annealing temperature for 30 s and 72 ◦C for 2.5 min, and a final extension at 72 ◦C for 10 min. The PCR results were verified on a 1% agarose gel (Figure 4). The PCR products of clusters were excised from the gels, cloned and sequenced at GATC Biotech (Konstanz, Germany) according to standard protocols.

4.6 FISH Procedure

FISH analysis was performed to further confirm the physical existence of the HOR units in the genome. Root tips were pre-treated in 0,002 M 8-hydroxyquinolin for 3 h in dark and fixed in 3:1 (v/v) 100% ethanol:acetic acid. The fixed root meristems were thoroughly washed in water and enzyme buffer (10 mM citrate buffer at pH 4.6) and partially digested in 0,3% (w/v) cytohelicase, pectolyase and cellulase (Sigma) at 37 ◦C for 3 h followed by washes in water [27]. The material, in a water drop, was carefully transferred onto a grease-free microscope slide and the cells were spread according to the technique of Pijnacker and Ferwerda [59] with modifications as previously described [60].

FISH experiments were performed with clones CficCl-61-40 X-1 and CacuCl-1-117 C-2 as probes labelled with Cy3 (Amersham, Amersham, Buckinghamshire, UK) and biotin (Roche, Basel, Switzerland) according to a standard oligolabeling protocol [61]. For evaluation of probe-specific chromosomal pattern probes were hybridized simultaneously to chromosomes of C. acuminatum, C. bryoniifolium, C. ficifolium, C. iljinii, C. pamiricum, C. suecicum and C. vulvaria (Figure 5, supplementary data 5). FISH was performed on ThermoBrite programmable temperature-controlled slide processing system at 63 ◦C for 3 h. Slides were stained with DAPI and mounted in antifade mountant (Vector Laboratories, Burlingame, CA, USA) and were examined and photographed on Zeiss Axio Imager.Z2 microscope system. Chromosome measurements were obtained by the analysis of metaphase plates using the computer application MicroMeasure version 3.3 [62].

5. CONCISE SUMMARY

Utilization of the RE process for whole genome shotgun analysis Illumina sequencing of genomes from seven diploid plant species across divergent lineages enabled the identification of three distinct evolutionary trajectories of satellite DNA (satDNA) families: (i) concerted evolution characterized by mutation and recombination events (most conserved); (ii) concerted evolution exhibiting a trend towards increased complexity and length of the satellite monomer (HOR formation); and (iii) non-concerted evolution, marked by low levels of homogenization and multidirectional trends.

REFERENCES

1. Bennetzen, J.L. The structure and evolution of angiosperm nuclear genomes. Curr. Opin. Plant Biol. 1998, 1, 103–108. [CrossRef]
2. Maumus, F.; Quesneville, H. Ancestral repeats have shaped epigenome and genome composition for millions of years in Arabidopsis thaliana. Nat. Commun. 2014, 5, 4104. [CrossRef] [PubMed]
3. Elder, J.F.; Turner, B.J. Concerted evolution of repetitive DNA sequences in eukaryotes. Q. Rev. Biol. 1995, 70, 297–320. [CrossRef] [PubMed]
4. Garrido-Ramos, M.A. Satellite DNA: An Evolving Topic. Genes 2017, 8, 230. [CrossRef] [PubMed]
5. Biscotti, M.A.; Olmo, E.; Heslop-Harrison, J.S. Repetitive DNA in eukaryotic genomes. Chromosome Res. 2015, 23, 415–420. [CrossRef] [PubMed]
6. Wei, K.H.-C.; Lower, S.E.; Caldas, I.V.; Sless, T.J.; Barbash, D.A.; Clark, A.G. Variable rates of simple satellite gains across the Drosophila phylogeny. Mol. Biol. Evol. 2018, 35, 925–941. [CrossRef]

7. Satovic´, E.; Vojvoda Zeljko, T.; Luchetti, A.; Mantovani, B.; Plohl, M. Adjacent sequences disclose potential for intra-genomic dispersal of satellite DNA repeats and suggest a complex network with transposable elements. BMC Genom. 2016, 17, 997. [CrossRef]

8. Charlesworth, B.; Sniegowski, P.; Stephan, W. The evolutionary dynamics of repetitive DNA in eukaryotes. Nature 1994, 371, 215–220. [CrossRef]

9. Raskina, O.; Barber, J.C.; Nevo, E.; Belyayev, A. Repetitive DNA and chromosomal rearrangements: Speciation-related events in plant genomes. Cytogenet. Gen. Res. 2008, 120, 351–357. [CrossRef]

10. Emadzade, K.; Jang, T.S.; Macas, J.; Kovarˇík, A.; Novák, P.; Parker, J.; Weiss-Schneeweiss, H. Differential amplification of satellite PaB6 in chromosomally hypervariable Prospero autumnale complex (Hyacinthaceae). Ann. Bot. 2014, 114, 1597–1608. [CrossRef]

11. Dodsworth, S.; Chase, M.W.; Kelly, L.J.; Leitch, I.J.; Macas, J.; Novák, P.; Piednoël, M.; Weiss-Schneeweiss, H.; Leitch, A.R. Genomic repeat abundances contain phylogenetic signal. Syst. Biol. 2015, 64, 112–126. [CrossRef]

12. Martienssen, R.A. Maintenance of heterochromatin by RNA interference of tandem repeats. Nat. Genet. 2003, 35, 213–214. [CrossRef] [PubMed]

13. Kloc, A.; Martienssen, R. RNAi, heterochromatin and the cell cycle. Trends Genet. 2008, 24, 511–517. [CrossRef] [PubMed]

14. Mehrotra, S.; Goyal, V. Repetitive sequences in plant nuclear DNA: Types, distribution, evolution and function. Genom. Proteom. Bioinform. 2014, 12, 164–171. [CrossRef] [PubMed]

15. Garrido-Ramos, M.A. SatDNA in plants: More than just rubbish. Cytogenet. Genome Res. 2015, 146, 153–170. [CrossRef] [PubMed]

16. Meštrovic´, N.; Mravinac, B.; Pavlek, M.; Vojvoda-Zeljko, T.; Šatovic´, E.; Plohl, M. Structural and functional liaisons between transposable elements and satellite DNAs. Chromosome Res. 2015, 23, 583–596. [CrossRef] [PubMed]

17. Plohl, M.; Meštrovic´, N.; Mravinac, B. Satellite DNA evolution. Genome Dyn. 2012, 7, 126–152.

18. Salser, W.; Bowen, S.; Browne, D.; el-Adli, F.; Fedoroff, N.; Fry, K.; Heindell, H.; Paddock, G.; Poon, R.; Wallace, B.; et al. Investigation of the organization of mammalian chromosomes at the DNA sequence level. Fed. Proc. 1976, 35, 23–35.

19. Dover, G. Molecular drive. Trends Genet. 2002, 18, 587–589. [CrossRef]

20. Plohl, M.; Luchetti, A.; Mestrovic, N.; Mantovani, B. Satellite DNAs between selfishness and functionality: Structure, genomics and evolution of tandem repeats in centromeric (hetero) chromatin. Gene 2008, 409, 72–82. [CrossRef]
21. Samoluk, S.S.; Robledo, G.; Bertioli, D.; Seijo, J.G. Evolutionary dynamics of an at-rich satellite DNA and its contribution to karyotype differentiation in wild diploid Arachis species. Mol. Genet. Genom. 2017, 292, 283–296. [CrossRef] [PubMed]
22. Ugarkovic, D.; Plohl, M. Variation in satellite DNA profiles-causes and effects. EMBO J. 2002, 2, 5955–5959. [CrossRef]
23. Novák, P.; Neumann, P.; Macas, J. Graph-based clustering and characterization of repetitive sequences in next-generation sequencing data. BMC Bioinform. 2010, 11, 378. [CrossRef] [PubMed]
24. Chu, G.-L.; Mosyakin, S.L.; Clemants, S.E. Chenopodiaceae. In Flora of China. Volume 5: Ulmaceae through Basellaceae; Wu, Z., Raven, P.H., Hong, D., Eds.; Missouri Botanical Garden Press: St. Louis, MI, USA, 2003; pp. 351–414.
25. Habibi, F.; Vít, P.; Rahiminejad, M.; Mandák, B. Towards a better understanding of the C. album aggregate in the Middle East: A karyological, cytometric and morphometric investigation. J. Syst. Evol. 2018, 56, 231–242. [CrossRef]
26. Mandák, B.; Krak, K.; Vít, P.; Pavlíková, Z.; Lomonosova, M.N.; Habibi, F.; Lei, W.; Jellen, E.N.; Douda, J. How genome size variation is linked with evolution within Chenopodium sensu lato. Perspect. Plant Ecol. Evol. System. 2016, 23, 18–32. [CrossRef]
27. Mandák, B.; Krak, K.; Vít, P.; Lomonosova, M.N.; Belyayev, A.; Habibi, F.; Wang, L.; Douda, J.; Storchova, H. Hybridization and polyploidization within the Chenopodium album aggregate analyzed by means of cytological and molecular markers. Mol. Phylogenet. Evol. 2018, 129, 189–201. [CrossRef]
28. Gao, D.; Schmidt, T.; Jung, C. Molecular characterization and chromosomal distribution of species-specific repetitive DNA sequences from Beta corolliflora, a wild relative of sugar beet. Genome 2000, 43, 1073–1080. [CrossRef]
29. Kolano, B.; Gardunia, B.W.; Michalska, M.; Bonifacio, A.; Fairbanks, D.; Maughan, P.J.; Coleman, C.E.; Stevens, M.R.; Jellen, E.N.; Maluszynska, J. Chromosomal localization of two novel repetitive sequences isolated from the Chenopodium quinoa Willd. Genome 2011, 54, 710–717. [CrossRef]
30. Benson, G. Tandem repeats finder: A program to analyze DNA sequences. Nucleic Acids Res. 1999, 27, 573–580. [CrossRef]
31. Gogarten, J.P.; Kibak, H.; Dittrich, P.; Taiz, L.; Bowman, E.J.; Bowman, B.J.; Manolson, M.F.; Poole, R.J.; Date, T.; Oshima, T.; et al. Evolution of the Vacuolar

H+-Atpase: Implications for the Origin of Eukaryotes. Proc. Natl. Acad. Sci. USA 1989, 86, 6661–6665. [CrossRef]

32. Iwabe, N.; Kuma, K.; Hasegawa, M.; Osawa, S.; Miyata, T. Evolutionary Relationship of Archaebacteria, Eubacteria, and Eukaryotes Inferred from Phylogenetic Trees of Duplicated Genes. Proc. Natl. Acad. Sci. USA 1989, 86, 9355–9359. [CrossRef] [PubMed]
33. Kadereit, G.; Hohmann, S.; Kadereit, J.W. A synopsis of Chenopodiaceae subfam. Betoideae and notes on the taxonomy of Beta. Willdenowia 2006, 36, 9–19. [CrossRef]
34. Koukalova, B.; Moraes, A.P.; Renny-Byfield, S.; Matyasek, R.; Leitch, A.; Kovarik, A. Fall and rise of satellite repeats in allopolyploids of Nicotiana over c. 5 million years. New Phytol. 2009, 186, 148–160. [CrossRef] [PubMed]
35. Plohl, M.; Petrovic′, V.; Luchetti, A.; Ricci, A.; Satovic′, E.; Passamonti, M.; Mantovani, B. Long-term conservation vs. high sequence divergence: The case of an extraordinarily old satellite DNA in bivalve mollusks. Heredity 2009, 104, 543–551. [CrossRef] [PubMed]
36. Willard, H.F.; Waye, J.S. Hierarchical order in chromosome-specific human alpha satellite DNA. Trends Genet. 1987, 3, 192–198. [CrossRef]
37. Gallagher, D.S.; Modi, W.S.; Ivanov, S. Concerted Evolution and Higher-Order Repeat Structure of the 1.709 (Satellite IV) Family in Bovids. J. Mol. Evol. 2004, 58, 460–465. [CrossRef] [PubMed]
38. Adega, F.; Chaves, R.; Guedes-Pinto, H.; Heslop-Harrison, J.S. Physical organization of the 1.709 satellite IV DNA family in Bovini and Tragelaphini tribes of the Bovidae: Sequence and chromosomal evolution. Cytogenet. Genome Res. 2006, 114, 140–146. [CrossRef]
39. Macas, J.; Navrátilová, A.; Koblížková, A. Sequence homogenization and chromosomal localization of VicTR-B satellites differ between closely related Vicia species. Chromosoma 2006, 115, 437–447. [CrossRef]
40. Jarvis, D.E.; Ho, Y.S.; Lightfoot, D.J.; Schmöckel, S.M.; Li, B.; Borm, T.J.; Ohyanagi, H.; Mineta, K.; Michell, C.T.; Saber, N.; et al. The genome of Chenopodium quinoa. Nature 2017, 542, 307–312. [CrossRef]
41. Belyayev, A.; Paštová, L.; Fehrer, J.; Josefiová, J.; Chrtek, J.; Mráz, P. Mapping of Hieracium (Asteraceae) chromosomes with genus-specific satDNA elements derived from next-generation sequencing data. Plant Syst. Evol. 2018, 304, 387–396. [CrossRef]
42. Henikoff, S.; Ahmad, K.; Malik, H.S. The centromere paradox: Stable inheritance with rapidly evolving DNA. Science 2001, 293, 1098–1102. [CrossRef] [PubMed]

43. Belyayev, A.; Josefiová, J.; Jandová, M.; Krak, K.; Mandák, B. Transposable elements dynamics in the evolution of Chenopodium album aggregate. in preparation.

44. Kejnovský, E.; Michalovova, M.; Steflova, P.; Kejnovska, I.; Manzano, S.; Hobza, R.; Kubat, Z.; Kovarik, J.; Jamilena, M.; Vyskot, B. Expansion of microsatellites on evolutionary young Y chromosome. PLoS ONE 2013, 8, e45519. [CrossRef] [PubMed]

45. Li, X.-M.; Lee, B.S.; Mammadov, A.C.; Koo, B.C.; Mott, I.W.; Wang, R.R.-C. CAPS markers specific to Eb, Ee, and R genomes in the tribe Triticeae. Genome 2007, 50, 400–411. [CrossRef] [PubMed]

46. Luchetti, A.; Marini, M.; Mantovani, B. Non-concerted evolution of the RET76 satellite DNA family in Reticulitermes taxa (Insecta, Isoptera). Genetica 2006, 128, 123–132.

47. Groom, Q.J. Piecing together the biogeographic history of Chenopodium vulvaria L. using botanical literature and collections. Peer J. 2015, 3, e723. [CrossRef] [PubMed]

48. Mayr, E. Populations Species and Evolution: An Abridgment of Animal Species and Evolution; Belknap Press: Cambridge, UK, 1970.

49. Grant, V. Plant Speciation, 2nd ed.; Columbia University Press: New York, NY, USA, 1981.

50. Husband, B.C. Chromosomal variation in plant evolution. Am. J. Bot. 2004, 91, 621–625. [CrossRef]

51. Belyayev, A. Bursts of transposable elements as an evolutionary driving force. J. Evol. Biol. 2014, 27, 2573–2584. [CrossRef]

52. Vít, P.; Krak, K.; Trávníc˘ek, P.; Douda, J.; Lomonosova, M.N.; Mandák, B. Genome size stability across Eurasian Chenopodium species (Amaranthaceae). Bot. J. Linn. Soc. 2016, 182, 637–649. [CrossRef]

53. Novák, P.; Neumann, P.; Pech, J.; Steinhaisl, J.; Macas, J. RepeatExplorer: A Galaxy-based web server for genome-wide characterization of eukaryotic repetitive elements from next-generation sequence reads. Bioinformatics 2013, 29, 792–793. [CrossRef]

54. Noe, L.; Kucherov, G. YASS: Enhancing the sensitivity of DNA similarity search. Nucleic Acids Res. 2005, 33, W540–W543. [CrossRef]

55. Vinga, S.; Almeida, J. Alignment-free sequence comparison-a review. Bioinformatics 2003, 19, 513–523. [CrossRef] [PubMed]

56. Edgar, R.C. MUSCLE: A multiple sequence alignment method with reduced time and space complexity. BMC Bioinform. 2003, 5, 113.

57. Kalendar, R.; Tselykh, T.; Khassenov, B.; Ramanculov, E.M. Introduction on using the FastPCR software and the related Java web tools for PCR, in silico PCR, and oligonucleotide assembly and analysis. Met. Mol. Biol. 2017, 1620, 33–64. [CrossRef]

58. Kumar, S.; Stecher, G.; Li, M.; Knyaz, C.; Tamura, K. MEGA X: Molecular Evolutionary Genetics Analysis across Computing Platforms. Mol. Biol. Evol. 2018, 35, 1547–1549. [CrossRef] [PubMed]

59. Pijnacker, L.P.; Ferwerda, M.A. Giemsa C-banding of potato chromosomes. Can. J. Genet. Cytol. 1984, 26, 415–419. [CrossRef]

60. Belyayev, A.; Raskina, O.; Nevo, E. Chromosomal distribution of reverse transcriptase containing retroelements in two Triticeae species. Chromosome Res. 2001, 9, 129–136. [CrossRef] [PubMed]

61. Feinberg, A.P.; Vogelstein, B. A technique for radiolabeling DNA restriction endonuclease fragments to high specific activity. Anal. Biochem. 1983, 132, 6–13. [CrossRef]

62. Reeves, A. MicroMeasure: A new computer program for the collection and analysis of cytogenetic data. Genome 2001, 44, 439–443. [CrossRef]

8

Chapter

THE COMPLEX FUNCTION OF PHEDOF12-1 IN MODULATING FLOWERING TIME AND RESPONSES TO ABIOTIC STRESS IN MOSO BAMBOO (PHYLLOSTACHYS EDULIS)

1. INTRODUCTION: AN OVERVIEW

DNA binding with one finger (Dof) transcription factors (TFs) constitute a family of transcription factors particular to plants. The proteins typically comprise 50–52 highly conserved amino acids, featuring a C2C2-type zinc-finger motif at the N-terminal region [1]. Dof transcription factors are extensively dispersed throughout the plant world. The cDNA sequence of Dof was initially derived from Zea mays [2]. Since that time, several Dofs have been cloned from diverse plant species [3–5]. Previous research suggest that Dof proteins regulate various biological processes, including seed germination, floral organ abscission, hormone signaling, and cell cycles. In Arabidopsis, DAG1 and DAG2 facilitate seed germination [6,7], DOF6 functions as a negative regulator of seed germination and interacts with TCP14 [8], while AtDOF4.7 is involved in the transcriptional regulation of floral organ abscission by influencing cell wall hydrolase gene expression [9]. Furthermore, certain Dof genes (AtDof2.4, AtDof5.8, and AtDof5.6/HCA2) are expressed during the initial stages of vascular cell formation [10]. OsDof3 in rice is implicated in the expression regulated by gibberellin [11]. Additionally, Dof transcription factors, including maize Dof1 and Dof2, play a role in regulating carbon and nitrogen metabolism by modulating phosphoenolpyruvate carboxykinase (PEPCK), glutamine synthase (GS), and glutamate synthase (GLU) [7,12–16].Genetic and molecular studies have suggested that Dof transcription factors participate in different stresses, light responsiveness, and flowering regulation. In Brachypodium distachyon, BdCBF1, BdCBF2, and BdCBF3 contribute to cold, drought, and salt stresses by regulating downstream targets such as DEHYDRIN5.1 (Dhn5.1) and

COR genes [17]. Overexpressing SlCDF3 shows increased transgenic Arabidopsis drought and salt tolerance [18]. In Chinese cabbage, most BraDof genes are induced by cold, heat, high salinity, and drought stresses [19]. Moreover, Dof proteins are involved in photoperiod flowering. In Arabidopsis, cycling Dof factor-1 (CDF1) binds to the COSTANS (CO) and FLOWERING LOCUS T (FT) promoter regions to block transactivation of these two flowering genes, whereas this inhibition could be released based on the GIGANTEA-FLAVIN-BINDING, KELCH REPEAT, F-BOX1(GI-FKF1) complex-mediated degradation of CDF1 under long-day (LD) conditions [20]. In addition, CDF2, CDF3, and CDF5 repress flowering of Arabidopsis by decreasing the mRNA level of CO [21]. In rice, overexpressing OsDof12 promotes early flowering under LD conditions by upregulating the expression of Hd3a and OsMADS14 [22]. Although a large number of Dofs have been extensively studied in annual plants [23,24], the knowledge of Dofs in moso bamboo is limited.

Moso bamboo (Phyllostachys edulis) is a perennial species distinguished by an extended vegetative phase that culminates in synchronized flowering, subsequently resulting in extensive mortality. Investigating the mechanism of moso bamboo flowering timing is crucial and complex, making it challenging to identify the essential regulating gene. The proliferation of moso bamboo in its natural habitat is significantly jeopardized by diverse environmental factors, including drought, salinity, and cold temperatures, which critically restrict its growth and distribution, thereby impacting the yield and quality of winter shoots, as well as the subsequent year's bamboo yield and wood harvests in ensuing years [26–28]. Furthermore, recent investigations on Dofs predominantly focus on annual plants and are restricted in perennials. Consequently, investigating the function of Dofs in moso bamboo is essential, particularly regarding abiotic stress and flowering time. This study isolates a Dof gene (PheDof12-1) from moso bamboo, which is activated by cold, drought, salt, and gibberellin (GA3) stressors. This study investigates the functional characteristics of PheDof12-1 through ectopic expression in Arabidopsis for the first time. Transgenic Arabidopsis lines overexpressing homozygous PheDof12-1 exhibit early flowering under long-day (LD) conditions and demonstrate a pronounced diurnal binding pattern to the promoter sequence of PheCOL4. The findings offer novel insights into the role of the Dof transcription factor in regulating photoperiodic flowering timing and abiotic stress in moso bamboo.

2. RESULTS

2.1 Isolation and Analysis of PheDof12-1

Based on the moso bamboo genome database, PheDof12-1 was isolated from moso bamboo. The full-length CDS of PheDof12-1 is 1299 bp, encoding 432-amino acids, with predicted molecular weight (MW) and isoelectric point (pI) of 46.37 kDa and 8.32, respectively. Structure analysis showed that PheDof12-1 contains one intron and two exons (Figure 1A). The deduced proteins contain the conserved zf-Dof domain. Furthermore, phylogenetic analysis of PheDof12-1 and homologous proteins from other plants shows that PheDof12-1 and other Dofs from monocotyledons belong to the same clade (Figure 1B). The amino acid sequence of PheDof12-1 shows 83% and 84% identity with Brachypodium (XP_003558722) and rice (XP_015690912), respectively. This result was consistent with the findings in the stated phylogeny and classification of plants. All these proteins contain the conserved zf-Dof domain.

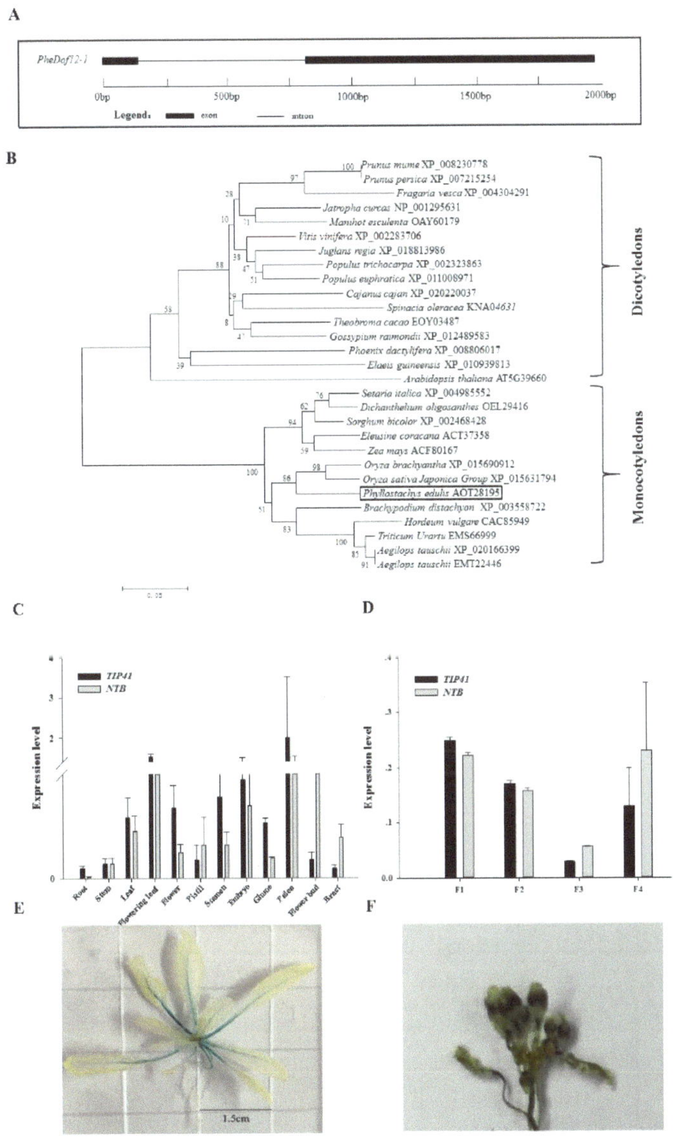

Figure 1. Characterization and preliminary expression analysis of PheDof12-1. (A) Gene structure of PheDof12-1. (B) Phylogenetic analysis of PheDof12-1 with other DNA binding with one finger (Dof) proteins. (C) qRT-PCR analysis of PheDof12-1 in

different tissues of moso bamboo. (D) Expression profile of PheDof12-1 in different flower developmental stages: F1: floral bud formation stage; F2: inflorescence growing stage; F3: blooming stage; F4: flowers are withered. (E) Glucuronidase (GUS) staining of ProPheDof12-1-GUS in transgenic Arabidopsis seedling. (F) GUS staining of ProPheDof12-1-GUS plants showing PheDof12-1 localization in flower and pollen.

2.2 Tissue-Specific Gene Expression

In order to analyze the expression of PheDof12-1 in different tissues (root, stem, leaf, flowering leaf, flower) and floral organs (pistil, stamen, embryo, glume, palea, flower bud, bract), RNA was isolated to perform qRT-PCR. The results show that the transcription level of PheDof12-1 in flowering leaf is significantly higher than in other tissues. In different flower organs, the expression of PheDof12-1 was highest in palea, and lowest in bract (Figure 1C). In developing flowers, PheDof12-1 had higher transcript accumulation at the floral bud formation stage (F1) (Figure 1D), and decreased gradually at flower development, which was consistent with the previously reported detection of PheDof1 at early stages of flower formation and development [29]. We further generated ProPheDof12-1-GUS transgenic lines, and glucuronidase (GUS) staining was detected in the vasculature of cotyledons and hypocotyls, true leaves, roots, flower, and pollen (Figure 1E,F). The results demonstrate that PheDof12-1 is expressed in different tissues and at different flower development stages, suggesting that it is dynamic during plant development and may play an important role in moso bamboo growth and development.

2.3 Expression Patterns of PheDof12-1 under Stress Treatments

Previous reports have shown that Dof TFs are involved in abiotic stress [30]. To determine the expression pattern of PheDof12-1 in moso bamboo under different stresses, we performed detailed qRT-PCR with TIP41 and NTB as internal reference genes. The results show that PheDof12-1 was responded to cold, drought, and salt stresses. In drought stress, PheDof12-1 was induced and upregulated at each time point, and levels of transcripts in leaves and stems were slightly elevated, but a sharp increase occurred after 1 h in roots, peaking at 70.9-fold. This implies that PheDof12-1 is induced and has a positive function in response to drought stress (Figure 2A–C). In cold treatment using NTB as a reference gene, the expression of PheDof12-1 rapidly increased in leaves, reaching 86.1-fold at 24 h (Figure 2I). Regarding salt treatment, the maximum increase was observed at 12 h, reaching 12.5-fold in leaves when TIP41 was used as the reference gene (Figure 2F), but the expression level was first induced and then decreased in roots. To further investigate the functions of PheDof12-1, we initially analyzed the effects of gibberellin A3 (GA3) and abscisic acid (ABA) on its expression (Figure 2J,K). In GA3 stress, the transcription level of PheDof12-1 was induced and upregulated at almost every time point, peaking at 15.0-fold at 24 h. Under ABA treatment, the translation level of PheDof12-1 initially decreased and then increased, was lowest at 6 h, dropping to undetectable levels, and reached a peak at 48 h. All of these data indicate that PheDof12-1 takes part in the hormones and different abiotic stresses of moso bamboo.

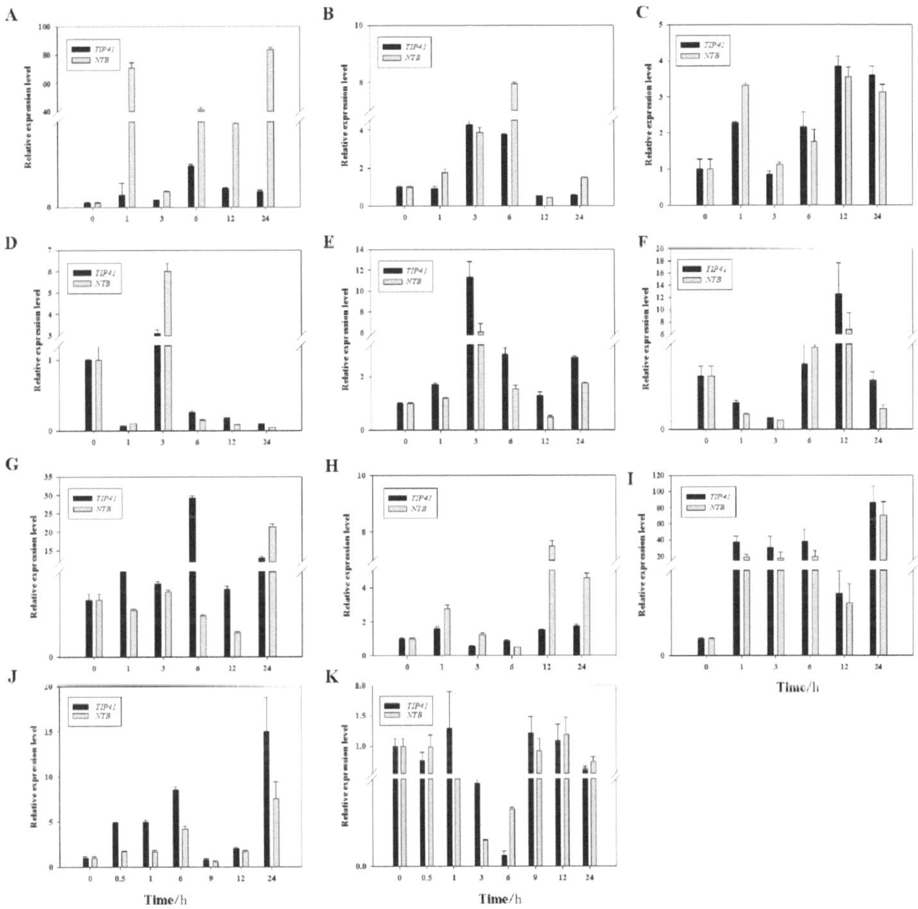

Figure 2. Relative expression of PheDof12-1 in different tissues of moso bamboo under drought: (A) root, (B) stem, (C) leaf; salt: (D) root, (E) stem, (F) leaf; under cold: (G) root, (H) stem, (I) leaf; and under (J) gibberellin A3 (GA3) and (K) abscisic acid (ABA) treatments.

2.4 Overexpression of PheDof12-1 Promotes Early Flowering in Arabidopsis

In order to verify the subcellular localization of PheDof12-1, we further amplified its coding region and fused it to the N-terminal of the eGFP vector. The subcellular localization assay indicated that PheDof12-1 was localized in the nucleus, in accordance with its function as a transcription factor (Figure 3B). To study the genetic functions of PheDof12-1, we transformed it in Arabidopsis. The overexpressed plants showed an early flowering phenotype under LD conditions (Figure 3A), whereas PheDof12-1 overexpression had no effect on flowering time under SD conditions (not shown). The flowering time was about 10 days earlier than wild-type, and the number of rosette leaves of overexpressed lines was smaller than that of wild Arabidopsis (Figure 3C). We further investigated the transcription levels of FT, SUPPRESSOR OF OVEREXPRESSION OF CONSTANS1 (SOC1), AGAMOUS-LIKE 24 (AGL24), FLOWERING LOCUS C (FLC), and SHORT VEGETATIVE PHASE (SVP) in the T3 generation to ascertain the downstream effects of this construct. FT, SOC1, and AGL24 were upregulated, while FLC and SVP expression were rather low compared with wild-type (Figure 3D). These data suggest that PheDof12-1 might regulate flowering by controlling the expression of FT, SOC1, AGL24, FLC, and SVP.

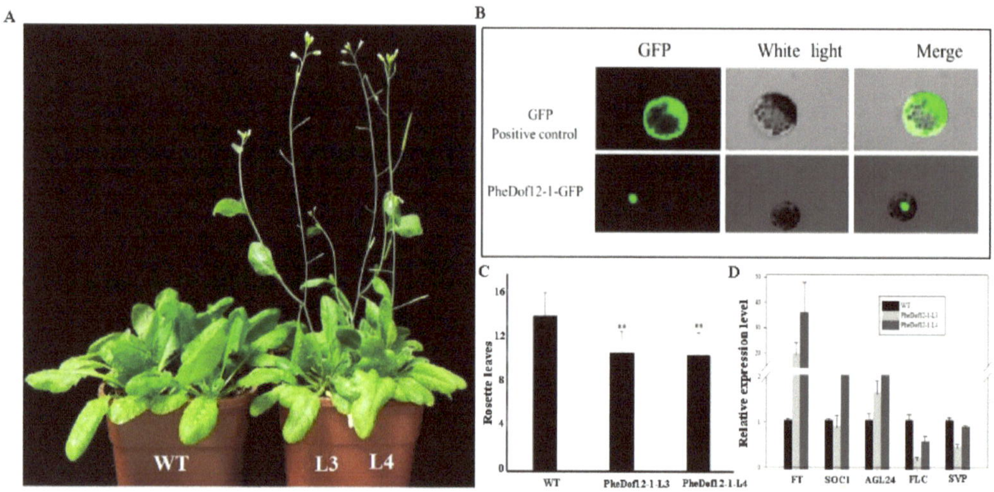

Figure 3. Analysis of an early flowering phenotype by overexpression of PheDof12-1 in Arabidopsis.Phenotypes of overexpressing PheDof12-1 transgenic lines (L3, L4) and wild-type (WT) plants as control under long-day (LD) conditions. (B) Subcellular localization of PheDof12-1. (C) Flowering time scored as number of rosette leaves at flowering of wild-type and transgenic plants under LD conditions. (D) Transcription levels of FT, SOC1, AGL24, FLC, and SVP in wild-type and transgenic plants. Arabidopsis Actin was used as the internal reference gene. Error bars indicate standard deviations. Asterisks indicate statistically significant difference between wild-type and transgenic plants ($p < 0.01$ by Student's t-test).

2.5 PheDof12-1 Interacts with Photoperiod-Related Regulators

In Arabidopsis, CDFs are transcriptional repressors that bind to CO and FT promoters to repress their transcription [20]. To explore whether PheDof12-1 can form heterodimers with other proteins, an interaction prediction was performed using STRING (https://stringdb.org/) based on the interaction network of rice orthologous genes. As shown in Figure 4E, PheDof12-1 interacted with 10 identified proteins. Among them, the B-box protein (PH01004196G0130), Dof transcription factor (PH01001184G0160), grain size gene (PheSLR1) [31], photoperiodic flowering response gene (PH01002431G0090) [32], and drought-induced protein (PH01000199G0750) were identified, suggesting that PheDof12-1 may be involved in growth and development, photoperiodic response, and abiotic stress.

CDF1, CDF2, CDF3, and CDF5 had high mRNA levels at the beginning of the light period in Arabidopsis [21], and CDFs displayed a similar expression pattern in Populus [33]. So, we detected the expression patterns under photoperiod treatments. The results show that PheDof12-1 was similarly expressed under both LD and SD conditions. The transcription level of PheDof12-1 decreased with the increased light time, reaching the minimum value before dark (Figure 4A,B), with high mRNA levels at the beginning of the light period, which was consistent with the expression patterns of CDFs in Arabidopsis and Populus. The highly similar expression pattern of CDFs in Populus, Arabidopsis, and moso bamboo suggests a functional conservation.

CO and CO-like (COL) proteins are members of the B-box family, playing a central role in the photoperiod response pathway by mediating between the circadian clock and the floral integrators [34]. CDFs are transcriptional repressors that bind to the CO promoter to repress its transcription [20]. PheDof12-1 interacted with B-box proteins by interaction prediction; moreover, PheDof12-1 and PheCOL4 had similar expression patterns under photoperiod treatments (Figure 4C,D), suggesting that PheDof12-1 may interact with PheCOL4 in moso bamboo. To examine whether the PheDof12-1 protein regulated PheCOL4 expression by directly binding to the promoter region, the PheCOL4 promoter sequence was investigated. We performed a targeted yeast one hybrid (Y1H) assay using PheDof12-1, and PheCOL4 was inserted upstream of the reporter plasmid pHIS2 and cotransfected into the yeast cells with the AD-PheCOL12-1 effector plasmid. The binding of PheCOL12-1 and the promoter of PheCOL4 was indicated by the growth of transfected yeast cells on a nutrient-deficient medium (synthetic dextrose (SD)/-Trp-Leu-His) plus 3-amino-1, 2, 4-triazole (3-AT) and 5-bromo-4-chloro-3-indoxyl-α-D-galactopyranoside (X-α-Gal). The results show that all transformants tested were found to grow well on the SD/-Leu/-Trp medium when transferred onto SD/-Trp/-Leu/-His/3-AT/X-α-Gal plates for 3 days; only the yeast cells of AD-PheDof12-1 + pHIS2-PheCOL4 vectors and the positive control grew strong and turned blue (Figure 4F). This result suggests that PheDof12-1 could bind to the promoter of PheCOL4 and regulate PheCOL4 expression in moso bamboo.

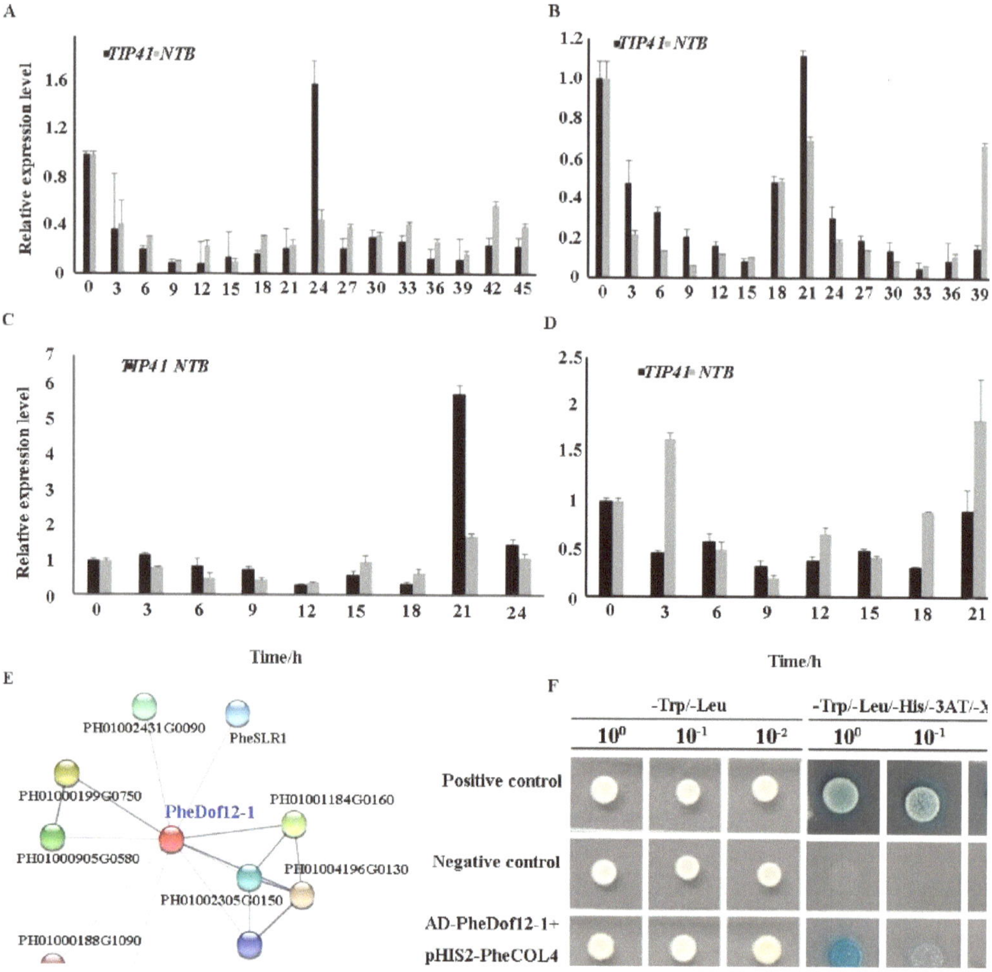

Figure 4. PheDof12-1 protein binds to the promoter region of PheCOL4. Relative expression of PheDof12-1 under (A) LD and (B) SD conditions. Transcription level of PheCOL4 under (C) LD and(D) SD conditions. (E) Interaction network of PheDof12-1 in moso bamboo. Colored balls (protein nodes) in the network were used as a visual aid to indicate different input proteins and predicted interactions. Enlarged protein nodes indicate the availability of 3D protein structure information. Gray lines connect proteins that are associated by recurring text mining evidence. (F) Yeast one-hybrid (Y1H) assay for AD-PheDof12-1 and pHIS2-PheCOL4. The reporter pHIS2 vector carrying the corresponding fragment and the effector AD-PheDof12-1 vector were cotransfected into yeast Y187 cells. Growth of the transfected yeast cells on a 3-AT and X-α-Gal medium indicates that the PheDof12-1 protein can bind to the PheCOL4 promoter.

3. DISCUSSION

Moso bamboo is a perennial plant characterized by rapid growth and a long vegetative stage that lasts for decades or even longer before flowering [25]. Dof proteins are a group of plant-specific TFs that are involved in diverse plant-specific biological processes [16]. In addition, recent research on Dofs is mainly in annual plants, and is limited in perennials. Therefore, researching the roles of Dofs in moso bamboo is necessary. In this study, a Dof gene, PheDof12-1, is identified from moso bamboo as a nucleus-localized transcription factor that contains typical zf-dof domains.

In recent decades, reports have indicated that Dof transcription factors are involved in stress response. In Arabidopsis, the expression level of AtCDF3 is upregulated by cold, drought, high salinity, and ABA treatment [30], and overexpression of 35S::SlCDF1 and 35S::SlCDF3 increases Arabidopsis's tolerance to salt and drought stresses [18]. In wheat, TaDof14 and TaDof15 are significantly induced under drought treatment [35]. Previous research has suggested that drought or other environmental stresses are functional in the flowering stage of bamboo, and the transcription levels of Dof genes are upregulated in drought stress [36]. In addition, studying the tolerance of PheDof12-1 will help to characterize moso bamboo cultivars such as salt, cold, and drought tolerance. In this study, PheDof12-1 exhibited differential expression patterns under the conditions of drought, cold, salt, and ABA and GA3 treatments. Through the drought, cold, salt, and GA3 stresses, the expression pattern of PheDof12-1 is basically upregulated in roots, stems, and leaves, indicating that it might participate in abiotic stress and hormone pathways, which is consistent with previous reports [36,37]. The results provide a better understanding of the stress tolerance of PheDof12-1 in moso bamboo.

Hd1/CO and Hd3a/FT are conserved genetic pathways that regulate photoperiodic flowering between rice and Arabidopsis by their genomic comparison [38]. In Arabidopsis, CDF1–CDF3 are suggested to participate in photoperiodic flowering [39]. JcDof3 is a circadian clock regulated gene involved in the regulation of flowering time in Jatropha curcas [40]. In rice, OsDof12 and CDF1 belong to the same group [41], and overexpression of OsDof12 resulted in early flowering by increasing the expression of Hd3a and OsMADS14 under LD conditions [22]. PheDof12-1 is the homologous gene of OsDof12, and Dof-Hd3a-MADS-flowering may play an important role in moso bamboo flowering [36]. Therefore, we researched the function of PheDof12-1 in flowering time by ectopic expression in Arabidopsis for the first time, and the transgenic lines overexpressing PheDof12-1 show earlier flowering than the wild-type plants under LD conditions. In addition, FT, SOC1, and AGL24 are upregulated and FLC and SVP are downregulated in the transgenic lines. FT promotes flowering [42], which is activated by CO in the phloem [43]. SOC1 is a core regulator of flowering in Arabidopsis, which can interact with SVP and AGL24 proteins, but SVP and AGL24 have opposite effects on flowering time, acting as floral repressor and inducer, respectively [44]. FLC encodes a MADS domain-containing transcription factor that acts as an inhibitor of flowering [45]. This leads us to suspect that PheDof12-1 promotes flowering time by regulating FT, SOC1, AGL24, FLC, and SVP directly or indirectly, suggesting that it might retain some function in the control of flowering time through similar molecular mechanisms to those observed when expressed in Arabidopsis.

Diurnal oscillation of the transcription levels of CDFs has been reported in Arabidopsis and other species [21,23]. In Arabidopsis, CDF1–CDF3 and CDF5 show maximum expression at the beginning of the light period, decreasing to a minimum between 16 and 20 h, then rising again during dawn [21]. In tomato, SlCDF1 and SlCDF3 exhibit maximum expression at the beginning of the day, while SlCDF2, SlCDF4, and SlCDF5 exhibit maximum levels during the night [18]. In rice, OsDof12 is strongly inhibited by dark treatment [22]. In the study, PheDof12-1 exhibited significantly diurnal expression patterns with high mRNA levels at the beginning of the light period under LD and SD conditions, supporting the assumption that it is a true homologue of the Arabidopsis CDFs. In Arabidopsis, CDFs can bind to the CO promoter to repress its transcription [20], and PttCDF3 can bind directly to the PttCO2 promoter in Populus [33]. In moso bamboo, the diurnal expression pattern of PheCOL4 is consistent with PheDof12-1, and Y1H analysis shows that PheDof12-1 binds directly to the promoter of PheCOL4. These results support the hypothesis that flowering regulator CO, a target of CDFs, is controlled precisely [21], which is similar to the situation in Arabidopsis and Populus.

4. MATERIALS AND METHODS

4.1 Plant Materials and Treatments

Moso bamboo seeds were harvested from Guilin in the Guangxi Zhuang Autonomous Region, China. Seedlings were grown in an illumination incubator under long-day conditions (16 h light/8 h dark) at day/night temperatures of 25/18 ◦C, and watered with Hoagland nutrient solution. For drought and salt stress, the seedlings were watered with 50% Hoagland's solution with 20% polyethylene glycol 6000 (PEG 6000) and 250 mM NaCl. For low temperature treatment, the plants were transferred to a growth chamber at 4 ◦C, and plant leaf, stem, and root tissues were collected [46]. For abscisic acid (ABA) and gibberellin A3 (GA3) treatments, the seedlings were watered with 200 μM ABA [47] and 200 μM GA3 solution [48]. To detect the transcriptional level of PheDof12-1 in photoperiod treatments, leaves were collected for analysis from plants exposed to LD (16 h light/8 h dark) and SD (16 h light/8 h dark) treatments [21]. All samples were immediately frozen in liquid nitrogen and stored at -80 ◦C until further analysis.

4.2 Bioinformatic Analysis

The sequences were downloaded from BambooGDB (http://forestry.fafu.edu.cn/db/ PhePacBio/) [49]. Molecular weight (MW) and isoelectric point (pI) were analyzed using ProtParam (http://web.expasy.org/protparam/) [50]. The structure was shown using Gene Structure Display Server software (http://gsds1.cbi.pku.edu.cn/index.php) [51]. To search the database, the Basic Local Alignment Search Tool (BLAST) network service from the National Center for Biotechnology Information (NCBI) web server was applied. Homologue alignment was obtained using Clustal 1.83, and a phylogenetic tree was constructed by MEGA6.0 [52] using the following parameters: NJ method, complete deletion, and bootstrap with 1000 replicates.

4.3 Vector Construction and Plant Transformation

The subcellular localization was performed by transfecting GFP-tagged PheDof12-1 into Arabidopsis sheath protoplasts [53] (Supplementary Table S1). The full-length cDNA of PheDof12-1 was fused in frame with the GFP cDNA and ligated between the CaMV 35 S promoter and the nopaline synthase terminator. The fluorescence signals were examined using a confocal laser scanning microscope (Leica Microsystems, Wiesler, Germany).

The full-length coding sequence of PheDof12-1 was cloned into the pCAMBIA 2300 vector under the control of the modified CaMV 35S promoter (Supplementary Table S1). The pCAMBIA 2300-PheDof12-1 vector was introduced into Agrobacterium umefaciens strain GV3101 for Arabidopsis transformation in the Col-0 background by the floral dipping method [54]. Putative transgenic plants were screened on 50% Murashige and Skoog (MS) solid medium supplemented with 50 mg/L kanamycin, and homozygous T3 or T4 seeds were used.

In order to analyze the spatial expression patterns of PheDof12-1, a 2 kb region upstream of the PheDof12-1 transcription start site was cloned and fused to the pCAMBIA2391Z vector to generate the ProPheDof12-1-GUS reporter, which was transformed into wild-type (WT) plants (Supplementary Table S1). For GUS staining, ProPheDof12-1-GUS transgenic plants were used as previously reported [55].

4.4 Gene Expression Analysis

Total RNA was extracted from the frozen samples using Trizol reagent (Invitrogen, Carlsbad, CA, USA) according to the manufacturerts instructions and treated with DNase I (TaKaRa, Tokyo, Japan) to remove genomic DNA contamination. Then, for each sample, the first-strand cDNA was synthesized using a PrimeScript™ RT Reagent Kit (TaKaRa). The expression profiles of PheDof12-1 in different tissues, abiotic stress, and photoperiod treatments were analyzed by quantitative RT-PCR (qRT-PCR). TIP41 and NTB were used as internal housekeeping genes [56]. The qRT-PCR reactions were carried out using a Light Cycler 480 System (Roche, Basel,

Switzerland) and a SYBR Premix EX TaqTMkit (Roche, Mannheim, Germany). All reactions were performed in triplicate, both technical and biological, and data were analyzed using the Roche manager software. The primer sequences are listed in Supplementary Table S1.

4.5 Yeast One-Hybrid Assay

To perform the Y1H assay, the full length of PheDof12-1 was cloned into the pGADT7-Rec2 bait vector, and the promoter sequence of PheCOL4 was cloned into the pHIS2 prey vector (Supplementary Table S1). The lithium acetate method was used to transform into the Y187 strain. The transformed yeast cells were selected on SD/-Trp/-Leu and SD/-Trp/-Leu/-His/-3AT/X-α-Gal plates at 30 ◦C for 3–5 days.

5. CONCISE SUMMARY

In conclusion, this study presents novel insights into the role of Dof transcription factors in moso bamboo, identifying PheCOL12-1 as a pivotal element with diverse functions associated with abiotic stress and the developmental processes governing the shift from the vegetative to the reproductive phase under long-day conditions. PheCOL12-1 is a nucleus-localized transcription factor that modulates regulators associated with photoperiodism. These findings enhance our comprehension of the functional roles of Dof proteins in regulating abiotic stress and flowering time, while also presenting a significant candidate gene for investigating the molecular regulatory mechanisms of moso bamboo flowering.

REFERENCES

1. Yanagisawa, S. The Dof family of plant transcription factors. Trends Plant Sci. 2002, 7, 555–560. [CrossRef]
2. Yanagisawa, S.; Izui, K. Molecular cloning of two DNA-binding proteins of maize that are structurally different but interact with the same sequence motif. J. Biol. Chem. 1993, 268, 16028–16036. [PubMed]
3. Plesch, G.; Ehrhardt, T.; Muellerroeber, B. Involvement of TAAAG elements suggests a role for Dof transcription factors in guard cell-specific gene expression. Plant J. 2001, 28, 455–464. [CrossRef] [PubMed]
4. Tanaka, M.; Takahata, Y.; Nakayama, H.; Nakatani, M.; Tahara, M. Altered carbohydrate metabolism in the storage roots of sweetpotato plants overexpressing the SRF1 gene, which encodes a Dof zinc finger transcription factor. Planta 2009, 230, 737–746. [CrossRef] [PubMed]
5. Wen, C.L.; Cheng, Q.; Zhao, L.; Mao, A.; Yang, J.; Yu, S.; Weng, Y.; Xu, Y. Ientification and characterisation of Dof transcription factors in the cucumber genome. Sci. Rep. 2016, 6, 23072–23083. [CrossRef]

6. Papi, M.; Sabatini, S.; Bouchez, D.; Camilleri, C.; Costantino, P.; Vittorioso, P. Identification and disruption of an Arabidopsis zinc finger gene controlling seed germination. Genes Dev. 2000, 14, 28–33. [CrossRef] [PubMed]

7. Gualberti, G.; Papi, M.; Bellucci, L.; Ricci, I.; Bouchez, D.; Camilleri, C.; Costantino, P.; Vittorioso, P. Mutations in the Dof zinc finger genes DAG2 and DAG1 influence with opposite effects the germination of Arabidopsis seeds. Plant Cell 2002, 14, 1253–1263. [CrossRef] [PubMed]

8. Rueda-Romero, P.; Barrero-Sicilia, C.; Gomez-Cadenas, A.; Carbonero, P.; OnateSanchez, L. Arabidopsis thaliana DOF6 negatively affects germination in non-after-ripened seeds and interacts with TCP14. J. Exp. Bot. 2012, 63, 1937–1949. [CrossRef]

9. Wei, P.C.; Tan, F.; Gao, X.Q.; Zhang, X.Q.; Wang, G.Q.; Xu, H.; Li, L.J.; Chen, J.; Wang, X.C. Overexpression of AtDOF4. 7, an Arabidopsis DOF family transcription factor, induces floral organ abscission deficiency in Arabidopsis. Plant Physiol. 2010, 153, 1031–1045. [CrossRef]

10. Guo, Y.; Qin, G.; Gu, H.; Qu, L.J. Dof5. 6/HCA2, a Dof transcription factor gene, regulates interfascicular cambium formation and vascular tissue development in Arabidopsis. Plant Cell 2009, 21, 3518–3534. [CrossRef]

11. Washio, K. Functional dissections between GAMYB and Dof transcription factors suggest a role for protein-protein associations in -the gibberellin-mediated expression of the RAmy1A gene in the rice aleurone. Plant Physiol. 2003, 133, 850 863. [CrossRef]

12. Yanagisawa, S.; Sheen, J. Involvement of maize Dof zinc finger proteins in tissue-specific and light-regulated gene expression. Plant Cell 1998, 10, 75 89. [CrossRef] [PubMed]

13. Yanagisawa, S. Dof1 and Dof2 transcription factors are associated with expression of multiple genes involved in carbon metabolism in maize. Plant J. 2000, 21, 281–288. [CrossRef] [PubMed]

14. Rueda-López, M.; Crespillo, R.; Cánovas, F.M.; Avila, C. Differential regulation of two glutamine synthetase genes by a single Dof transcription factor. Plant J. 2008, 56, 73–85. [CrossRef] [PubMed]

15. Kurai, T.; Wakayama, M.; Abiko, T.; Yanagisawa, S.; Aoki, N.; Ohsugi, R. Introduction of the ZmDof1 gene into rice enhances carbon and nitrogen assimilation under low-nitrogen conditions. Plant Biotechnol. J. 2011, 9, 826–837. [CrossRef] [PubMed]

16. Yanagisawa, S.; Akiyama, A.; Kisaka, H.; Uchimiya, H.; Miwa, T. Metabolic engineering with Dof1 transcription factor in plants: Improved nitrogen

assimilation and growth under low-nitrogen conditions. Proc. Natl. Acad. Sci. USA 2004, 101, 7833–7838. [CrossRef] [PubMed]

17. Ryu, J.Y.; Hong, S.Y.; Jo, S.H.; Woo, J.C.; Lee, S.; Park, C.M. Molecular and functional characterization of cold responsive C-repeat binding factors from Brachypodium distachyon. BMC Plant Biol. 2014, 14, 15. [CrossRef]
18. Corrales, A.R.; Nebauer, S.G.; Carrillo, L.; Fernández-Nohales, P.; Marqués, J.; Renau-Morata, B.; Granell, A.; Pollmann, S.; Vicente-Carbajosa, J.; Molina, R.V.; et al. Characterization of tomato Cycling Dof Factors reveals conserved and new functions in the control of flowering time and abiotic stress responses. J. Exp. Bot. 2014, 65, 995–1012. [CrossRef]
19. Ma, J.; Li, M.Y.; Wang, F.; Tang, J.; Xiong, A.S. Genome-wide analysis of Dof family transcription factors and their responses to abiotic stresses in Chinese cabbage. BMC Genom. 2015, 16, 33. [CrossRef]
20. Song, Y.H.; Smith, R.W.; To, B.J.; Millar, A.J.; Imaizumi, T. FKF1 conveys timing information for CONSTANS stabilization in photoperiodic flowering. Science 2012, 336, 1045–1049. [CrossRef]
21. Fornara, F.; Panigrahi, K.C.; Gissot, L.; Sauerbrunn, N.; Rühl, M.; Jarillo, J.A.; Coupland, G. Arabidopsis DOF transcription factors act redundantly to reduce CONSTANS expression and are essential for a photoperiodic flowering response. Dev. Cell 2009, 17, 75–86. [CrossRef] [PubMed]
22. Li, D.; Yang, C.; Li, X.; Gan, Q.; Zhao, X.; Zhu, L. Functional characterization of rice OsDof12. Planta 2009, 229, 1159–1169. [CrossRef] [PubMed]
23. Iwamoto, M.; Higo, K.; Takano, M. Circadian clock- and phytochrome-regulated Dof-like gene, Rdd1, is associated with grain size in rice. Plant Cell Environ. 2009, 32, 592–603. [CrossRef] [PubMed]
24. Ahmad, M.; Rim, Y.; Chen, H.; Kim, J.K. Functional characterization of Arabidopsis Dof transcription factor AtDof4.1. Russ. J. Plant Physiol. 2013, 60, 116–123. [CrossRef]
25. Lin, X.C.; Chow, T.Y.; Chen, H.H.; Liu, C.C.; Chou, S.J.; Huang, B.L.; Kuo, C.I.; Wen, C.K.; Huang, L.C.; Fang, W. Understanding bamboo flowering based on large-scale analysis of expressed sequence tags. Genet. Mol. Res. 2010, 9, 1085–1093. [CrossRef] [PubMed]
26. Wu, M.; Liu, H.; Han, G.; Cai, R.; Pan, F.; Xiang, Y. A moso bamboo WRKY gene PeWRKY83 confers salinity tolerance in transgenic Arabidopsis plants. Sci. Rep. 2017, 7, 11721. [CrossRef] [PubMed]
27. Chen, J.; Shafi, M.; Li, S.; Wang, Y.; Wu, J.; Ye, Z.; Peng, D.; Yan, W.; Liu, D. Copper induced oxidative stresses, antioxidant responses and phytoremediation

potential of moso bamboo (Phyllostachys pubescens). Sci. Rep. 2015, 5, 13554. [CrossRef]

28. Zhang, P.; Wang, J.; Zhang, H. Measures of water management and increasing drought resistance of moso forests in Anji County, Zhejiang Province. World Bamboo Rattan 2008, 6, 23–24. [CrossRef]

29. Ge, W.; Zhang, Y.; Cheng, Z.; Hou, D.; Li, X.; Gao, J. Main regulatory pathways, key genes and microRNAs involved in flower formation and development of moso bamboo (Phyllostachys edulis). Plant Biotechnol. J. 2017, 15, 82–96. [CrossRef]

30. Corrales, A.R.; Carrillo, L.; Lasierra, P.; Nebauer, S.G.; Dominguez-Figueroa, J.; Renau-Morata, B.; Pollmann, S.; Granell, A.; Molina, R.V.; Vicente-Carbajosa, J.; et al. Multifaceted role of cycling DOF factor 3 (CDF3) in the regulation of flowering time and abiotic stress responses in Arabidopsis. Plant Cell Environ. 2017, 40, 748–764. [CrossRef]

31. Sun, L.; Li, X.; Fu, Y.; Zhu, Z.; Tan, L.; Liu, F.; Sun, X.; Sun, X.; Sun, C. GS6, a member of the GRAS gene family, negatively regulates grain size in rice. J. Integr. Plant Biol. 2013, 55, 938–949. [CrossRef] [PubMed]

32. Murakami, M.; Ashikari, M.; Miura, K.; Yamashino, T.; Mizuno, T. The evolutionarily conserved OsPRR quintet: Rice pseudo-response regulators implicated in circadian rhythm. Plant Cell Physiol. 2003, 44, 1229–1236. [CrossRef] [PubMed]

33. Ding, J.; Böhlenius, H.; Rühl, M.G.; Chen, P.; Sane, S.; Zambrano, J.A.; Zheng, B.; Eriksson, M.E.; Nilsson, O. GIGANTEA-like genes control seasonal growth cessation in Populus. New Phytol. 2018, 218, 1491–1503. [CrossRef] [PubMed]

34. Searle, L.; Coupland, G. Induction of flowering by seasonal changes in photoperiod. Embo J. 2004, 23, 1217–1222. [CrossRef] [PubMed]

35. Shaw, L.M.; McIntyre, C.L.; Gresshoff, P.M.; Xue, G.P. Members of the Dof transcription factor family in Triticum aestivum are associated with light-mediated gene regulation. Funct. Integr. Genom. 2009, 9, 485–498. [CrossRef] [PubMed]

36. Gao, J.; Zhang, Y.; Zhang, C.; Qi, F.; Li, X.; Mu, S.; Peng, Z. Characterization of the floral transcriptome of moso bamboo (Phyllostachys edulis) at different flowering developmental stages by transcriptome sequencing and RNA-seq analysis. PLoS ONE 2014, 9, e98910. [CrossRef] [PubMed]

37. Cheng, Z.; Hou, D.; Liu, J.; Li, X.; Xie, L.; Ma, Y.; Gao, J. Characterization of moso bamboo (Phyllostachys edulis) Dof factors in floral development and abiotic stress responses. Genome 2018, 61, 151–156. [CrossRef] [PubMed]

38. Izawa, T.; Takahashi, Y.; Yano, M. Comparative biology comes into bloom: Genomic and genetic comparison of flowering pathways in rice and Arabidopsis. Curr. Opin. Plant Biol. 2003, 6, 113–120. [CrossRef]

39. Imaizumi, T.; Schultz, T.F.; Harmon, F.G.; Ho, L.A.; Kay, S.A. FKF1 F-box protein mediates cyclic degradation of a repressor of CONSTANS in Arabidopsis. Science 2005, 309, 293–297. [CrossRef]

40. Yang, J.; Yang, M.F.; Zhang, W.P.; Chen, F.; Shen, S.H. A putative flowering-time-related Dof transcription factor gene, JcDof3, is controlled by the circadian clock in Jatropha curcas. Plant Sci. 2011, 181, 667–674. [CrossRef]

41. Lijavetzky, D.; Carbonero, P.; Vicentecarbajosa, J. Genome-wide comparative phylogenetic analysis of the rice and Arabidopsis Dof gene families. BMC Evol. Biol. 2003, 3, 17. [CrossRef] [PubMed]

42. Kardailsky, I.; Shukla, V.K.; Ahn, J.H.; Dagenais, N.; Christensen, S.K.; Nguyen, J.T.; Chory, J.; Harrison, M.J.; Weigel, D. Activation tagging of the floral inducer FT. Science 1999, 286, 1962–1965. [CrossRef] [PubMed]

43. An, H.; Roussot, C.; Suárez-López, P.; Corbesier, L.; Vincent, C.; Piñeiro, M.; Hepworth, S.; Mouradov, A.; Justin, S.; Turnbull, C.; et al. CONSTANS acts in the phloem to regulate a systemic signal that induces photoperiodic flowering of Arabidopsis. Development 2004, 131, 3615–3626. [CrossRef] [PubMed]

44. Lee, J.; Oh, M.; Park, H.; Lee, I. SOC1 translocated to the nucleus by interaction with AGL24 directly regulates LEAFY. Plant J. 2010, 55, 832–843. [CrossRef] [PubMed]

45. Michaels, S.D.; Amasino, R.M. FLOWERING LOCUS C encodes a novel MADS domain protein that acts as a repressor of flowering. Plant Cell 1999, 11, 949–956. [CrossRef] [PubMed]

46. Wu, H.; Lv, H.; Li, L.; Liu, J.; Mu, S.; Li, X.; Gao, J. Genome-Wide Analysis of the AP2/ERF Transcription Factors Family and the Expression Patterns of DREB Genes in Moso Bamboo (Phyllostachys edulis). PLoS ONE 2015, 10, e0126657. [CrossRef] [PubMed]

47. Wang, H.; Zhao, S.; Gao, Y.; Yang, J. Characterization of Dof transcription factors and their responses to osmotic stress in Poplar (Populus trichocarpa). PLoS ONE 2017, 12, e0170210. [CrossRef]

48. Huang, R.; Li, L.; Liu, J.; Gao, H.; Li, X.P. Sequence of PheWRKY9-1 gene in Phyllostachys edulis and analysis of related expression. Mol. Plant Breed. 2018, 14, 4569–4575. [CrossRef]

49. Wang, T.; Wang, H.; Cai, D.; Gao, Y.; Zhang, H.; Wang, Y.; Lin, C.; Ma, L.; Gu, L. Comprehensive profiling of rhizome-associated alternative splicing and

alternative polyadenylation in moso bamboo (Phyllostachys edulis). Plant J. 2017, 91, 684–699. [CrossRef]

50. Gasteiger, E.; Gattiker, A.; Hoogland, C.; Ivanyi, I.; Appel, R.D.; Bairoch, A. ExPASy: The proteomics server for in-depth protein knowledge and analysis. Nucleic Acids Res. 2003, 31, 3784–3788. [CrossRef]
51. Guo, A.Y.; Zhu, Q.H.; Chen, X.; Luo, J.C. GSDS:a gene structure display server. Yi Chuan 2007, 29, 1023–1026. [CrossRef] [PubMed]
52. Tamura, K.; Stecher, G.; Peterson, D.; Filipski, A.; Kumar, S. MEGA6: Molecular evolutionary genetics analysis version 6.0. Mol. Biol. Evol. 2013, 30, 2725–2729. [CrossRef]
53. Yoo, S.D.; Cho, Y.H.; Sheen, J. Arabidopsis mesophyll protoplasts: A versatile cell system for transient gene expression analysis. Nat. Protoc. 2007, 2, 1565–1572. [CrossRef]
54. Clough, S.J.; Bent, A.F. Floral dip: A simplified method for Agrobacterium-mediated transformation of Arabidopsis thaliana. Plant J. 1998, 16, 735–743. [CrossRef] [PubMed]
55. Jefferson, R.A.; Kavanagh, T.A.; Bevan, M.W. GUS fusions: Betaglucuronidase as a sensitive and versatile gene fusion marker in higher plants. Embo J. 1987, 6, 3901–3907. [CrossRef]
56. Fan, C.; Ma, J.; Guo, Q.; Li, X.; Wang, H.; Lu, M. Selection of reference genes for quantitative Real-Time PCR in bamboo (Phyllostachys edulis). PLoS ONE 2013, 8, e56573. [CrossRef] [PubMed]

9 Chapter

IDENTIFICATION AND CHARACTERIZATION OF THE EXO70 GENE FAMILY IN POLYPLOID WHEAT AND ASSOCIATED SPECIES

1. INTRODUCTION: AN OVERVIEW

The exocyst complex is an evolutionarily conserved octameric tethering component that facilitates the fusing of post-Golgi secretory vesicles with the plasma membrane and is integral to exocytosis. EXO70 is a crucial component of the exocyst complex and is extensively distributed across yeast, animals, and plants [3]. In yeast and mammals, the EXO70 gene exists as a single copy, whereas plants possess numerous copies, with the number of EXO70 members varying from 21 to 47 in species such as potatoes (Symphytum tuberosum), Arabidopsis, Populus trichocarpa, and rice. The EXO70s of terrestrial plants likely arose from three ancestral EXO70 genes, categorizing them into three groups: EXO70.1, EXO70.2, and EXO70.3. They have been independently duplicated further in the moss, lycophyte, and angiosperm lineages, as well as in the succeeding lineage-specific multiplications represented by nine subgroups (EXO70A-EXO70I) [4,7–9].

The function of EXO70 has been extensively studied in yeast, and mammals [3,10–15]. In plants, EXO70s have been proven to play diverse roles in regulating plant growth and coping with adverse biotic/abiotic stresses. In Arabidopsis, EXO70A1 has been implicated in a wide range of developmental processes, including the differentiation of tracheary elements [16,17], the development of seed coat, root hair and stigmatic papillae [18], the recycling of the auxin efflux carrier proteins (PIN1 and PIN2) [19] and the formation of the Casparian strip [20]. EXO70C1 and EXO70C2 regulated the polarized growth and maturation of the pollen tube [21,22]. EXO70H4 regulates trichome cell wall maturation by mediating the secretion and accumulation of callose and silica [23,24]. The rice OsEXO70A1 is necessary for vascular bundle differentiation and assimilation of mineral nutrients [5]. The legumes EXO70J7, EXO70J8 and EXO70J9 are members of an atypical subgroup of EXO70 proteins (EXO70J) that regulate leaf senescence and nodule formation [25].

In Nicotiana benthamiana, silencing all the paralogue genes in subgroups EXO70A (six), C (three), D (four) and G (six) resulted in a smaller leaf phenotype [6].

Evidence has accumulated for the critical role of EXO70s in plant-pathogen interactions or responses to abiotic stresses. In N. benthamiana, the silencing of two EXO70B paralogues led to increased susceptibility to Phytophthora infestans [6]. Three of the 23 members of the Arabidopsis EXO70 gene family (EXO70B1, EXO70B2 and EXO70H1) have been proven to be involved in plant immunity [26]. AtEXO70B1 and AtEXO70B2 belong to the same subgroup and are both involved in plant immunity, of which AtEXO70B1 a negative regulator, while AtEXO70B2 is a positive regulator. AtEXO70B1 underwent autophagic transport, and the loss-of-function of exo70B1 led to reduced numbers of internalized autophagosomes, accumulation of salicylic acid (SA), and finally, ectopic hypersensitive responses and enhanced resistance to several pathogens. AtEXO70B1's regulation of disease resistance, either by interacting with TIR-NBS2, a truncated version of the classical nucleotide binding (NB) domain and a leucine-rich repeat (LRR)-containing (NLR) intracellular immune receptor-like protein [27–29], or by interacting with RIN4, a well-known regulator of pathogen-associated molecular pattern (PAMP)-triggered immunity (PTI) [29]. AtEXO70B2 regulated innate immunity via interacting with a negative PTI regulator, AtPUB22, which mediated the ubiquitination and degradation of AtEXO70B2 and contributed to PTI. The exo70B2 mutants showed aberrant papillae with halos and were susceptible to different PAMPs and pathogens [30,31]. AtEXO70B1 and AtEXO70B2 also contribute to the abiotic stress response; both were positive regulators of stomatal movement. The response to mannitol (drought) treatments is in either an abscisic acid (ABA)-dependent or -independent manner [32,33]. AtEXO70H1 is a homolog of AtEXO70B2 and is also involved in plant immunity [31]. Three of the 47 EXO70 members of rice (OsEXO70E1, OsEXO70F2 and OsEXO70F3) were reported to participate in plant immunity [5]. OsEXO70E1 is attributed to planthopper resistance by interacting with a broad resistance protein, Bph6. Interaction of the two proteins increased exocytosis and blocked the feeding of a planthopper by cell wall thickening at the infection sites [34]. The importance of EXO70 in plant immunity was also shown by the fact that some of the EXO70s were targets of the secreted effectors of the plant pathogen. Both OsEXO70F2 and OsEXO70F3 were targets of the Magnaporthe oryzae effector AVR-Pii, and OsEXO70F3 was proven to play an important role in Pii-dependent resistance by interacting with AVR-Pii [35].

Accumulated evidence has shown that a large number of EXO70s exist in plants; however, only a few have had their biological roles elucidated [5,35,36]. Due to the huge genome size and complexity [37], the knowledge of the EXO70 gene family from the Triticeae species is rather limited. In the last five years, the genome sequences of wheat and its ancestor species have been released, which makes genome-wide identification of a gene family in the Triticeae species feasible [38–43].

Haynaldia villosa L. (2n = 2x = 14, VV) is a diploid wild progenitor of wheat. Prior research indicated that H. villosa is a significant genetic reservoir containing numerous superior features, including resistance to various wheat diseases and tolerance to abiotic stresses [44–46]. This work identifies many members of the EXO70 gene family by examining the available genome sequences of Triticeae species. Targeted primer pairs are developed, EXO70s are extracted from H. villosa, and their prospective activities are clarified by expression profiles derived from in silico analysis and quantitative RT-PCR (qRT-PCR). The results gained will enhance our understanding of the development and diversification of EXO70s within Triticeae species, as well as their possible functions in plant immunology and responses to abiotic stressors..

2. RESULTS

2.1 Identification and Phylogenetic Relationship Analysis of the EXO70 Gene Family in Triticeae Species

In total, 200 EXO70 genes were identified from the public database of five Triticeae species. Among them, there were 26 each from T. urartu, Ae. Tauschii and H. vulgare; 47 from T. dicoccoides; and 75 from common wheat (T. aestivum), respectively. Fifteen EXO70s from H. villosa were obtained by homology cloning (Figure 1a). The evolutionary relationship of the above 215 Triticeae EXO70s, along with 22 from Brachypodium distachyon, 41 from rice and 23 from Arabidopsis, were phylogenetically analyzed (Figure 1b, Table S1). These EXO70s were divided into three major groups, EXO70.1, EXO70.2 and EXO70.3, which were further assigned to nine subgroups, from EXO70A to EXO70I, according to a phylogenetic tree (Figure 1b). The EXO70A subgroup belongs to the group EXO70.1, in which 43 (14.28%) EXO70s were included; the EXO70B, C, D, E, F, H, and I subgroups belong to the group EXO70.2, in which 229 (76.08%) EXO70s were included; the EXO70G subgroup belongs to the group EXO70.3, in which 29 (9.63%) EXO70s were included. The EXO70I subgroup has the most members (74, 24.58%), followed by EXO70F (48, 15.95%) and EXO70A (43, 14.28%) (Figure 1c). Based on the subgroups and genome allocation, the wheat EXO70s were designated [7,47]. For example, the EXO70B1 from T. dicoccoides located on chromosome 1A was assigned TdEXO70B1-1A. In Arabidopsis, the EXO70I subgroup is missing, whereas the EXO70I subgroup in Triticeae species appeared to be the most divergent. However, our analysis led to a new insight: the EXO70I subgroup belongs to EXO70.2, rather than EXO70.3 from other species [36] (Figure 1b,c).

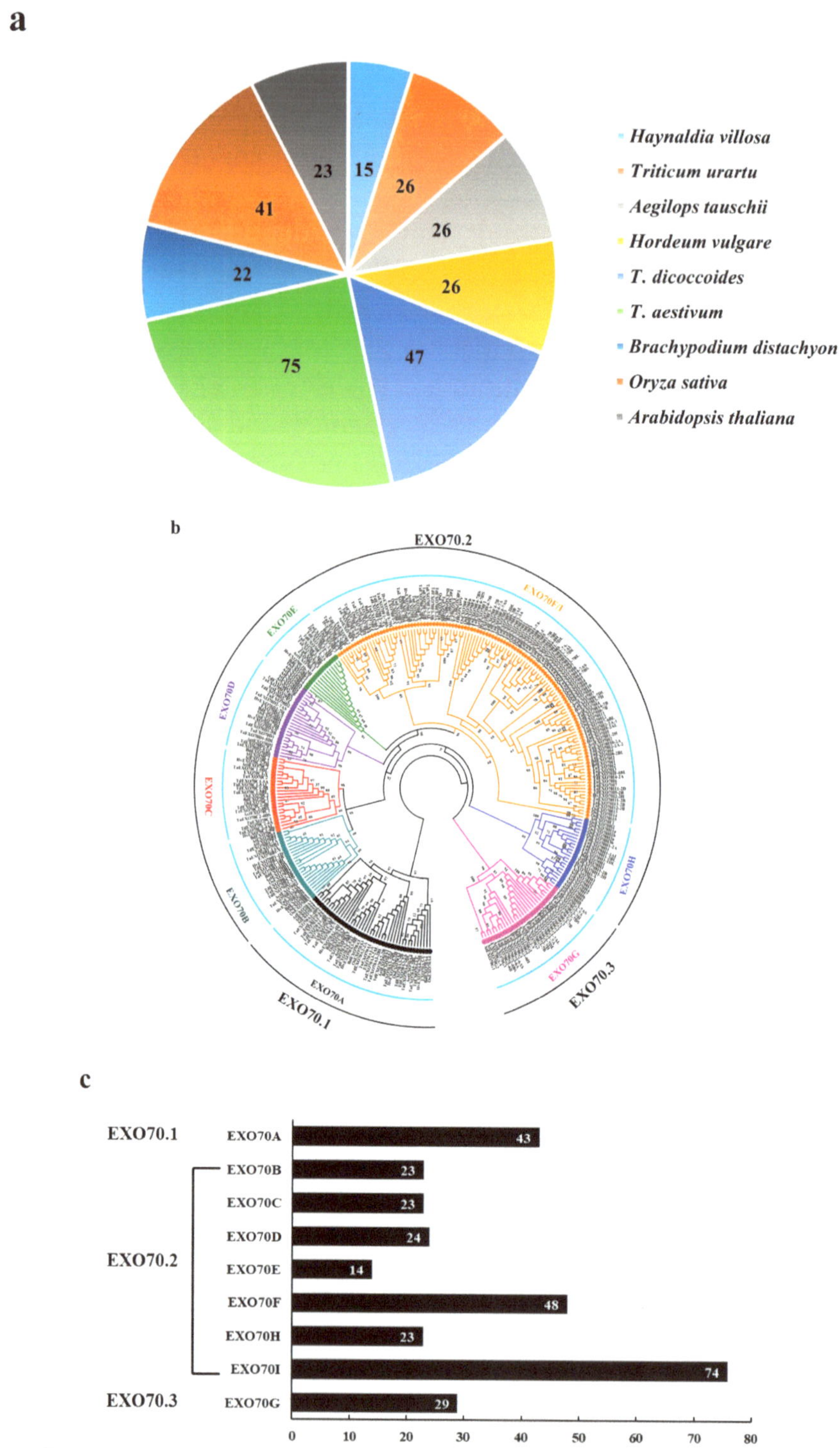

Figure 1. The number and phylogenetic relationships of EXO70 family genes from T. aestivum, T. urartu, Ae. tauschii, T. dicoccoides, H. vulgare, A. thaliana, Oryza sativa, Brachypodium distachyon and H. villosa.(a) The total number of EXO70

gene family in nine species. (b) The phylogenetic tree of nine Triticeae species. Species abbreviations: Ta, Triticum aestivum; Tu, Triticum urartu; Aet, Aegilops tauschii; Td, Triticum dicoccoides; Hv, Hordeum vulgare; At, Arabidopsis thaliana; Bd, Brachypodium distachyon; Os, Oryza sativa; -V, Haynaldia villosa. (c) The number of the EXO70 gene family from nine species in each of the subgroups. The horizontal/longitudinal coordinate axis represents the number of genes and different subgroups, respectively.

The 15 cloned EXO70s from H. villosa were designated as EXO70A1-V to EXO70I1-V according to the phylogenetic relationship to wheat EXO70s. They belong to nine subgroups, three each to EXO70A and F, two each to EXO70D and G and one each to EXO70B, C, E, H and I (Figure 1). Their CDS length ranges from 801 bp (EXO70H1-V) to 2007 bp (EXO70C1-V) and their isoelectric point varies from 4.52 (EXO70B1-V) to 10.19 (EXO70G1-V) (Table 1).

Table 1. EXO70 genes cloned from H. villosa.

Number	Name	ORF (bp)	AA (aa)	DL (aa)	PI	MW (KD)
1	EXO70A1-V	1914	637	268–623	7.71	71.48
2	EXO70A2-V	1944	647	273–631	8.78	73.07
3	EXO70A3-V	882	294	1–283	8.83	32.61
4	EXO70B1-V	1626	541	180–534	4.52	59.74
5	EXO70C1-V	2007	668	287–654	5.72	73.9
6	EXO70D1-V	1650	549	152–532	6.98	61.2
7	EXO70D2-V	1644	547	152–525	7.00	61.02
8	EXO70E1-V	1815	605	233–580	5.04	68.93
9	EXO70F1-V	1323	441	52–424	5.46	49.88
10	EXO70F2-V	1542	514	147–504	5.03	57.62
11	EXO70F3-V	1431	477	108–466	5.60	52.91
12	EXO70G1-V	807	268	1–232	10.19	30.95
13	EXO70G2-V	1437	478	85–439	9.42	53.8
14	EXO70H1-V	801	268	1–238	8.66	28.65
15	EXO70I1-V	1455	484	126–480	5.61	42.48

Abbreviations: ORF, open reading frame; AA, amino acids; DL, PFam03081 domain location; PI, protein isoelectric point; MW, protein molecular weight.

DNAMan was used to explore the amino acid sequence feature of H. villosa EXO70s. The pfam03081 domain at the C-terminus was relatively conservative, shown by the fact that all of the EXO70s had the same amino acid in the 11 sites (in the red rectangle in Figure 2a). R programming language was used to visualize the sequence similarities. The same subgroups shared more common sequences, e.g., EXO70D1-V and EXO70D2-V shared about 80% similarity, while different subgroups had low sequence similarities, e.g., EXO70H1-V and EXO70I1-V only had 21.46% similarity (Figure 2b).

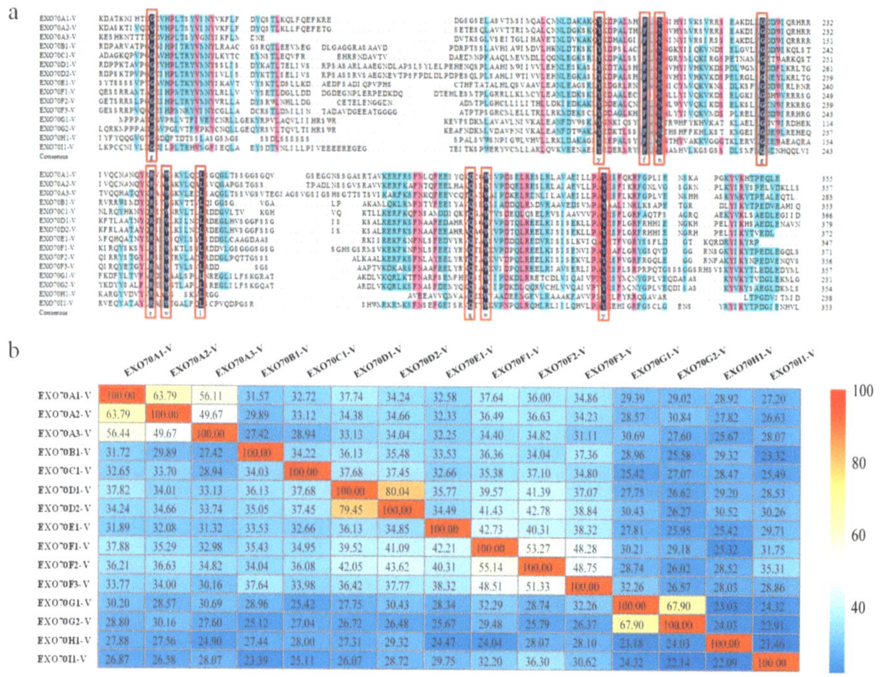

Figure 2. The amino acid sequence feature analysis of EXO70 gene in H. villosa. (a) The amino acid sequence of pfam03081 domain for each EXO70 was selected for multiple sequence alignment analysis by DNAMan. The 11 identical amino acids were indicated in red frame. The three colors of black, red and blue represent the level of similarity of amino acids, from high to low. (b) Sequence similarity analysis using the R programming language. The color scale bar represents sequence similarity between different genes. Red and yellow indicate that the sequence similarity was greater than 80% and 60%, respectively. Blue indicates that the sequence similarity was less than 40%.

2.2 The Number of EXO70s in Genomes of Triticeae Species

The diploid species B. distachyon and Arabidopsis have 22 and 23 EXO70s, respectively, while diploid rice has 41, nearly twice as many. We identified 22 EXO70s each in A and D of common wheat and B of the tetraploid wheat, 25 in A of tetraploid wheat, 26 in D of Ae. Tauschii and H of barley and 29 EXO70s in B of common wheat. For the A genome, tetraploid has three more EXO70s than hexaploidy wheat; for the B genome, hexaploidy has seven more than tetraploid wheat; for the D genome, Ae. tauschii has four more EXO70s than hexaploidy wheat. In common wheat, the chromosomes 5A, 5B and 5D had the same number of EXO70s, belonging to the same gene type and having the same gene order. The chromosomes 7A, 7B and 7D also have the same number of EXO70s; however, their gene type and gene order were not completely the same. For example, TaEXO70D is only present on 7A and 7B, but not on 7D; TaEXO70G1 is only present on 7A and 7D, but not 7B. The gene number and order for EXO70s in the remaining five homoeologous were also different among corresponding homoeologous chromosomes, mainly due to their difference for subgroup EXO70I (Tables 2 and S2).

Table 2. Number of EXO70 from different species in each of the chromosomes.

Chromosome	T aestivum			T. dicoccoides		Ae. tauschii	H. vulgare	Total
	A	B	D	A	B	D	H	
Chr.1	1	1	3	1	1	1	1	9
Chr.2	5	8	4	8	6	6	7	44
Chr.3	4	5	3	4	3	4	4	27
Chr.4	1	3	2	2	2	2	2	14
Chr.5	3	3	3	2	2	3	3	19
Chr.6	1	2	0	1	1	1	1	7
Chr.7	7	7	7	7	7	9	7	51
Total	22	29	22	25	22	26	26	171
*Unknow		2					1	3

* The genes that were assigned to unknown chromosome.

The three analyzed diploid Triticea species, H. vulgare, T. uratu and Ae. tauschii, each had 26 EXO70s and had the same number as the EXO70B, C, D and H subgroups. They had the least EXO70Hs and the most EXO70Is. T. uratu only had one EXO70G, while the other two species had three. However, T. uratu had more EXO70I than the other two species (Table 3). The variation among different diploid species may be due either to the quality of the genome assembly or the duplication or deletion during evolution. The tetraploid T. dicoccoides and hexaploid T. aestivum have 47 and 75 EXO70s, which are about twice and three times the number in diploid species, respectively. In addition, the number of EXO70s from each subgroup is also exactly (EXO70C, H) or almost (EXO70A, B, D, E, F, I) 2 or 3 times that of the diploid species (Table 3). This indicated that, in polyploid wheat, the increased number of EXO70s is mostly due to the genome polyploidization.

We only identified 15 EXO70s in diploid H. villosa, which is much less than the average number for other diploid species. H. villosa had the same number of subgroups D and H, but fewer members of subgroups B, C, E, F and I. This is due either to the lack of the sequence of H. villosa, or the divergence of H. villosa from other Triticea species (Table 3).

Table 3. Numbers of EXO70 paralogs encoded by the surveyed genomes in total and individual subgroups.

Genome	Total Number	Subgroup								
		A	B	C	D	E	F	G	H	I
H. villosa (VV)	15	3	1	1	2	1	3	2	1	1
H. vulgare (HH)	26	4	2	2	2	2	5	3	1	5
T. urartu (AA)	26	3	2	2	2	1	5	1	1	9
Ae. tauschii (DD)	26	4	2	2	2	1	4	3	1	7
T. dicoccoides (AABB)	22–25 (47)	3(6)	2(4)	2(4)	2(4)	1(2)	4(8)	3(6)	1(2)	7(11)
T. aestivum (AABBDD)	22–29 (75)	4(12)	2(5)	2(6)	2(5)	1(3)	5(13)	3(6)	1(3)	14 (22)
B. distachyon (Bd)	22	4	2	2	2	1	4	3	1	3
Oryza sativa	41	4	3	2	2	1	5	3	5	16
Arabidopsis thaliana	23	3	2	2	3	2	1	2	8	0
Total	238(301)	32 (43)	18 (23)	17 (23)	19 (24)	11 (14)	36 (48)	23 (29)	20 (23)	64 (74)
Numbers in brackets indicate number of copies for polyploid genomes.										

2.3 The Chromosomal Distribution of EXO70s in Triticeae Species

The identified 174 EXO70s from four Triticeae species (T. aestivum, Ae. tauschii, T. dicoccoides and H. vulgare) were assigned to corresponding chromosomes (including three on unknown chromosome). With the exception of wheat 6D, all chromosomes have at least one EXO70 gene (Figure 3). The EXO70s were not evenly distributed on difference chromosomes or in different homologous groups. In total, there were 9, 44, 27, 14, 19, 7 and 51 EXO70s in the homologous groups 1 to 7, with group 7 having the most EXO70s (29.31%), followed by group 2 (25.29%) (Table 2). The EXO70s in group 7 was dispersely distributed along the chromosome, while those in group 2 were clutered at the dital region of the long arm (Figure 3). We found that nine groups EXO70s were conversely present on the same homoeologous groups of the four species, such as the B1 in group1, F2/3 in group 2, E1 and F3/2 in group 3, C2/C1 in group 4, CI and D1 in group 5, F1 and F4 in group 7 (Table S2, Figure 3).

For homologous chromosomes from different genomes of wheat and its ancestral/related species, most of the corresponding EXO70 orthologs were present at the syntenic genome regions. For example, EXO70B1s were on the long chromosome arm of homoeologous group 1 and EXO70E1 on group 3 chromosomes (Figure 3).

There are some exceptions. The EXO70G1 is present on all group 6 chromosomes except for sub-genome 6D of the hexaploidy wheat. The EXO70H1 is located on chromosomes 2A, 2B of common wheat and T. dicoccoides, 2D of Ae. tauschii and 2H of barley and on 1D of common wheat (Figure 3). We found that the EXO70C2 is on the 4BS of common wheat and T. dicoccoides (Figure 3a,b), on the 4DS of common wheat and Ae. tauschii (Figure 3a,d), but on the 4AL of common wheat and T. dicoccoides (Figure 3a,b). This also supports the presence of an inter-arm translocation of 4A during the evolution from diploid to tetraploid wheat [48,49].

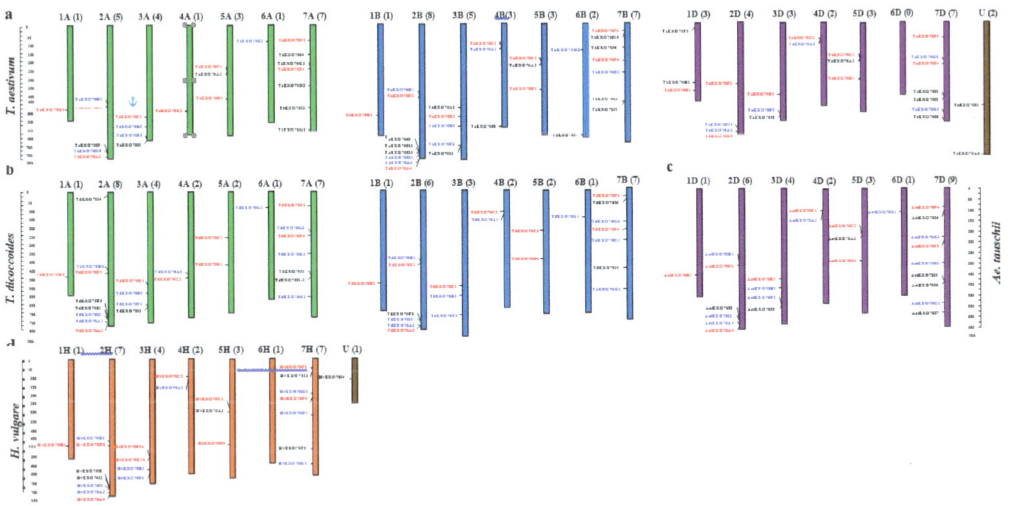

Figure 3. The chromosomal distribution of the EXO70 gene family in four Triticeae species. (a–d) represent T. aestivum, T. dicoccoides, A. tauschii and H. vulgare, respectively. Chromosome numbers are indicated at the top of each bar and the number in parentheses corresponds to the number of EXO70 genes present on that chromosome. The name of each gene is to the left of each chromosome. Gene names labeled with red, blue, or black indicate that they are conserved in four species, missing in one due to incomplete data, or missing in more than two, respectively.

2.4 The Diversification of Gene Structure of Triticeae EXO70s

The C-terminal PFam03081 domain, which may determine the function or structure of the proteins, is a specific characteristic of the EXO70 superfamily [47]. All the predicted 200 and 15 homologous cloned EXO70 proteins possessed such a domain; however, their amino acid sequence length is different for different EXO70s, varying from 103aa to 669aa and with an average length of 345aa (Table S1).

The exon–intron structure of the 200 Triticeae EXO70s was visualized by using the online Gene Structure Display Server and compared among different subgroups or within each individual subgroup. The exon-intron structures for most of the EXO70s within the same subgroups were relatively conserved among Triticeae species, was similar to in Arabidopsis or rice [7,36]. Compared with other subgroups, the subgroup EXO70A with 30 genes has the most introns on average, e.g., TuEXO70A3 has the most introns (20) and HvEXO70A4-2H has the fewest introns (five) (Figure 4a). The 170 EXO70 genes in the remaining eight subgroups only have about one intron on average. Among them, 83 genes (41.50%) are intronless, including eight EXO70Bs, seven EXO70Cs, nine EXO70Ds, one EXO70Es, 18 EXO70sF, 12 EXO70Gs, five EXO70Hs and 23 EXO70sI; 60 genes (30.00%) have one intron (e.g., AetEXO70B1 and TaEXO70C1-5BL), 19 genes (9.50%) have two introns (e.g., TuEXO70C2 and TaEXO70G2-6BS) and eight genes (4.00%) have three to six introns (e.g., AetEXO70I3 and TaEXO70D1-5AL). We also observed that genes that have a closer phylogenetic relationship in the same subgroup have a similar gene structure; however, within the same subgroup, some genes showed quite different gene structure. For instance, TaEXO70D1-5AL has six introns in EXO70D (Figure 4d) and the three copies of TaEXO70I5 and AetEXO70I3 have five introns, while the other genes in the same subgroup (EXO70D or EXO70I) only have one or two introns (Figure 4i). The most diversified gene structure of the EXO70A subgroup may be a clue that their diversified biological role is different from that of other subgroups.

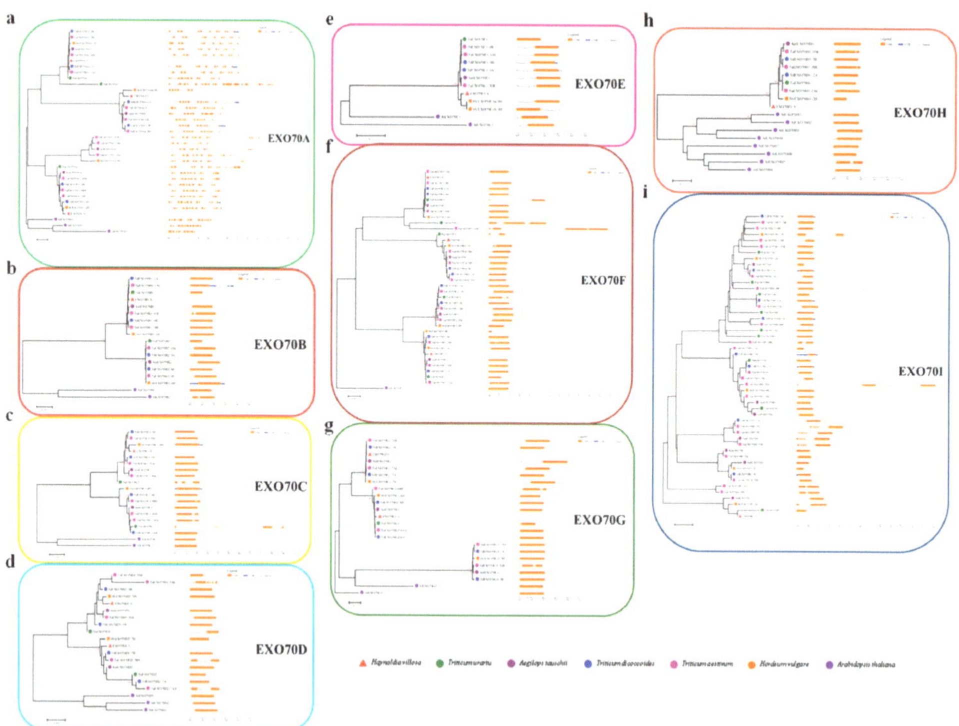

Figure 4. Phylogenetic analysis and exon–intron structures of EXO70 gene family in common wheat and related Triticeae species. The phylogenetic analysis was performed using the sequences of the conserved domain of EXO70; proteins were aligned by ClustalW, constructed by MEGA6 using the N-J method, with

1000 bootstrap replicates; the branch length scale bar indicates the evolutionary distance. The left column identifies subgroups and is marked with different alternating background tones to make subgroup identification easier. Introns and exons are represented by black lines and colored boxes, respectively.

2.5 The Expression Pattern of TaEXO70 Genes

The expression patterns of different members in a gene family will help us predict their potential biological roles. To elucidate the potential roles of the identified EXO70s, their expression in different tissues or in responses to various biotic and abiotic stresses was investigated by in silico expression profiling or qRT-PCR analysis. The expression patterns of wheat TaEXO70s in different tissues (root, stem and leaf of seeding stage), under two biotic stresses (stripe rust pathogen CYR31 and powdery mildew pathogen E09) and two abiotic stresses (drought and heat) [50,51] were first investigated using the wheat RNA-seq data from the publicly available databases. The expression level was measured as tags per million (TPM). To facilitate the portraits of transcript abundance, we assume the expression was high if TPM ≥ 2.5; moderate if 2.5 > TPM ≥ 1.5; low if 1.5 >TPM > 0; and undetectable if TPM = 0. All 75 TaEXO70 genes exhibited significantly diverse expression patterns (Figure 5). Their expression can be classified into five groups. The group "e" includes 23 TaEXO70s (30.67%). Their transcript abundance was undetectable; among them 11 were from subgroup I and six were from subgroup C (Figure 5e). The group "d" includes 20 genes (26.67%). They had detectable but weak transcript abundance; among them six were from subgroup A and six were from subgroup I (Figure 5d). The group "c" has six genes (8.00%), whose expression is high in roots and stems, whereas it does not respond to the four stresses (Figure 5c). The group "b" has 15 genes (20.00%). These genes were found to be negligibly to moderately expressed in the three tissue types and in response to one or two stresses. Among them, nine were from the subgroup F (Figure 5b). The group "a" has ten genes (13.33%). Their expression was generally high, either in different tissues or in response to the four stresses and none of them were from the subgroup I (Figure 5a).

Six genes were upregulated in response to both biotic and abiotic stresses. TaEXO70E1-3B was the only gene that was highly expressed in three tissues and upregulated in response to all four stresses. Its homologs on 3AL TaEXO70E1-3AL displayed moderate expression in three organs and were only responsive to Bgt inoculation and drought stress. The TaEXO70D1-5DL and TaEXO70F3-3B also displayed moderate expression in three organs and stresses. The TaEXO70H1-1DL expression was higher in root/leaf tissue and showed upregulation under both Pst infection and heat treatment. The TaEXO70B1-1BL/1DL expression was higher in root/stem tissue, and it was downregulated in response to Bgt and Pst inoculation but upregulated by drought treatment. TaEXO70A3-2AL/2BL/2DL had a similar expression pattern, being expressed in three tissues (stems > roots >

leaves) and downregulated by Pst inoculation but upregulated by both drought and heat treatment. TaEXO70D2-7BS and TaEXO70G1-7DL were both only responsive to biotic stresses; however, they showed opposite patterns, with TaEXO70D2-7BS upregulated and TaEXO70G1-7DL downregulated. Twelve genes were only responsive to one of the four stresses; five were only upregulated in response to Bgt or Pst inoculation; and seven were only upregulated in response to drought or heat stress. Conservation roles were observed for all three homolog genes from different genomes, such as TaEXO70A3(2AL/2BL/2DL) and TaEXO70F2(2AL/2BL/2DL).

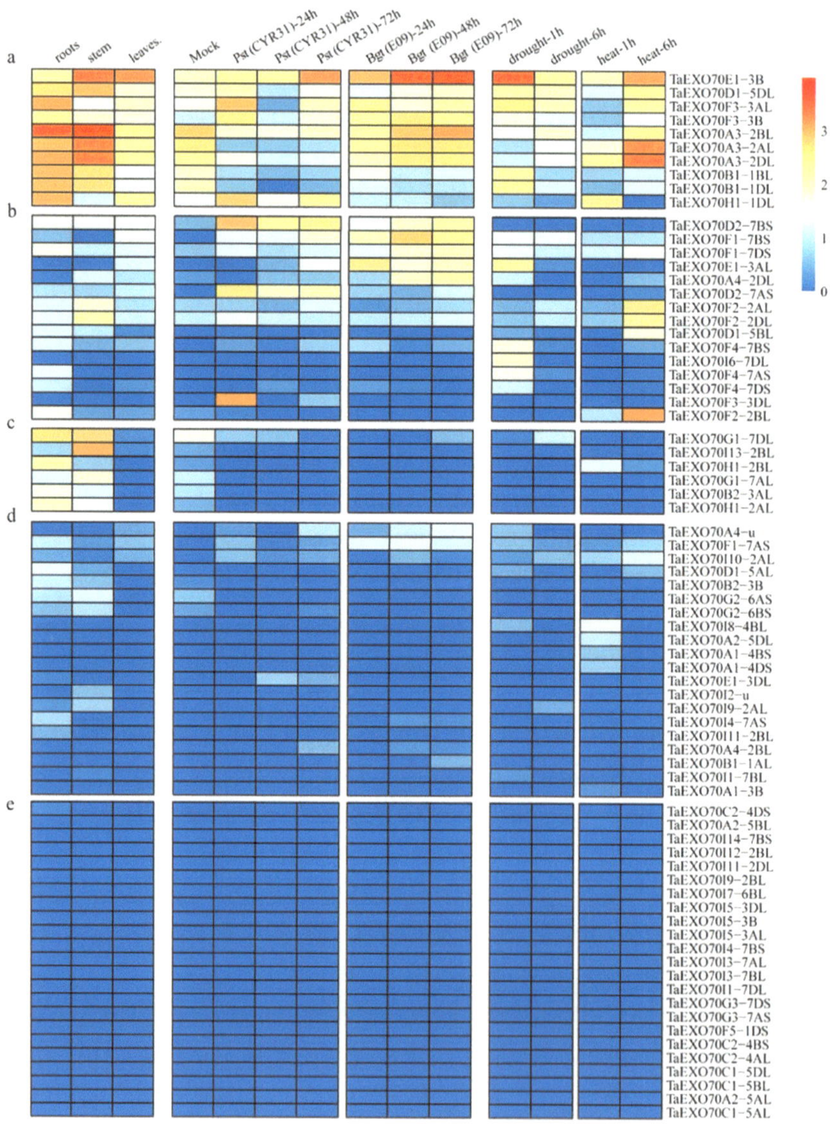

Figure 5. Heat map of the expression profiling of wheat EXO70 genes in different tissues and under various stresses. The color scale bar represents the expression values of the genes. (a–e): Genes with different expression types. Abbreviations: Bgt, powdery mildew; Pst: Stripe rust.

2.6 Differential Expression of 15 EXO70 Genes from H. villosa

The expression patterns of 15 EXO70-Vs in different tissues and in response to different stresses or treatments in H. villosa were investigated by qRT-PCR. Different expression patterns were observed for the analyzed genes (Figure 6). Six of the genes (EXO70D1-V, EXO70A2-V, EXO70F3-V, EXO70H1-V, EXO70I1-V and EXO70G2-V) were not differentially expressed in all three tissues. Four genes (EXO70B1-V, EXO70G1-V, EXO70A1-V, EXO70C1-V) showed more abundant transcript in stem; seven genes (EXO70E1-V, EXO70A3-V, EXO70B1-V, EXO70G1-V, EXO70F1-V, EXO70D2-V and EXO70F2-V) showed higher expression level in leaves (Figure 6a).

EXO70A2-V showed a similar expression level in all the three tissues; however, its expression was significantly increased in response to Bgt inoculation and treatments by chitin, flg22, heat stress, phytohormones, or H2O2. EXO70G1-V showed a similar expression level in the stems and leaves; its expression was also significantly increased by Bgt inoculation and treatment by chitin, flg22, four phytohormones, or H2O2.. The expression of EXO70H1-V was strongly upregulated by Bgt inoculation and cold stress, and moderately upregulated by chitin, Flg22, ET and heat stress. EXO70I1-V only responded to abiotic stresses (drought, salt and cold) and ABA treatment. EXOG2-V was responsive to flg22 treatment and drought stress, and EXO70A3-V was only responsive to heat stress (Figure 6b,c). The divergence of expression patterns of different gene members indicated their clear-cut roles in the adaptation of H. villosa to various environments or stresses.

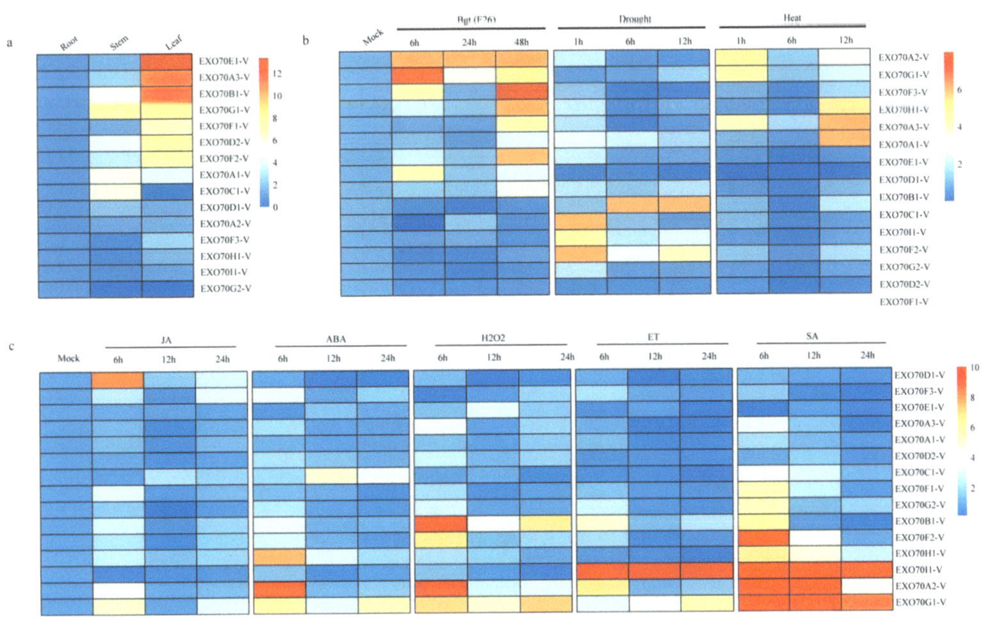

Figure 6. Heat map of the expression profiling of wheat and H. villosa EXO70 genes in tissues and in response to biotic/abiotic stress, phytohormones and H2O2 treatments. (a) Tissue-specific expression pattern of 15 EXO70-V genes in

H. villosa. (b) Expression levels of EXO70-V genes in biotic/abiotic stresses of H. villosa. (c) Expression profiling of EXO70-V genes in response to phytohormones and H2O2 treatments. The scale bar showing expression level of the genes. Abbreviations: Bgt, powdery mildew; SA, salicylic acid; MeJA, methyl jasmonate; ET, ethephon; ABA, abscisic acid; H2O2 hydrogen peroxide.

2.7 The Subcellular Localization of EXO70s from H. villosa

Knowledge of the subcellular localization of a plant protein can help us predict its potential role in the biological process. The subcellular localization of EXO70-Vs was investigated by transiently expressing the construct into leaves of Nicotiana tabacum via Agrobacterium method. Eleven EXO70-Vs generated fluorescence signals. Compared with the relatively even distribution of GFP signals in the cell (Figure 7a), the 11 confusion proteins had distinct localization patterns. EXO70A1, A3 and F1-V displayed weak signals on the plasma membrane (PM) (Figure 7b–d), while the PM signals for EXO70C1-V and EXO70D2-V were more intensive (Figure 7e,f). EXO70B1, E1 and F3-V displayed signals both in the PM and the nucleus (Figure 7g–i). EXO70D1-V and EXO70F2-V also produced signals in the PM; in addition, they also had small and discrete spot signals in the PM (Figure 7j,k). EXO70I1-V was the only one with no continuous PM localized signal; however, we observed discrete punctate signals along the PM (Figure 7l).

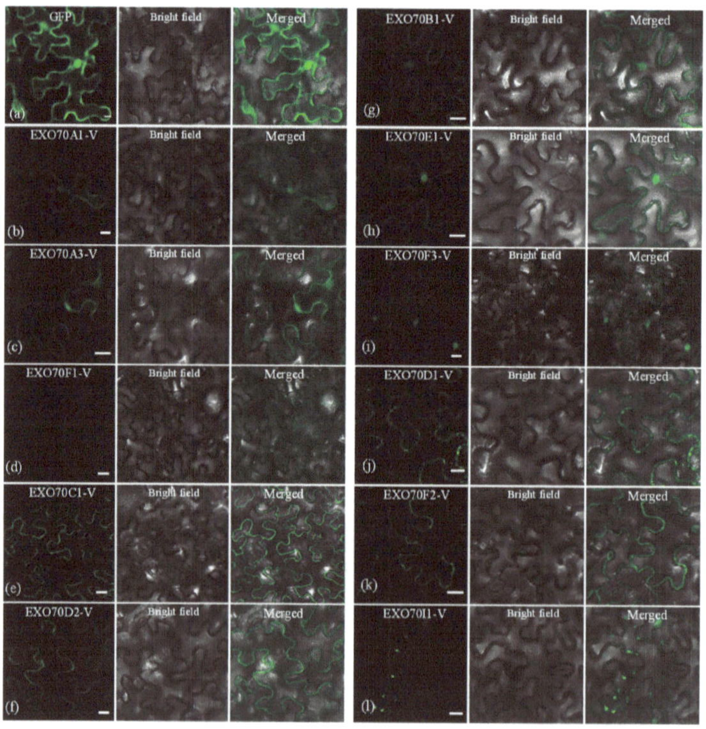

Figure 7. Subcellular localization of H. villosa EXO70-GFP-Vs proteins. H. villosa EXO70-GFP-Vs proteins were transiently expressed in N. benthamiana

leaves carried out with injection of Agrobacterium and examined by a confocal microscope. Green fluorescence was observed 48 h after infection. Bar = 20 μm. (a) The empty GFP vector was used as the control. The green channel shows that GFP signals were localized in the nucleus and cytoplasmic and plasma membranes. (b-l) The subcellular localization pattern of EXO70A1-V to EXO70I1-V, respectively.

3. DISCUSSION

3.1 Evolutionary Relationship of the EXO70 Gene Family in Wheat and Its Relatives

The evolutionary relationships of the EXO70 gene family (between wheat, T. urartu, Ae. tauschii, T. dicoccoides, H. vulgare and H. villosa) have been speculated about based on the total number, classification, chromosomal distribution and structure. The surveyed diploid species of seven chromosome pairs except for H. villosa all possessed 26 EXO70 genes, which suggested this gene family appeared before the divergence among Triticum species [52]. Allohexaploid wheat originated from two hybridizations between three diploid progenitors approximately 2.5–4.5 million years ago [53]. The number of EXO70 genes in tetraploid and common wheat (a total of 47 and 75, respectively) is approximately twice and three times as many as diploids, implying they have undergone one and two rounds of polyploidization events [54]. Although the polyploidization event induced rapid and extensive genetic and epigenetic changes in the genome which were related to a large range of molecular and physiological adjustment [55] as well as a significant loss of gene family members upon domestication [53], by comparing with diploid species, the EXO70 gene family did not go through a wide range of expansion or diminution in tetraploid wheat and allohexaploid wheat. This deduction is also supported by their chromosomal location analysis, which showed that 73.1% of genes have good collinearity among wheat, Ae. tauschii, T. dicoccoides and H. vulgare, and most of the same type of orthologous genes maintain the relative order of their ancestral genes (Figures 3 and 4). As for the small difference in the number of genes on individual subgroups among wheat and its relatives, this may be because of the quality of the genome assembling or the gene duplication of subgroup EXO70I.

Phylogenetic analysis showed that all three groups (EXO70.1, EXO70.2 and EXO70.3) and nine subgroups (EXO70A to EXO70I) are represented in each of the six Triticeae species. A similar gene structure was found in the same subgroup; subgroup EXO70A consists of multiple introns, while the other eight subgroups had fewer or were intronless. The variable intron numbers confirmed the classifications of the EXO70 genes. Additionally, EXO70 subgroups diversified before the divergence within polyploid wheat and related species during the evolutionary process of the EXO70 gene family; however, no new groups/subgroups have emerged.

EXO70I members were most represented in wheat and its five relatives (57), as well as in rice (16), but not in Arabidopsis [6,36]. EXO70I belongs to EXO70.2, not EXO70.3, which is different from what was found in previous studies [1,7,36]. This suggests that the EXO70I subgroup arose before the evolutionary divergence of rice from other Triticeae crops and disappeared during the evolution of Arabidopsis. The EXO70I subgroup underwent rapid divergence, producing a large number of members; this event can probably be explained by unequal cross-over or segmental chromosomal duplication [56,57]. During long-term natural selection, numerous EXO70 genes diverged and evolved in order to respond to various conditions. The study of Arabidopsis showed that the duplicated gene loss process is non-random; those involved in DNA repair are more likely to be lost, while genes involved in signal transduction and transcription have been preferentially retained [56]. Therefore, the function of EXO70I subgroup in Arabidopsis was inclined to responses to DNA repair, and in grass species may participate in signal transduction. Research on a larger range of species is needed to figure out whether the EXO70I branch is unique to monocotyledons.

3.2 Diversification of Subcellular Localization Pattern of the EXO70 in H. villosa

Protein subcellular localization analysis provided important clues to their specialized biological functions [58]. The diversification of EXO70-V subcellular localization patterns implies functional differentiation. Except for EXO70I1-V, all EXO70 genes showed plasma membrane (PM) signals. The EXO70D1/F2-V-GFP locates to the PM merged with some small, discrete punctate. At the same time, EXO70I1-V-GFP only gave rise to smaller fluorescent discrete punctate along with PM, which are similar to AtEXO70E2 in Arabidopsis protoplasts, which was a maker of a novel double-membraned structure termed EXPO (exocyst-positive organelles). AtEXO70E2 was involved in unconventional protein secretion for cytosolic proteins that lack a signal peptide, because of its ability to recruit several other exocyst complex subunits [8,9,59,60]. Therefore, EXO70D1, F2 and I1-V might have the ability to recruit different partners, then form various complexes to execute different biological functions. AtEXO70A1 was distributed in different patterns in different systems' cytosol; it showed up in the nucleus and numerous small punctate structures in the BY-2 cell [36], at the apex of growing tobacco pollen tubes [61] and is strongly present in the cell plate [62]. In the study, EXO70A1/A3-V showed a weak PM signal, while EXO70A2-V was characterized by mis-localizations. This was probably because AtEXO70A1 took part in a different vesicular transport process. Moreover, despite both AtEXO70B2 and AtEXO70H1 participating in the interaction between plants and pathogens, the signal of EXO70B2-GFP was mainly found in the cytoplasm, while EXO70H1-GFP was in vesicle-like structures in

Nicotiana benthamiana leaf [31]. In our analysis, EXO70B1-V was present in the PM and nucleus (Figure 7g). It is likely that they went through different action sites to take part in the process of disease resistance. An exocyst is a tethering factor that mediates secretory vesicles to the plasma membrane before SNARE-mediated fusion [63,64]. EXPOs deliver cytosolic proteins to the cell surface [65,66] and therefore all of those were related to PM. The results explained why most genes had the PM location pattern.

3.3 Function Conserve or Differentiation of the EXO70 Gene Family in Common Wheat and H. villosa

An orthologous gene is one that diverged after evolution to give rise to different species; this gene generally maintains a similar function to that of the ancestral gene that it evolved from [67]. In our study, some EXO70 orthologous genes from H. villosa exhibited a similar expression pattern to common wheat. For example, EXO70A3-V and TaEXO70A3 showed a high expression level under heat stress; EXO70B1-V was preferred to TaEXO70B1-3AL, which was induced by Bgt at a late stage (48 h), but not by drought and heat stress; EXO70E1-V and TaEXO70E1-3B were upregulated by Bgt treatment; the expression of EXO70H1-V and TaEXO70H1-1DL was increased in response to heat (Figures 5 and 6). Therefore, it is reasonable to presume that some EXO70 genes from H. villosa may have a similar function to the corresponding EXO70 genes from common wheat. In plants, AtEXO70B1, AtEXO70B2, AtEXO70H1 and OsEXO70E1 are known for their roles in innate immunity [6,27–31, 34]. In the literature, genes such as TaEXO70B1/B2 (with their homologous alleles), EXO70B1-V, TaEXO70E1-3B, EXO70E1-V, TaEXO70H1-1DL and EXO70H1-V were induced by Pst/Bgt treatment. Thus, we hypothesize that those genes also play an important role in plant defense responses and it is worth conducting further study to prove their function.

The long-term evolutionary fate of paralogous genes will still be determined by functions, with the genes that appeared to be sub-functionalized or neo-functionalized probably having higher rates of gene birth because of the increased adaptability. In contrast, the functional redundancy gene is unlikely to be stably maintained in the genome [55,57,68,69]. Paralogous/orthologous genes may diverge in expression to achieve more complex control of the same genetic network, balancing the relationship between internal growth and external environmental stimuli so that they "pay" the least and get the most [70]. In Arabidopsis, EXO70C1/C2 were involved in pollen development and mainly localized pollen related tissue [21,22]. In common wheat, six TaEXO70C genes had an undetectable expression level in roots/stem/leaves (Figure 5a), but EXO70C1-V showed a moderate expression level in the stem and responded to drought and ABA treatment (Figure 6). Studies have shown that ABA has contributed to osmotic stress tolerance by regulating stomatal aperture and guard cells [71]. Therefore, EXO70C1-V perhaps plays an important role in drought tolerance. In rice, OsExo70F3 interacts with AVR-Pii and plays a crucial role in triggered immunity [35]. In our research, including the copy number, 13 TaEXO70F members were identified and EXO70F1-V, EXO70F2-V and EXO70F3-V were cloned from H. villosa. Expression studies have revealed that 11/13 common wheat varieties were induced by stress treatment (Figure 5), and three EXO70-V genes showed clearly diverse expression patterns, of which EXO70F1-V was only induced by SA, EXO70F2-V in response to drought, H2O2 and SA at an early stage, and EXO70F3-V in response to Bgt treatment at a late stage (48 h) (Figure 6). Research shows that both SA and H2O2 function as a key regulator against pathogens and stress tolerance [72–74]. Thus, we can infer the function of EXO70F genes not only in plant defense responses but also in abiotic stress.

In N. benthamiana, EXO70D and EXO70G mainly affect the size of the leaf [6]. In wheat, TaEXO70D2-7BS and TaEXO70G1/G2 (with their homeoalleles) were induced by Pst and Bgt treatments. EXO70D1-V was upregulated by Bgt and MeJA treatments, EXO70G1-V had an increased expression level under phytohormones and H2O2. treatments, while EXO70G2-V was induced by drought and SA. MeJA is important for regulating the growth of plants and promotes plant resistance of various stresses [75]. This might suggest that EXO70D1-V plays an important role in plant growth and defense responses EXO70G are multifunctional. Of 22 EXO70I genes from common wheat, only five genes had distinct inducible expression. For instance, TaEXO70I6-7DL and TaEXO70I8-4BL were upregulated by drought and heat, respectively. EXO70I1-V was induced by ET and SA and maintained a high expression level. ET and SA regulate many diverse metabolic and developmental processes in plants, such as seed germination, abiotic stress response and pathogen defense [76,77]. Thus, we hypothesize that EXO70I1-V might play a vital role in growth or against multiple stresses. EXO70 genes provide diverse expression patterns in different tissues and stresses, implying that EXO70 genes may play an essential role in plant adaptation to a complicated and changeable environment.

4. MATERIALS AND METHODS

4.1 Plant Materials

H. villosa (genome VV, accession no. 91C43), from the Cambridge Botanical Garden, Cambridge, UK, was used for gene cloning and expression analysis. Powdery mildew susceptible variety Sumai 3 was used for propagation of fresh spores of powdery mildew isolate E26. Nicotiana benthamiana plants were used for subcellular localization analysis. All the materials were grown in a greenhouse under a 14 h light/10 h dark cycle at 24 ◦C/18 ◦C, with 70% relative air humidity.

4.2 Plant Treatments

The seedlings of H. villosa were grown in liquid or soil until the three-leaf stage. For heat shock or drought stress treatment, the plants were transferred to 42 □C conditions, or dipped into 20% PEG 6000 and leaves were sampled at 0, 1, 6, or 12 h after treatment. For powdery mildew treatment, the plant was inoculated with pathogen isolate E26 and the leaf tissues were sampled at 0, 24, 48 and 72 h after inoculation. For phytohormones and H2O2 treatments, the plants were sprayed with 5 mmol salicylic acid (SA), 0.1 mmol methyl jasmonate (MeJA), 0.1 mmol ethephon (ET), 0.2 mmol abscisic acid (ABA) and 7 mmol hydrogen peroxide H2O2 respectively and all leaf tissues were collected at 0, 6, 12, or 24 h after spraying [78]. All the samples were rapidly frozen in liquid nitrogen, then stored in an ultra-freezer (-80 °C) before use.

4.3 RNA Isolation and Real-Time PCR Analysis

Total RNA was extracted using a Trizol Reagent kit (Invitrogen, CA, USA) according to the manufacturer's instructions and analyzed by gel electrophoresis. The first-strand cDNA was synthesized with random oligonucleotides using the HiScript® II Reverse Transcriptase system (Vazyme, Nanjing, China). qRT-PCR was carried out in a total volume of 20 μL containing 2 μL of cDNA, 0.4 μL gene-specific primers (10 μM), 10 μL SYBR Green Mix and 7.2 μL of RNase free ddH2O, using the Roche LightCycler480 Real-time System (Roche, Basel, Swiss Confederation). The expression was represented in the form of relative fold change using the 2-ΔΔCT method [79]. Differentially expressed genes between each two samples pair were defined as two-fold up-regulated or two-fold down-regulated genes. Primers used for qRT-PCR are designed by Primer3 (Table S3). Three biological replications were performed. Heat map analysis of the expression data was performed using heat map drawing software MeV (version No. 4.7, Institute for Genomic Research, MD, USA)

4.4 Identification of EXO70 Gene Families in Wheat and Related Triticeae Species

We searched for the keywords 'exocyst subunit exo70 family protein' in the annotated proteins database of Hordeum vulgare (HH, 2n = 2x = 14, accession No. FJWB02000000) and obtained entries containing gene ID and protein sequences [42]. Then we performed a Conserved Domains (CD) search (https://www.ncbi.nlm.nih.gov/Structure/cdd/wrpsb.cgi) and reserved the protein sequences that contain the typical pfam03081 domain [47].

The identified EXO70 protein sequences of H. vulgare were used as query sequences to blast (E-value ≤ 10-10) against protein database of the other species, including Triticum urartu (AA, 2n = 2x = 14, accession No. NMPL02000000) [39], T. aestivum

(AABBDD, 2n = 6x = 42, accession No. NMPL02000000) [80], Brachypodium distachyon (BdBd, 2n = 2x = 14, accession No. ADDN03000000) [81], Oryza sativa Japonica (2n = 2x = 24, accession No. AP008207–AP008218) [82], T. dicoccoides (AABB, 2n = 4x = 28, accession No. FXXJ01000000) [43] and Aegilops tauschii (DD, 2n = 2x = 14, accession No. AOCO02000000) [83]. After removing the redundant gene sequences for each the species, the alignment hits were validated by performing a CD search as described above.

4.5 Cloning and Protein Sequences Analysis of EXO70 Genes from Haynaldia villosa

According to the sequences obtained from the database of H. vulgare, primers (Table S3) for cloning the full-length cDNA of the EXO70 gene from Haynaldia villosa were designed with online software Primer3 designing tool (v. 0.4.0, University of California, USA) [84] (Table S3). Mixed root, stem and leaf tissue cDNA of H. villosa served as a template for the isolation. This was performed at 95 ◦C for 30 s, followed by 35 cycles of 95 ◦C for 15 s, 58 ◦C for 15 s or 30 s and 72 ◦C for 3 min and then by 5 min at 72 ◦C in Phanta Max Super-Fidelity DNA polymerase (Vazyme, Nanjing, China). Before subcloning into their destination vectors, the PCR-amplified cDNA products were first cloned into the pTOPO-Blunt Vector (Aidlab, Beijing, China) as per the manufacturer's instructions. Multiple sequence alignments were conducted using DNAMan (Lynnon Corporation, Quebec, QC, Canada) software. The sequences similarity was visualized using the R programming language.

4.6 Subcellular Localization Assay

The ORFs of EXO70-V genes (without stop codon) were amplified from the pTOPO-Blunt Vector, then inserted into the pCambia1305-GFP vector, which contains a green fluorescent protein (GFP) reporter gene driven by the CaMV 35S promoter, using homologous cloning technology as per the manufacturer's instructions (Vazyme, Nanjing, China) (Table S3). Then it was introduced into Agrobacterium tumefaciens (strain GV3101) bacteria by a freeze–thaw procedure and grown in Luria-Bertani (LB) medium at 28 ◦C for 2 or 3 d.

Agrobacterium tumefaciens (strain GV3101) bacteria containing fusion constructed were grown in Luria-Bertani (LB) medium with both rifampicin and kanamycin (0.05 µg/mL) at 28 ◦C overnight. The bacterial cells were centrifuged and resuspended in an infiltration solution (10 mM MES pH 5.6, 0.1 mM Acetosyringone, 10 mM MgCl2) to a final OD600 = 1.5. Bacterial suspensions were infiltrated into five- to six-week growing stage leaves of N. benthamiana by depressing the plunger of a 1-mL disposable needleless syringe into the abaxial side of leaves [85,86]. The fluorescence signals were observed 48–60 h after injection and images were captured using a confocal laser scanning microscope (LSM780; Carl Zeiss, Jena, Germany) according to the methods described by Wang et al. [87].

4.7 Phylogenetic Analysis of EXO70 Gene Family

Multiple sequence alignment was conducted by ClustalW which was integrated in Mega v6.0 [88]. Phylogenetic analysis was performed through online software PhyML 3.0 [89] using maximum-likelihood method with default parameter [90]. EXO70 proteins are rather diverse at their N-terminal and could not be aligned reliably, so we only used the conserved domain proteins to construct the phylogenetic tree and removed five genes where the length of the domain is fewer than 200 amino acids for further analysis.

4.8 Chromosomal Distribution and Exon-Intron Structure Analysis

Chromosomal information of predicted EXO70 genes from each species was obtained after using cDNA sequences as a query sequence blasted to the genomic sequence to determine their chromosomal locations. Then we drew their locations onto the physical map of each chromosome using MapInspect tool (http://mapinspect.software.informer.com/).

The gff3 files of each species was downloaded from the Ensembl Plants FTP server (http://plants. ensembl.org/index.html) on 20 January 2018 for exon–intron structure analysis; the image of the exon–intron structure was obtained using the online Gene Structure Display Server (last accessed date on 20 January 2018) with the gff3 files for each species [91]. The corresponding evolutionary tree were constructed by Mega v6.0 [88], all sequences were aligned by ClustalW using the default parameters [92,93], used the Neighbor-Joining method with the pairwise deletion option, Poisson correction and bootstrap analysis conducted with 1000 replicates [26,94]

4.9 RNA-seq Expression Analysis

Publicly available RNA-seq data were retrieved from the expVIP [50] were used to analyze the expression pattern of predicted wheat EXO70 genes in different tissues and stresses. The tissue-specific expression data were compiled from three wheat tissues (leaf, stem, root) collected from Chinese Spring at seeding development. The biotic stress expression data included two diseases (stripe rust and powdery mildew pathogen) and were collected from disease-resistant wheat varieties N9134 (at 7 days seedling stage). The abiotic stress expression data, including drought and heat treatments, were collected from heat-resistant wheat cultivar TAM107 (at 7 days seedling stage). The relative expression of each TaEXO70 gene in different tissues and stresses was presented as a heat map, which was constructed by the heat map drawing software MeV (version No. 4.7, Institute for Genomic Research, MD, USA).

REFERENCES

1. Cvrckova, F.; Grunt, M.; Bezvoda, R.; Hala, M.; Kulich, I.; Rawat, A.; Zarsky, V. Evolution of the land plant exocyst complexes. Front. Plant Sci. 2012, 3, 159. [CrossRef] [PubMed]
2. He, B.; Guo, W. The exocyst complex in polarized exocytosis. Curr. Opin. Cell Biol. 2009, 21, 537–542. [CrossRef] [PubMed]
3. Ma, W.; Wang, Y.; Yao, X.; Xu, Z.; An, L.; Yin, M. The role of Exo70 in vascular smooth muscle cell migration. Cell. Mol. Biol. Lett. 2016, 21, 20. [CrossRef] [PubMed]
4. Elias, M. The exocyst complex in plants. Cell Biol. Int. 2003, 27, 199–201. [CrossRef]
5. Tu, B.; Hu, L.; Chen, W.; Li, T.; Hu, B.; Zheng, L.; Lv, Z.; You, S.; Wang, Y.; Ma, B.; et al. Disruption of OsEXO70A1 Causes Irregular Vascular Bundles and Perturbs Mineral Nutrient Assimilation in Rice. Sci. Rep. 2015, 5, 18609. [CrossRef] [PubMed]
6. Du, Y.; Overdijk, E.J.R.; Berg, J.A.; Govers, F.; Bouwmeester, K. Solanaceous exocyst subunits are involved in immunity to diverse plant pathogens. J. Exp. Bot. 2018, 69, 655–666. [CrossRef]
7. Synek, L.; Schlager, N.; Elias, M.; Quentin, M.; Hauser, M.T.; Zarsky, V. AtEXO70A1, a member of a family of putative exocyst subunits specifically expanded in land plants, is important for polar growth and plant development. Plant J. 2006, 48, 54–72. [CrossRef]
8. Wang, J.; Ding, Y.; Wang, J.; Hillmer, S.; Miao, Y.; Lo, S.W.; Wang, X.; Robinson, D.G.; Jiang, L. EXPO, an exocyst-positive organelle distinct from multivesicular endosomes and autophagosomes, mediates cytosol to cell wall exocytosis in Arabidopsis and tobacco cells. Plant Cell 2010, 22, 4009–4030. [CrossRef]
9. Ding, Y.; Wang, J.; Chun Lai, J.H.; Ling Chan, V.H.; Wang, X.; Cai, Y.; Tan, X.; Bao, Y.; Xia, J.; Robinson, D.G.; et al. Exo70E2 is essential for exocyst subunit recruitment and EXPO formation in both plants and animals. Mol. Biol. Cell 2014, 25, 412–426. [CrossRef]
10. Dellago, H.; Loscher, M.; Ajuh, P.; Ryder, U.; Kaisermayer, C.; Grillari-Voglauer, R.; Fortschegger, K.; Gross, S.; Gstraunthaler, A.; Borth, N.; et al. Exo70, a subunit of the exocyst complex, interacts with SNEV(hPrp19/hPso4) and is involved in pre-mRNA splicing. Biochem. J. 2011, 438, 81–91. [CrossRef]
11. Ren, J.; Guo, W. ERK1/2 regulate exocytosis through direct phosphorylation of the exocyst component Exo70. Dev. Cell 2012, 22, 967–978. [CrossRef]

12. Zhao, Y.; Liu, J.; Yang, C.; Capraro, B.R.; Baumgart, T.; Bradley, R.P.; Ramakrishnan, N.; Xu, X.; Radhakrishnan, R.; Svitkina, T.; et al. Exo70 generates membrane curvature for morphogenesis and cell migration. Dev. Cell 2013, 26, 266–278. [CrossRef] [PubMed]

13. Zuo, X.; Zhang, J.; Zhang, Y.; Hsu, S.C.; Zhou, D.; Guo, W. Exo70 interacts with the Arp2/3 complex and regulates cell migration. Nat. Cell Biol. 2006, 8, 1383–1388. [CrossRef] [PubMed]

14. Liu, J.; Yue, P.; Artym, V.V.; Mueller, S.C.; Guo, W. The role of the exocyst in matrix metalloproteinase secretion and actin dynamics during tumor cell invadopodia formation. Mol. Biol. Cell 2009, 20, 3763–3771. [CrossRef] [PubMed]

15. Xiao, L.; Zheng, K.; Lv, X.; Hou, J.; Xu, L.; Zhao, Y.; Song, F.; Fan, Y.; Cao, H.; Zhang, W.; et al. Exo70 is an independent prognostic factor in colon cancer. Sci. Rep. 2017, 7, 5039. [CrossRef] [PubMed]

16. Li, S.; Chen, M.; Yu, D.; Ren, S.; Sun, S.; Liu, L.; Ketelaar, T.; Emons, A.M.; Liu, C.M. EXO70A1-mediated vesicle trafficking is critical for tracheary element development in Arabidopsis. Plant Cell 2013, 25, 1774–1786. [CrossRef] [PubMed]

17. Vukasinovic, N.; Oda, Y.; Pejchar, P.; Synek, L.; Pecenkova, T.; Rawat, A.; Sekeres, J.; Potocky, M.; Zarsky, V. Microtubule-dependent targeting of the exocyst complex is necessary for xylem development in Arabidopsis. New Phytol. 2017, 213, 1052–1067. [CrossRef]

18. Kulich, I.; Cole, R.; Drdova, E.; Cvrckova, F.; Soukup, A.; Fowler, J.; Zarsky, V. Arabidopsis exocyst subunits SEC8 and EXO70A1 and exocyst interactor ROH1 are involved in the localized deposition of seed coat pectin. New Phytol. 2010, 188, 615–625. [CrossRef] [PubMed]

19. Drdova, E.J.; Synek, L.; Pecenkova, T.; Hala, M.; Kulich, I.; Fowler, J.E.; Murphy, A.S.; Zarsky, V. The exocyst complex contributes to PIN auxin efflux carrier recycling and polar auxin transport in Arabidopsis. Plant J. 2013, 73, 709–719. [CrossRef] [PubMed]

20. Kalmbach, L.; Hematy, K.; De Bellis, D.; Barberon, M.; Fujita, S.; Ursache, R.; Daraspe, J.; Geldner, N. Transient cell-specific EXO70A1 activity in the CASP domain and Casparian strip localization. Nat. Plants 2017, 3, 17058. [CrossRef]

21. Lai, K.S. Analysis of EXO70C2 expression revealed its specific association with late stages of pollen development. Plant Cell Tissue Organ Cult. 2015, 124, 209–215. [CrossRef]

22. Synek, L.; Vukasinovic, N.; Kulich, I.; Hala, M.; Aldorfova, K.; Fendrych, M.; Zarsky, V. EXO70C2 Is a Key Regulatory Factor for Optimal Tip Growth of Pollen. Plant Physiol. 2017, 174, 223–240. [CrossRef] [PubMed]

23. Chen, C.; Liu, M.; Jiang, L.; Liu, X.; Zhao, J.; Yan, S.; Yang, S.; Ren, H.; Liu, R.; Zhang, X. Transcriptome profiling reveals roles of meristem regulators and polarity genes during fruit trichome development in cucumber (Cucumis sativus L.). J. Exp. Bot. 2014, 65, 4943–4958. [CrossRef] [PubMed]

24. Kulich, I.; Vojtikova, Z.; Glanc, M.; Ortmannova, J.; Rasmann, S.; Zarsky, V. Cell wall maturation of Arabidopsis trichomes is dependent on exocyst subunit EXO70H4 and involves callose deposition. Plant Physiol. 2015, 168, 120–131. [CrossRef] [PubMed]

25. Wang, Z.; Li, P.; Yang, Y.; Chi, Y.; Fan, B.; Chen, Z. Expression and Functional Analysis of a Novel Group of Legume-specific WRKY and Exo70 Protein Variants from Soybean. Sci. Rep. 2016, 6, 32090. [CrossRef] [PubMed]

26. Yang, Y.; Zhou, Y.; Chi, Y.; Fan, B.; Chen, Z. Characterization of Soybean WRKY Gene Family and Identification of Soybean WRKY Genes that Promote Resistance to Soybean Cyst Nematode. Sci. Rep. 2017, 7, 17804. [CrossRef] [PubMed]

27. Kulich, I.; Pecenkova, T.; Sekeres, J.; Smetana, O.; Fendrych, M.; Foissner, I.; Hoftberger, M.; Zarsky, V. Arabidopsis exocyst subcomplex containing subunit EXO70B1 is involved in autophagy-related transport to the vacuole. Traffic 2013, 14, 1155–1165. [CrossRef]

28. Stegmann, M.; Anderson, R.G.; Westphal, L.; Rosahl, S.; McDowell, J.M.; Trujillo, M. The exocyst subunit Exo70B1 is involved in the immune response of Arabidopsis thaliana to different pathogens and cell death. Plant Signal. Behav. 2013, 8, e27421. [CrossRef]

29. Zhao, T.; Rui, L.; Li, J.; Nishimura, M.T.; Vogel, J.P.; Liu, N.; Liu, S.; Zhao, Y.; Dangl, J.L.; Tang, D. A truncated NLR protein, TIR-NBS2, is required for activated defense responses in the exo70B1 mutant. PLoS Genet. 2015, 11, e1004945. [CrossRef]

30. Stegmann, M.; Anderson, R.G.; Ichimura, K.; Pecenkova, T.; Reuter, P.; Zarsky, V.; McDowell, J.M.; Shirasu, K.; Trujillo, M. The ubiquitin ligase PUB22 targets a subunit of the exocyst complex required for PAMP-triggered responses in Arabidopsis. Plant Cell 2012, 24, 4703–4716. [CrossRef]

31. Pecenkova, T.; Hala, M.; Kulich, I.; Kocourkova, D.; Drdova, E.; Fendrych, M.; Toupalova, H.; Zarsky, V. The role for the exocyst complex subunits Exo70B2 and Exo70H1 in the plant-pathogen interaction. J. Exp. Bot. 2011, 62, 2107–2116. [CrossRef] [PubMed]

32. Seo, D.H.; Ahn, M.Y.; Park, K.Y.; Kim, E.Y.; Kim, W.T. The N-Terminal UND Motif of the Arabidopsis U-Box E3 Ligase PUB18 Is Critical for the Negative Regulation of ABA-Mediated Stomatal Movement and Determines Its Ubiquitination Specificity for Exocyst Subunit Exo70B1. Plant Cell 2016, 28, 2952–2973. [CrossRef] [PubMed]

33. Hong, D.; Jeon, B.W.; Kim, S.Y.; Hwang, J.U.; Lee, Y. The ROP2-RIC7 pathway negatively regulates light-induced stomatal opening by inhibiting exocyst subunit Exo70B1 in Arabidopsis. New Phytol. 2016, 209, 624–635. [CrossRef] [PubMed]

34. Guo, J.; Xu, C.; Wu, D.; Zhao, Y.; Qiu, Y.; Wang, X.; Ouyang, Y.; Cai, B.; Liu, X.; Jing, S.; et al. Bph6 encodes an exocyst-localized protein and confers broad resistance to planthoppers in rice. Nat. Genet. 2018, 50297–50306. [CrossRef] [PubMed]

35. Fujisaki, K.; Abe, Y.; Ito, A.; Saitoh, H.; Yoshida, K.; Kanzaki, H.; Kanzaki, E.; Utsushi, H.; Yamashita, T.; Kamoun, S.; et al. Rice Exo70 interacts with a fungal effector, AVR-Pii, and is required for AVR-Pii-triggered immunity. Plant J. 2015, 83, 875–887. [CrossRef] [PubMed]

36. Chong, Y.T.; Gidda, S.K.; Sanford, C.; Parkinson, J.; Mullen, R.T.; Goring, D.R. Characterization of the Arabidopsis thaliana exocyst complex gene families by phylogenetic, expression profiling, and subcellular localization studies. New Phytol. 2010, 185, 401–419. [CrossRef]

37. Marcussen, T.; Sandve, S.R.; Heier, L.; Spannagl, M.; Pfeifer, M.; Jakobsen, K.S.; Wulff, B.B.; Steuernagel, B.; Mayer, K.F.; Olsen, O.A. Ancient hybridizations among the ancestral genomes of bread wheat. Science 2014, 345, 1250092. [CrossRef]

38. Zimin, A.V.; Puiu, D.; Hall, R.; Kingan, S.; Clavijo, B.J.; Salzberg, S.L. The first near-complete assembly of the hexaploid bread wheat genome, Triticum aestivum. GigaScience 2017, 6, gix097. [CrossRef]

39. Ling, H.Q.; Ma, B.; Shi, X.; Liu, H.; Dong, L.; Sun, H.; Cao, Y.; Gao, Q.; Zheng, S.; Li, Y.; et al. Genome sequence of the progenitor of wheat A subgenome Triticum urartu. Nature 2018, 557, 424–428. [CrossRef]

40. Bettgenhaeuser, J.; Krattinger, S.G. Rapid gene cloning in cereals. Theor. Appl. Genet. 2018. [CrossRef]

41. Luo, M.C.; Gu, Y.Q.; Puiu, D.; Wang, H.; Twardziok, S.O.; Deal, K.R.; Huo, N.; Zhu, T.; Wang, L.; Wang, Y.; et al. Genome sequence of the progenitor of the wheat D genome Aegilops tauschii. Nature 2017, 551, 498–502. [CrossRef] [PubMed]

42. Mascher, M.; Gundlach, H.; Himmelbach, A.; Beier, S.; Twardziok, S.O.; Wicker, T.; Radchuk, V.; Dockter, C.; Hedley, P.E.; Russell, J.; et al. A chromosome conformation capture ordered sequence of the barley genome. Nature 2017, 544, 427–433. [CrossRef] [PubMed]

43. Avni, R.; Nave, M.; Barad, O.; Baruch, K.; Twardziok, S.O.; Gundlach, H.; Hale, I.; Mascher, M.; Spannagl, M.; Wiebe, K.; et al. Wild emmer genome architecture and diversity elucidate wheat evolution and domestication. Science 2017, 357, 93–97. [CrossRef] [PubMed]

44. Xing, L.; Qian, C.; Cao, A.; Li, Y.; Jiang, Z.; Li, M.; Jin, X.; Hu, J.; Zhang, Y.; Wang, X.; et al. The Hv-SGT1 gene from Haynaldia villosa contributes to resistances towards both biotrophic and hemi-biotrophic pathogens in common wheat (Triticum aestivum L.). PLoS ONE 2013, 8, e72571. [CrossRef]

45. Cao, A.; Xing, L.; Wang, X.; Yang, X.; Wang, W.; Sun, Y.; Qian, C.; Ni, J.; Chen, Y.; Liu, D.; et al. Serine/threonine kinase gene Stpk-V, a key member of powdery mildew resistance gene Pm21, confers powdery mildew resistance in wheat. Proc. Natl. Acad. Sci. USA 2011, 108, 7727–7732. [CrossRef] [PubMed]

46. Xing, L.; Hu, P.; Liu, J.; Witek, K.; Zhou, S.; Xu, J.; Zhou, W.; Gao, L.; Huang, Z.; Zhang, R.; et al. Pm21 from Haynaldia villosa Encodes a CC-NBS-LRR that Confers Powdery Mildew Resistance in Wheat. Mol. Plant 2018, 11, 874–878. [CrossRef] [PubMed]

47. Li, S.; van Os, G.M.; Ren, S.; Yu, D.; Ketelaar, T.; Emons, A.M.; Liu, C.M. Expression and functional analyses of EXO70 genes in Arabidopsis implicate their roles in regulating cell type-specific exocytosis. Plant Physiol. 2010, 154, 1819–1830. [CrossRef] [PubMed]

48. Miftahudin; Ross, K.; Ma, X.F.; Mahmoud, A.A.; Layton, J.; Milla, M.A.; Chikmawati, T.; Ramalingam, J.; Feril, O.; Pathan, M.S.; et al. Analysis of expressed sequence tag loci on wheat chromosome group 4. Genetics 2004, 168, 651–663. [CrossRef]

49. Hao, M.; Luo, J.; Zhang, L.; Yuan, Z.; Zheng, Y.; Zhang, H.; Liu, D. In situ hybridization analysis indicates that 4AL-5AL-7BS translocation preceded subspecies differentiation of Triticum turgidum. Genome 2013, 56, 303–305. [CrossRef]

50. Borrill, P.; Ramirez-Gonzalez, R.; Uauy, C. expVIP: A Customizable RNA-seq Data Analysis and Visualization Platform. Plant Physiol. 2016, 170, 2172–2186. [CrossRef]

51. Pearce, S.; Vazquez-Gross, H.; Herin, S.Y.; Hane, D.; Wang, Y.; Gu, Y.Q.; Dubcovsky, J. WheatExp: An RNA-seq expression database for polyploid wheat. BMC Plant Biol. 2015, 15, 299. [CrossRef] [PubMed]

52. Li, X.; Gao, S.; Tang, Y.; Li, L.; Zhang, F.; Feng, B.; Fang, Z.; Ma, L.; Zhao, C. Genome-wide identification and evolutionary analyses of bZIP transcription factors in wheat and its relatives and expression profiles of anther development related TabZIP genes. BMC Genom. 2015, 16, 976. [CrossRef] [PubMed]

53. Brenchley, R.; Spannagl, M.; Pfeifer, M.; Barker, G.L.; D'Amore, R.; Allen, A.M.; McKenzie, N.; Kramer, M.; Kerhornou, A.; Bolser, D.; et al. Analysis of the bread wheat genome using whole-genome shotgun sequencing. Nature 2012, 491, 705–710. [CrossRef] [PubMed]

54. Feldman, M.; Levy, A.A. Allopolyploidy–a shaping force in the evolution of wheat genomes. Cytogenet. Genome Res. 2005, 109, 250–258. [CrossRef] [PubMed]

55. Feldman, M.; Levy, A.A. Genome evolution in allopolyploid wheat—A revolutionary reprogramming followed by gradual changes. J. Genet. Genom. 2009, 36, 511–518. [CrossRef]

56. Blanc, G.; Wolfe, K.H. Functional divergence of duplicated genes formed by polyploidy during Arabidopsis evolution. Plant Cell 2004, 16, 1679–1691. [CrossRef] [PubMed]

57. Zhang, J. Evolution by gene duplication: An update. Trends Ecol. Evol. 2003, 18, 292–298. [CrossRef]

58. Sperschneider, J.; Catanzariti, A.M.; DeBoer, K.; Petre, B.; Gardiner, D.M.; Singh, K.B.; Dodds, P.N.; Taylor, J.M. LOCALIZER: Subcellular localization prediction of both plant and effector proteins in the plant cell. Sci. Rep. 2017, 7, 44598. [CrossRef]

59. Lin, Y.S.; Ding, Y.; Wang, J.; Shen, J.B.; Kung, C.H.; Zhuang, X.H.; Cui, Y.; Yin, Z.; Xia, Y.J.; Lin, H.X.; et al. Exocyst-Positive Organelles and Autophagosomes Are Distinct Organelles in Plants. Plant Physiol. 2015, 169, 1917–1932.

60. Robinson, D.G.; Ding, Y.; Jiang, L. Unconventional protein secretion in plants: A critical assessment. Protoplasma 2016, 253, 31–43. [CrossRef]

61. Hala, M.; Cole, R.; Synek, L.; Drdova, E.; Kulich, I.; Pecenkova, T.; Hochholdinger, F.; Cvrckova, F.; Fowler, J.; Zarsky, V. Exocyst complex functions in plant development. Comp. Biochem. Phys. A 2008, 150, S188–S189. [CrossRef]

62. Fendrych, M.; Synek, L.; Pecenkova, T.; Toupalova, H.; Cole, R.; Drdova, E.; Nebesarova, J.; Sedinova, M.; Hala, M.; Fowler, J.E.; et al. The Arabidopsis exocyst complex is involved in cytokinesis and cell plate maturation. Plant Cell 2010, 22, 3053–3065. [CrossRef] [PubMed]

63. Wu, B.; Guo, W. The Exocyst at a Glance. J. Cell Sci. 2015, 128, 2957–2964. [CrossRef] [PubMed]

64. Lurick, A.; Kummel, D.; Ungermann, C. Multisubunit tethers in membrane fusion. Curr. Biol. 2018, 28, R417–R420. [CrossRef] [PubMed]

65. Chi, Y.; Yang, Y.; Li, G.; Wang, F.; Fan, B.; Chen, Z. Identification and characterization of a novel group of legume-specific, Golgi apparatus-localized WRKY and Exo70 proteins from soybean. J. Exp. Bot. 2015, 66, 3055–3070. [CrossRef] [PubMed]

66. Ebine, K.; Ueda, T. Roles of membrane trafficking in plant cell wall dynamics. Front. Plant Sci. 2015, 6, 878. [CrossRef] [PubMed]

67. Koonin, E.V. Orthologs, paralogs, and evolutionary genomics. Annu. Rev. Genet. 2005, 39, 309–338. [CrossRef] [PubMed]

68. Panchy, N.; Lehti-Shiu, M.; Shiu, S.H. Evolution of Gene Duplication in Plants. Plant Physiol. 2016, 171, 2294–2316. [CrossRef]

69. Roulin, A.; Auer, P.L.; Libault, M.; Schlueter, J.; Farmer, A.; May, G.; Stacey, G.; Doerge, R.W.; Jackson, S.A. The fate of duplicated genes in a polyploid plant genome. Plant J. 2013, 73, 143–153. [CrossRef] [PubMed]

70. Liang, Z.; Schnable, J.C. Functional Divergence Between Subgenomes and Gene Pairs After Whole Genome Duplications. Mol. Plant. 2018, 11, 388–397. [CrossRef]

71. Fernando, V.C.D.; Schroeder, D.F. Role of ABA in Arabidopsis Salt, Drought, and Desiccation Tolerance. In Abiotic and Biotic Stress in Plants—Recent Advances and Future Perspectives; IntechOpen: London, UK, 2016; Chapter 22. Kumar, D. Salicylic acid signaling in disease resistance. Plant Sci. 2014, 228, 127–134. [CrossRef] [PubMed]

72. Maity, A.; Sharma, J.; Sarkar, A.; More, A.K.; Pal, R.K.; Nagane, V.P.; Maity, A. Salicylic acid mediated multi-pronged strategy to combat bacterial blight disease (Xanthomonas axonopodis pv. punicae) in pomegranate. Eur. J. Plant Pathol. 2018, 150, 923–937. [CrossRef]

73. You, J.; Chan, Z. ROS Regulation During Abiotic Stress Responses in Crop Plants. Front. Plant Sci. 2015, 6, 1092. [CrossRef] [PubMed]

74. Chung, H.S.; Howe, G.A. A critical role for the TIFY motif in repression of jasmonate signaling by a stabilized splice variant of the JASMONATE ZIM-domain protein JAZ10 in Arabidopsis. Plant Cell. 2009, 21, 131–145. [CrossRef] [PubMed]

75. Vlot, A.C.; Dempsey, D.A.; Klessig, D.F. Salicylic Acid, a multifaceted hormone to combat disease. Annu. Rev. Phytopathol. 2009, 47, 177–206. [CrossRef] [PubMed]

76. Bleecker, A.B.; Kende, H. Ethylene: A gaseous signal molecule in plants. Annu. Rev. Cell Dev. Biol. 2000, 16, 1–18. [CrossRef] [PubMed]

77. Zhu, Y.; Li, Y.; Fei, F.; Wang, Z.; Wang, W.; Cao, A.; Liu, Y.; Han, S.; Xing, L.; Wang, H.; et al. E3 ubiquitin ligase gene CMPG1-V from Haynaldia villosa L. contributes to powdery mildew resistance in common wheat (Triticum aestivum L.). Plant J. 2015, 84, 154–168. [CrossRef]

78. Livak, K.J.; Schmittgen, T.D. Analysis of relative gene expression data using real-time quantitative PCR and the 2-ΔΔCT Method. Methods 2001, 25, 402–408. [CrossRef]

79. Mayer, K.F.X.; Rogers, J.; Doležel, J.; Pozniak, C.; Eversole, K.; Feuillet, C.; Gill, B.; Friebe, B.; Lukaszewski, A.J.; Sourdille, P. A chromosome-based draft sequence of the hexaploid bread wheat (Triticum aestivum) genome. Science 2014, 345, 1251788.

80. International Brachypodium Initiative. Genome sequencing and analysis of the model grass Brachypodium distachyon. Nature 2010, 463, 763–768. [CrossRef]

81. Fujisawa, M.; Baba, T.; Nagamura, Y.; Nagasaki, H.; Waki, K.; Vuong, H.; Matsumoto, T.; Wu, J.Z.; Kanamori, H.; Katayose, Y. The map-based sequence of the rice genome. Nature 2005, 436, 793–800.

82. Zhao, G.; Zou, C.; Li, K.; Wang, K.; Li, T.; Gao, L.; Zhang, X.; Wang, H.; Yang, Z.; Liu, X. The Aegilops tauschii

83. genome reveals multiple impacts of transposons. Nat. Plants 2017, 3, 946–955. [CrossRef] [PubMed]

84. Andreas, U.; Ioana, C.; Triinu, K.; Jian, Y.; Brant, C.F.; Maido, R.; Steven, G.R. Primer3—New capabilities and interfaces. Nucleic Acids Res. 2012, 40, e115.

85. Chen, H.; Zou, Y.; Shang, Y.; Lin, H.; Wang, Y.; Cai, R.; Tang, X.; Zhou, J.M. Firefly luciferase complementation imaging assay for protein-protein interactions in plants. Plant Physiol. 2008, 146, 368–376. [CrossRef] [PubMed]

86. Wu, C.; Tan, L.; van Hooren, M.; Tan, X.; Liu, F.; Li, Y.; Zhao, Y.; Li, B.; Rui, Q.; Munnik, T.; et al. Arabidopsis EXO70A1 recruits Patellin3 to the cell membrane independent of its role as an exocyst subunit. J. Integr. Plant Biol. 2017, 59, 851–865. [CrossRef]

87. Wang, Z.; Cheng, J.; Fan, A.; Zhao, J.; Yu, Z.; Li, Y.; Zhang, H.; Xiao, J.; Muhammad, F.; Wang, H.; et al. LecRK-V, an L-type lectin receptor kinase in Haynaldia villosa, plays positive role in resistance to wheat powdery mildew. Plant Biotechnol. J. 2018, 16, 50–62. [CrossRef]

88. Tamura, K.; Stecher, G.; Peterson, D.; Filipski, A.; Kumar, S. MEGA6: Molecular Evolutionary Genetics Analysis version 6.0. Mol. Biol. Evol. 2013, 30, 2725–2729. [CrossRef]

89. Lefort, V.; Longueville, J.E.; Gascuel, O. SMS: Smart Model Selection in PhyML. Mol. Biol. Evol. 2017, 34, 2422–2424. [CrossRef]

90. Guindon, S.; Dufayard, J.F.; Lefort, V.; Anisimova, M.; Hordijk, W.; Gascuel, O. New Algorithms and Methods to Estimate Maximum-Likelihood Phylogenies: Assessing the Performance of PhyML 3.0. Syst. Biol. 2010, 59, 307–321. [CrossRef]

91. Hu, B.; Jin, J.; Guo, A.Y.; Zhang, H.; Luo, J.; Gao, G. GSDS 2.0: An upgraded gene feature visualization server. Bioinformatics 2015, 31, 1296–1297. [CrossRef]

92. Thompson, J.D.; Gibson, T.J.; Higgins, D.G. Multiple sequence alignment using ClustalW and ClustalX. Curr. Protoc. Bioinform. 2002. [CrossRef] [PubMed]

93. Larkin, M.A.; Blackshields, G.; Brown, N.P.; Chenna, R.; McGettigan, P.A.; McWilliam, H.; Valentin, F.; Wallace, I.M.; Wilm, A.; Lopez, R.; et al. Clustal W and Clustal X version 2.0. Bioinformatics 2007, 23, 2947–2948. [CrossRef] [PubMed]

94. Wang, M.; Yue, H.; Feng, K.; Deng, P.; Song, W.; Nie, X. Genome-wide identification, phylogeny and expressional profiles of mitogen activated protein kinase kinase kinase (MAPKKK) gene family in bread wheat (Triticum aestivum L.). BMC Genom. 2016, 17, 668. [CrossRef] [PubMed]

10
Chapter

MOLECULAR IDENTIFICATION AND CHARACTERIZATION OF HYDROXYCINNAMOYL TRANSFERASE IN CAMELLIA SINENSIS L.

1. INTRODUCTION: AN OVERVIEW

Following water, tea has been the most extensively consumed beverage globally for millennia due to its distinctive scent and flavor. Tea plants (Camellia sinensis L.) are rich in specific secondary metabolites, including polyphenolic chemicals, alkaloids, terpenoids, and amino acids, with the highest concentrations seen in tea plants [1]. Tea polyphenols constitute 30% of the dry mass of tea leaves. They can be approximately categorized into five groups: flavanols, flavonols, flavones, proanthocyanidins, and phenolic acids [2,3]. Hydroxycinnamoyl transferase (HCT) facilitates the transfer of hydroxycinnamoyl groups to receptor substrates, including shikimic acid, quinic acid, anthocyanins, flavanol glycosides, polyamines, and long-chain fatty acids. Plants subjected to environmental challenges can activate gene expression associated with the phenylpropanoid metabolic pathway to produce diverse secondary metabolites that confer resistance or adaptation to such stresses. The HCT in the phenylpropanoid pathway catalyzes shikimic acid and quinic acid, contributing to the upstream process of lignin production. Lignins in plant cell walls offer a physical barrier that safeguards polysaccharides in cell walls against microbial breakdown. Low temperature, elevated salinity, drought, mechanical injury, abscisic acid (ABA), salicylic acid (SA), and hydrogen peroxide can stimulate HcHCT expression in Hibiscus cannabinus. HcHCT enhances abiotic stress tolerance in plants [6]. In Cucumis sativus, HCT expression decreased following pectinase treatment. Furthermore, channeling the phenylpropanoid route to produce H-lignin resulted in the buildup of p-coumaraldehyde [7]. HCT is involved in the acylation of secondary metabolites and also modulates hypersensitive reactions (HRs) in plants. HCT1806 or HCT4918 in Zea mays interacts with Rp1-D21, which is translated from resistance genes, therefore limiting the development of HR. When pathogens invade plants, effectors released by the pathogens can alter the protein structure of HCT1806 or HCT4918, thereby affecting their interaction with Rp1-D21 and prompting the plants to initiate hypersensitive response (HR). This inhibits the dissemination of infections in localized cell necrosis [8]. HCT exhibits specificity for many substrates, including gentisate, 3-hydroxybenzoate, hydroxyanthranilate, and protocatechuate, and competes with shikimic acid or quinate acid for the enzyme's binding site, therefore yielding alternative acylation products [9–11]. This study aims to elucidate the molecular characteristics of HCT in tea plants by analyzing highly conserved domains in the amino acid sequences of HCT from Arabidopsis thaliana, Nicotiana tabacum, H. cannabinus, Theobroma cacao, and Fragaria vesca. We designed degenerate primers for polymerase chain reaction (PCR) and cloned the genetic sequence of CsHCT from a Chin-Shin Oolong tea plant. Quantitative real-time PCR (qRT-PCR) was employed to assess CsHCT expression levels in the tissues of tea plants and seedlings. The findings revealed a significant expression of CsHCT in the stem tissues of tea plants and seedlings. The transcription levels of CsHCT in tea plants across varying altitudes and seasons were quantified, revealing that CsHCT expression was significantly elevated at higher elevations and lower temperatures. Furthermore, an abiotic stress experiment demonstrated that low temperature, drought, and high salinity stressors elicited CsHCT transcription. Furthermore, the expression of CsHCT was elevated following ABA treatment. This study revealed that CsHCT may be implicated in the response to abiotic stress and ABA signaling pathways in tea plants..

2. METHODS

2.1 Plant Materials and Growth Conditions

This study used the tea [C. sinensis (L.) Kuntze] cultivar Chin-Shin Oolong in these experiments. These were obtained from a tea seed germination farm operated in Nantou in central Taiwan. The seedlings were 1–2 years old with a height of 40–50 cm. The tea plant (>10 years old) samples were obtained from a tea farmer in Nantou County. The samples were collected between 2015 and 2016 and consisted of buds and young leaf (YL), old leaf (OL), young stem (YS), and old stem (OS) tissues. They were obtained from high-mountain tea plantations in the Alishan area of Chiayi County. The tea plantations were located at altitudes between 700 and 1300 m, and all the tea plants were >10 years old.

2.2 Bioinformatics Analysis of the CsHCT Gene and Amino Acid Sequence

This study analyzed the highly conserved domains of the HCT protein sequence in A. thaliana, N. tabacum, H. cannabinus, T. cacao, and F. vesca, designed a degenerate primer, and used the cDNA of the Chin-Shin Oolong tea plant as the template to perform PCR for obtaining the gene fragment sequence of CsHCT. The SMARTerTM

RACE cDNA amplification kit (Clontech Laboratories, Inc., Mountain View, CA, USA) was used to expand 5 -end and 3 -end cDNA sequences. The full-length cDNA sequence of CsHCT was obtained after sequencing.

In the bioinformatics analysis conducted on the amino acid sequence of CsHCT, the ExPASy Translate tool (https://web.expasy.org/translate/) (access on July, 2016) was used for estimating the amino acid sequence translated by a nucleotide. The ExPASy Compute pI/Mw tool (https://web.expasy.org/compute_pi/) (access on July, 2016) was used to estimate the protein molecular weight and isoelectric point. Multiple sequence alignments were performed using the EMMA function of the EMBOSS explorer (http://www.bioinformatics.nl/ emboss-explorer/) (access on July, 2016) and the BLOSUM50 scoring matrix, and GeneDoc software was used to compare the results. Subsequently, Motif Scan (https://myhits. isb-sib.ch/cgi-bin/ motif_scan) (access on July, 2016) was used to predict the structural and functional protein regions. This study used the National Center for Biotechnology Information (https://www.ncbi.nlm.nih.gov/) (access on July, 2016) and Phytozome v10.3 (https://phytozome.jgi.doe.gov/pz/portal.html) (access on July, 2016) websites to obtain the protein sequences of clade Vb of BAHD (BEAT, benzylalcohol-O-acetyltransferase; AHCT, anthocyanin O-hydroxycinnamoyltransferase; HCBT, anthranilate N-hydroxycinnamoyl-benzoyltransferase; and DAT, deacetylvindoline 4-O-acetyltransferase) acyltransferase from A. thaliana, Oryza sativa, Populus trichocarpa, Coffea canephora, and H. cannabinus. Sequence alignment was performed using the ClustalW model. A phylogenetic tree was constructed using the MEGA6 software, after which statistical analysis was conducted through the neighbor joining method. The 1000 iterations of the tree algorithm were performed using the bootstrap method. SignalP (http://www.cbs.dtu.dk/ services/SignalP/) (access on July, 2016) was used to predict whether a protein signal peptide existed. The Hphob./Kyte and Doolittle method of the ExPASy ProtScale (https://web. expasy.org/protscale/) (access on July, 2016) was adopted for predicting whether the proteins were hydrophilic or hydrophobic. The subcellular localization of proteins was predicted using WoLF PSORT (https://www.genscript. com/wolf-psort.html) (access on July, 2016).

2.3 Abiotic Stress and Hormone Treatments on Tea Seedlings

The 1-year-old tea seedlings were treated with low temperature, high temperature, high salinity, and drought. Treatment conditions were as follows. The low- and high-temperature stresses were 5 ◦C and 35 ◦C, respectively. The seedlings were watered on optimum level and treated with the stresses for 12 h. Under the high-salinity stress, the seedlings were given 50 mL of 300 mM NaCl at 20 ◦C per day, whereas under the drought stress, they were not given additional water. The two treatments lasted for 5 days. In the control group, the seedlings were on optimum level watered at 20 ◦C for 5 days. After the treatments, samples were collected and preserved in a -80 ◦C environment for subsequent analysis. For hormone treatments, 100 μM solutions of ABA, SA, methyl jasmonate (MeJA), and 1-aminocyclopropane-1-carboxylic acid (ACC) solutions were prepared and sprayed on the YLs of the seedlings. After waiting for 6 h, the samples were collected and preserved in a -80 ◦C environment for subsequent analysis.

2.4 Extraction of Total RNA and qRT-PCR

In the experiment, 0.2 g of tea leaf samples was ground into powder in liquid nitrogen, and the total RNA was extracted using the Plant Total RNA Purification Kit (GeneMark, Taichung, Taiwan). The Moloney Murine Leukemia Virus (MMLV) first-strand synthesis kit (Gene DireX, Las Vegas, NV, USA) was used for reaction of 2 μg of total RNA. In the reaction solution, 1 μL of Oligo dT (1 μg/μL) was mixed with 1 μL of 10 mM dNTP to react for 10 min at 70 ◦C and for 5 min at 4 ◦C. After the reaction, 4 μL of 5 × reaction buffer, 2 μL of 0.1 M Dithiothreitol (DTT), 1 μL of diethylpyrocarbonate (DEPC) H2O, and 1 μL of MMLV reverse transcriptase were sequentially added into the solution for 1 h of reaction at 37 ◦C and 10 min of reaction at 65 ◦C, after which the reaction was terminated. Subsequently, 80 μL of DEPC H2O was added into a 0.2-mL microcentrifuge tube to perform qRT-PCR on the obtained cDNA.

For qRT-PCR, the cDNA was amplified using the CFX ConnectTM Real-Time System (Bio-Rad, Hercules, CA, USA), and data were analyzed using Bio-Rad CFX Manager 3.1. The 18S rRNA of the tea leaf samples were used as the internal control to normalize cDNA levels. The reaction conditions were as follows: 5 min at 94 ◦C; 15 s of 45 circulations at 94 ◦C, 60 ◦C, and 72 ◦C each; and finally, 10 min at 72 ◦C. Nonspecific products or primer dimers were identified based on their lower melting temperature than that of the specific amplicon. The primers used for qRT-PCR analyses were as follows: CsHCT forward sequence 5 -caaattaaccaaggaccaactcaac-3 and reverse sequence 5 -tgtaattgaccatgttcccatcttc-3 ; and 18S rRNA forward sequence 5 -ccgctggcaccttatgagaa-3 and reverse sequence 5 -tttcagccttgcgaccatact-3 . The qRT-PCR experiments were repeated at least 3 times each biologically independently, and the data shown are average values. Statistical analyses were performed using Statistical Analysis System (SAS) 9.4 software.

3. RESULTS

3.1 Bioinformatics Analysis of the CsHCT Gene and Amino Acid Sequences of C. sinensis L.

C. sinensis L. contains numerous polyphenolic compounds that provide multiple health benefits. To understand the role of HCT in the reaction of acylated flavonol glycosides, the CsHCT gene was cloned from the Chin-Shin Oolong tea plant. The

CsHCT gene has cDNA of length 1552 bp that includes 35-bp and 182-bp 5 and 3 untranslated regions (excluding a poly-A tail). The open reading frame includes 1311 nucleotide sequences, which can encode 436 amino acid sequences (GenBank accession number: MH271107).

Sequence alignment analysis was performed on the amino acid sequence of CsHCT of C. sinensis L. and the HCT sequences of A. thaliana (AtHCT), H. cannabinus (HcHCT), P. trichocarpa (PtHCT), and C. canephora (CcHCT). Using global alignment and the BLOSUM50 scoring matrix, the similarities between CsHCT and AtHCT, HcHCT, PtHCT, and CcHCT were found to be 79.5%, 81.5%, 81.9%, and 82.1%, respectively. CsHCT and the HCT of other higher plants exhibited the sequences HXXXD and DFGWG, which are the conserved sequences of BAHD acyltransferase (Figure 1). Motif scanning was then performed to predict the amino acid sequence of CsHCT, and the results indicated that the N-terminus of its protein possesses the predicted N-myristoylation, casein kinase II phosphorylation, and protein kinase C phosphorylation sites.

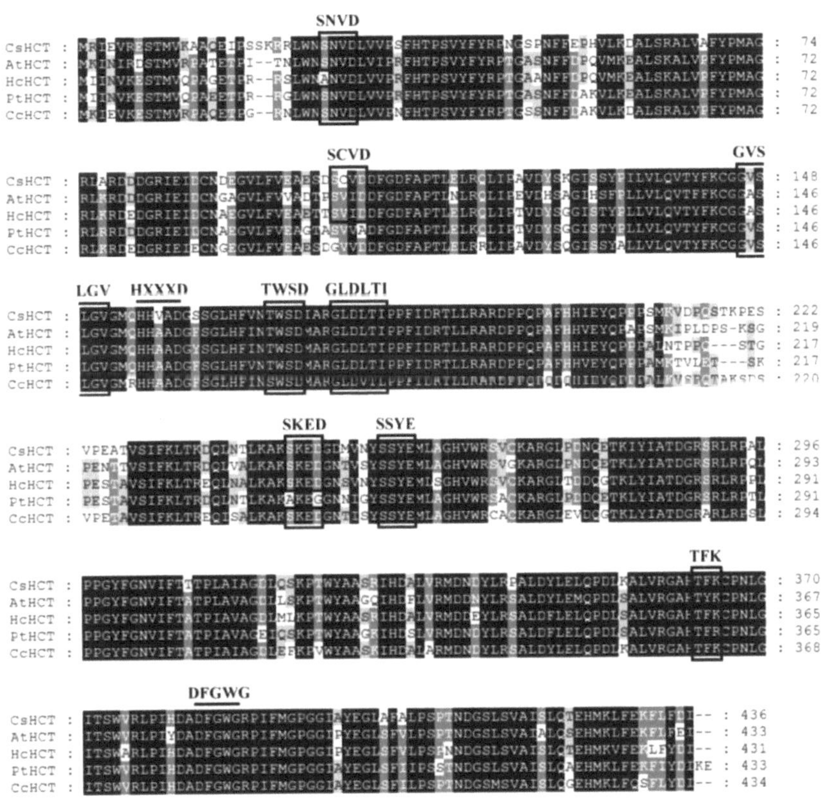

Figure 1. Alignment of deduced amino acid sequences of CsHCT with other putative HCTs. Black and gray shadings indicate conservation of 100% and at least 80%, respectively. Amino acid residues enclosed by squares correspond to consensus sequences of SXXD, SXXE, HXXXD, GVXXGV, TXXD, GLXXTI, DFGWG, etc. AtHCT,

hydroxycinnamoyl-CoA shikimate/quinate hydroxycinnamoyl transferase was from A. thaliana (NP_199704); HcHcT, putative hydroxycinnamoyl-CoA shikimate/quinate hydroxycinnamoyl transferase was from H. cannabinus (AFN85668); PtHCT, quinate O-hydroxycinnamoyl transferase/shikimate O-hydroxycinnamoyl transferase was from P. trichocarpa (ACC63882); and, CcHCT, hydroxycinnamoyl-CoA shikimate/quinate hydroxycinnamoyl transferase was from C. canephora (ABO47805). Alignment was performed using the ClustalW algorithm. Sequence identities with CsHCT were as follows: AtHCT, 79.5%; HcHCT, 81.5%; PtHCT, 81.9%; and, CcHCT, 82.1%.

Compute pI/Mw was used to analyze the amino acid sequence of CsHCT; its molecular weight and isoelectric point were predicted to be 48.53 kDa and 5.86, respectively. Analysis using SignalP failed to identify a signal peptide in CsHCT. The hydrophilicity and hydrophobicity of the protein were then analyzed using ProtScale, and the results demonstrated that the amino acid sequence did not possess an apparent hydrophobic end. The distribution of the hydropathy indices indicated that CsHCT is a hydrophilic protein (Figure 2A). Tuominen et al. (2011) reported that HCT is a member of clade Vb of the BAHD acyltransferase family. To understand the phylogenetic relationships between CsHCT and other clade Vb members, the MEGA6 software was used to perform neighbor joining. This generated a phylogenetic tree for the CsHCT of C. sinensis L. and the proteins of clade Vb members in A. thaliana, O. sativa, P. trichocarpa, C. canephora, and H. cannabinus. The results demonstrated that CsHCT and AtHCT had the closest phylogeny (Figure 2B); thus, they may have similar biochemical characteristics.

A

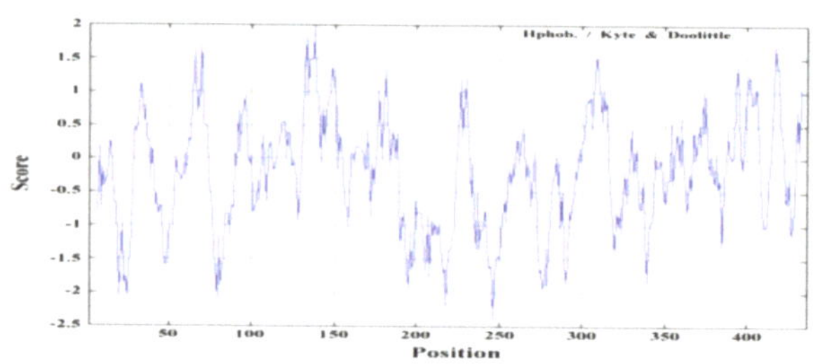

B

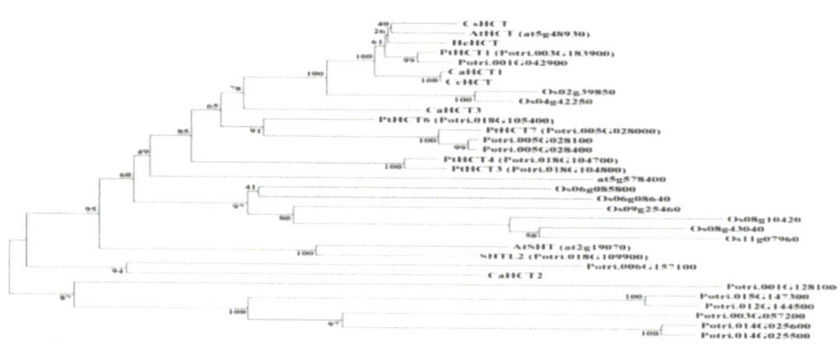

Figure 2. Hydropathy plot and phylogenetic tree analysis for CsHCT. (A) Hydropathy plot of CsHCT using the Kyte–Doolittle method with a window

size of 436 amino acids. The window position values indicated on the x-axis of the graph reveal the average hydropathy of the entire window, with the corresponding amino acids as the middle element of that window. Plots above 0 (zero) in the graph indicate hydrophobic regions in the protein, and those below 0 (zero) indicate hydrophilic regions. (B) Unrooted phylogram of members of the HCT protein family. Phylogenetic tree of AtSHT (at2g19070), AtHCT (at5g48930), and HXXXD-type acyl transferase family protein (at5g57840) proteins in Arabidopsis; transferase family protein (Os02g39850; Os04g42250; Os06g08580; Os06g08640; Os08g10420; Os08g43040; Os09g25460; and Os11g07960) in O. sativa; PtHCT1 (Potri.003G183900), PtHCT3 (Potri.018G104800), PtHCT4 (Potri.018G104700), PtHCT6 (Potri.018G105400), PtHCT7 (Potri.005G028000), SHTL2 (Potri.018G109900), Shikimate O-hydroxycinnamoyl transferase (Potri.005G028100 and Potri.005G028400), transferase family protein (Potri.006G157100; Potri.015G147300; Potri.003G057200; Potri.001G042900; Potri.001G128100; Potri.014G025500; and Potri.012G144500), anthranilate N-hydroxycinamoyl/benzoyltransferase-like protein (Potri.014G025600) in P. trichocarpa; CaHCT1 (CAJ40778), CaHCT2 (CAT00082), CaHCT3 (CAT00081), CcHCT (ABO77955) in C. canephora; and HcHCT (AFN85668) in H. cannabinus. The phylogenetic tree was constructed by the Neighbor Joining algorithm implemented in the MEGA 6 software package. The 1000 iterations of the tree algorithm were performed using the bootstrap method.

3.2 High Level of CsHCT Expression in the Stem Tissues of Tea Plants and Seedlings

To assess CsHCT expression in tissues of tea plants and seedlings, we selected YL, OL, YS, and OS tissues from 1-year-old tea seedlings and extracted the total RNA. Reverse transcriptase was then used to synthesize the first cDNA for qRT-PCR analysis. The results indicated that the CsHCT expression level was the highest in YS tissues, followed by OS tissues, buds, and leaves in tea seedlings (Figure 3A). The transcription level was the highest in OS tissues, followed by YS, YL, and OL tissues in tea plants (Figure 3B). Nonlignified YS tissues in tea seedlings exhibited the highest CsHCT expression level, but in tea plants, the CsHCT expression level was high in both OS and YS tissues. Moreover, CsHCT expression was evident in the buds and YL tissues of both tea plants and seedlings; however, expression levels remained lower than those in stem tissues (Figure 3).

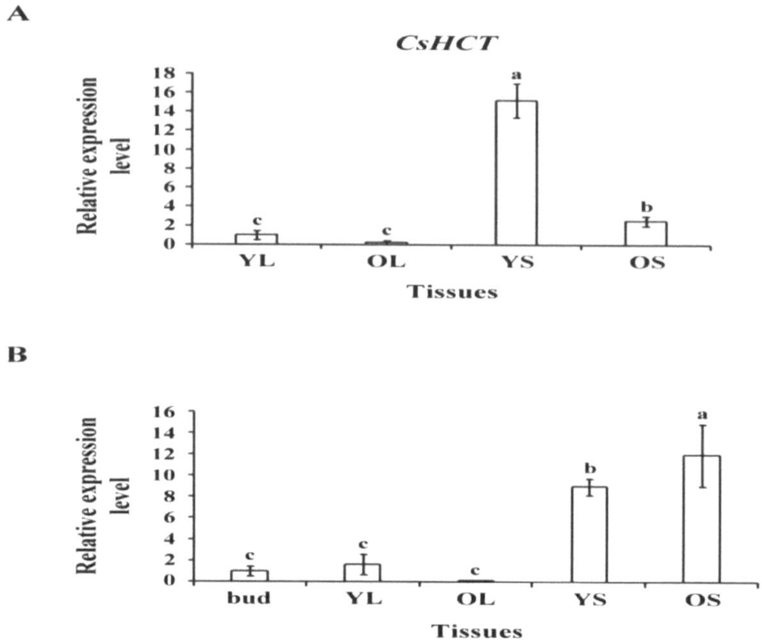

Figure 3. CsHCT transcription levels in various tissues of tea plants; qRT-PCR analysis of CsHCT transcription levels in various tissues of (A) tea seedlings and (B) tea plants. Total RNA was isolated from young leaf (YL), old leaf (OL), young stem (YS), old stem (OS), and buds. CsHCT transcription levels were determined. Relative amounts of transcripts were calculated and normalized to that of 18S rRNA. Values represent means ± SD from three biologically independent experiments. Values with different letters are significantly different at $p < 0.05$, according to a post hoc least significant difference (LSD) test.

3.3 CsHCT Transcription Levels in Oolong Tea Plants during Four Growing Seasons and at Different Altitudes

Temperature affects gene expression related to the secondary metabolism of plants, which regulates the generation of secondary metabolites and enables plants to adapt to environmental changes. To evaluate the influence of temperature on CsHCT expression, we used qRT-PCR to analyze its transcription levels in one-tip-two-leaf tissues of oolong tea plants during four seasons and compared their CsHCT expression. Specifically, the tea samples were collected from Zhushan, Nantou County, Taiwan in April, June, September, and December. The results showed that the CsHCT expression level of tea samples from December was the highest, and that of the tea samples from June was the lowest (Figure 4A). According to the average monthly temperature in Zhushan between 2015 and 2016 released by the Central Weather Bureau, CsHCT expression levels were negatively correlated with temperature (Figure 4B). CsHCT expression levels were high in low-temperature seasons.

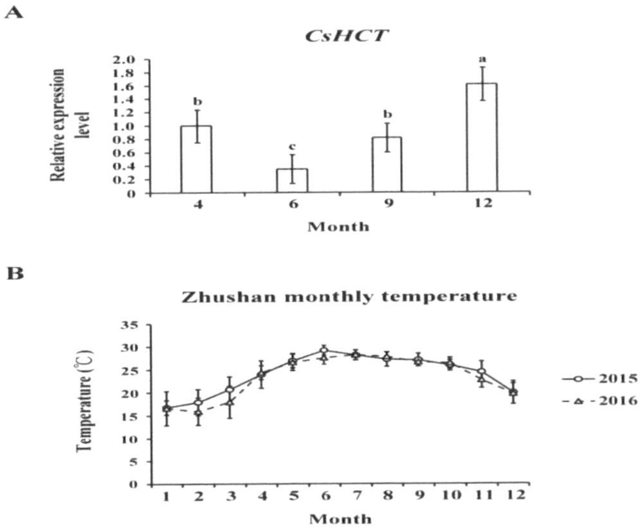

Figure 4. CsHCT transcription levels in four growing seasons. (A) The qRT-PCR analysis of CsHCT transcription levels in four growing seasons in the field. (B) Monthly average temperature in the Zhushan area. Total RNA was isolated from tissues of handpicked one-tip-two-leaf oolong tea samples collected in April, June, September, and December. Transcript levels of CsHCT were calculated and normalized to that of 18S rRNA. Values represent means ± SD from five biologically independent experiments. Values with different letters are significantly different at $p < 0.05$, according to a post hoc least significant difference (LSD) test.

Altitude influences temperature, humidity, and sunlight intensity. Temperature decreases as altitude increases. To further verify CsHCT expression, the tea samples collected from the Alishan area of Nantou County from altitudes of 700, 1000, and 1300 m and bud tissues were analyzed using qRT-PCR to determine CsHCT expression. The results demonstrated that the CsHCT expression level was the highest at 1300 m and decreased as the altitude decreased (Figure 5).

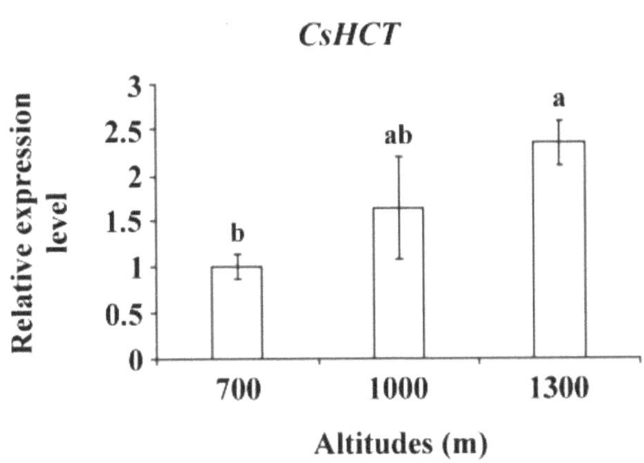

Figure 5. CsHCT expression levels at various altitudes; qRT-PCR analysis of CsHCT transcription levels in the Alishan area of Nantou County. Total RNA was

isolated from the tissues of handpicked one-tip-two-leaf oolong tea samples collected at altitudes of 700, 1000, and 1300 m. Transcription levels of CsHCT were calculated and normalized to that of 18S rRNA. Values represent means ± SD from three biologically independent experiments. Values with different letters are significantly different at $p < 0.05$, according to a post hoc LSD test.

3.4 Effects of Abiotic Stresses and Hormone Signaling on CsHCT Transcript Levels in Oolong Tea Seedlings

To investigate whether CsHCT expression is induced by abiotic stress, we exposed 1-year-old tea seedlings to stresses such as high temperature (35 ◦C), low temperature (5 ◦C), high salinity (300 mM NaCl), and drought; subsequently extracted the total RNA from one-tip-two-leaf tissues of oolong tea seedlings; and then performed qRT-PCR analysis. The results demonstrated that compared with the control group, CsHCT expression was higher at low-temperature stress and lower under high-temperature stress, thus verifying the association between CsHCT expression and temperature. Furthermore, CsHCT transcription levels also increased in response to the high-salinity and drought treatments (Figure 6A).

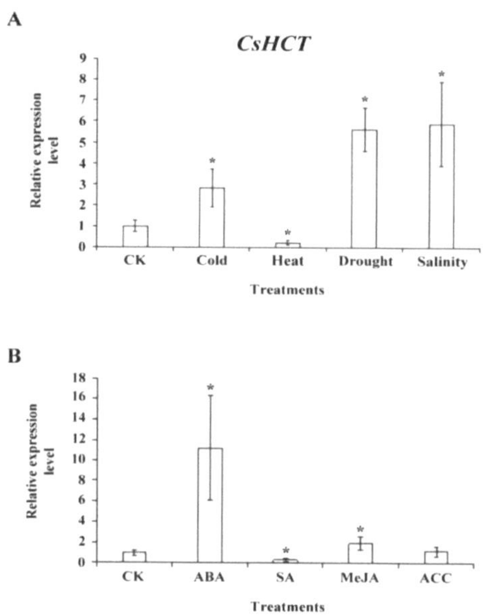

Figure 6. CsHCT expression levels under various abiotic stress and phytohormone treatment;RT-PCR analysis of CsHCT transcription levels in tea seedlings after (A) abiotic stress and (B) phytohormone treatment. Total RNA was isolated from YLs of tea seedlings after cold treatment (5 ◦C) for 12 h, heat treatment (35 ◦C) for 12 h, drought (no water) for 3 days, and salt treatment (300 mM NaCl) for 5 days. The phytohormone treatment included treatment with 100 μM ABA, SA, MeJA, and ACC for 6 h. CsHCT transcription levels were calculated and normalized to that of 18S rRNA. Values represent means ± SD from four biologically independent experiments. * $p < 0.05$, versus value in control check (CK) treatment (Student's t test).

When a plant is under biotic or abiotic stresses, stress-related hormone signaling initiates the plant's defense mechanisms and increases its stress tolerance. To assess whether CsHCT expression is induced by various stress hormone signals, we treated the 1-year-old tea seedlings with ABA, SA, MeJA, and the ethylene precursor ACC for 6 h and subsequently collected the one-tip-two-leaf tissues for analysis of CsHCT expression through qRT-PCR. The results indicated that CsHCT expression was induced under ABA and MeJA treatment and that the expression level was the highest in the ABA group. By contrast, the CsHCT expression level in the SA group was significantly lower than that in the control (Figure 6B). Accordingly, this study determined that CsHCT expression can be induced by abiotic stresses such as low temperature, high salinity, and drought, and inferred that CsHCT may be involved in the ABA signaling pathways.

4. DISCUSSION

Acyltransferase in higher plants can catalyze transfer of acyl group to donor substrate. Acyl esters are produced as a result of the acyl group transfer from the donor substrate to the acceptor substrate [12]. Acyltransferase can be divided into two protein families, namely BAHD acyltransferase and serine carboxypeptidase-like acyltransferase, according to different donor substrates. HCT is categorized into the clade Vb of BAHD acyltransferase and characterized by its ability to catalyze various substrates.

Amino acid analysis results demonstrated that the structure and function of the CsHCT and HCT sequences of other higher plants were highly conserved (Figure 1). The amino acid domain was highly conserved in clade Vb that possessed a particularly conserved sequence SXXDL in the BAHD acyltransferase family [13]. However, whether the amino acid domain affects the catalytic function of enzymes remains to be investigated and verified. BAHD acyltransferase acts mainly in the cytoplasm of plant cells. Its substrate, acyl coenzyme A thioesters, can perform biosynthesis in various cell organs and be transferred to the cytoplasm by the transporter on the cell membrane, which facilitates the catalytic reaction of the BAHD acyltransferase [14]. After analyzing the cellular localization of CsHCT, we inferred that CsHCT is primarily located in the cytoplasm, and the amino acid sequence does not possess a signal peptide or apparent hydrophobic end (Figure 2A). Therefore, this study inferred that CsHCT primarily reacts in the cytoplasm for catalytic reaction.

Our data determined that the CsHCT expression level in the YS tissues of tea seedlings was higher than that in the OS and bud tissues, whereas the CsHCT expression level in tea plants was higher in OS and YS tissues (Figure 3). Studies on Trifolium pretense have shown that HCT1 is primarily expressed in stem and flower tissues, whereas HCT2 is mainly expressed in leaf and flower tissues, which indicates that HCT1 and HCT2 have different catalytic functions in diverse plant tissues [15].

P. trichocarpa possesses seven PtrHCTs that can be expressed in the tissues of various plant parts and exhibit differences with respect to their relative performance. In particular, PtrHCT1 and PtrHCT6 are primarily expressed in stem tissues, whereas PtrHCT3 has a higher level of expression in leaf tissues [16]. In this study, nonlignified YS tissues of tea seedlings were found to contain a relatively large amount of CsHCT transcripts. The CsHCT expression level in YS tissues was eight times higher than that in OS tissues. This indicated that secondary metabolites and the expression of related biosynthesis genes in tissues vary according to the growth stages of C. sinensis L.

When plants are under environmental stresses, the expression of genes related to the biosynthesis of secondary metabolites is induced, which results in the generation and accumulation of compounds such as phenylpropanoid, flavonoids, and anthocyanins that can increase plants' tolerance to stresses [4]. Our results demonstrated that the CsHCT expression level in C. sinensis L. was relatively high in winter and at high altitudes (Figures 4 and 5), indicating that CsHCT has a high level of expression in low temperatures. The CsHCT expression level increased under low-temperature stress and decreased under high-temperature stress (Figure 6A). Thus, CsHCT expression is induced in low temperatures and may be involved in the defense pathways against low-temperature stress. Research demonstrated that HCT expression is regulated by biotic and abiotic stresses, thereby increasing stress tolerance in plants [6]. Our data indicated that CsHCT expression in C. sinensis L. can be induced with low-temperature, high-salinity, and drought stresses, and the expression level was particularly high with ABA treatment (Figure 6B).

Phytohormone ABA involved in stress tolerance in plants can be divided into ABA-dependent and ABA-independent signaling pathways. ABA-dependent pathways transmit signals through ABA, thereby activating downstream transcription factors such as the ABRE-binding factor/ABA-responsive element-binding protein, myelocytomatosis, and myeloblastosis to regulate plants' stress tolerance [17,18]. This study demonstrated that CsHCT expression was induced by the abiotic stresses of low temperature, high salinity, and drought as well as ABA treatment. The signals of the three stresses may be transmitted through ABA-dependent pathways and may affect the expression of transcription factors of genes associated with the regulation of secondary metabolism. In this study, our results demonstrated the relationship between CsHCT expression and hormone signaling in oolong tea plants and may help improve the quality and possible health benefits of tea in the future.

5. CONCISE SUMMARY:

Collectively, our findings suggest that CsHCT, cloned from the Chin-Shin Oolong tea plant, may play a role in the response of tea plants to abiotic stresses (such as low temperature, high salinity, and drought) and in various hormonal signaling pathways that influence the secondary metabolic pathways of tea plants.

REFERENCES

1. Chen, Y.J.; Kuo, P.C.; Yang, M.L.; Li, F.Y.; Tzen, J.T.C. Effects of baking and aging on the changes of phenolic and volatile compounds in the preparation of old Tieguanyin oolong teas. Food Res. Int. 2013, 53, 732–743. [CrossRef]
2. Park, J.S.; Kim, J.B.; Hahn, B.S.; Kim, K.H.; Ha, S.H.; Kim, J.B.; Kim, Y.H. EST analysis of genes involved in secondary metabolism in Camellia sinensis (tea), using suppression subtractive hybridization. Plant Sci. 2004, 166, 953–961. [CrossRef]
3. Lin, L.Z.; Chen, P.; Harnly, J.M. New phenolic components and chromatographic profiles of green and fermented teas. J. Agric. Food Chem. 2008, 56, 8130–8140. [CrossRef] [PubMed]
4. Dixon, R.A.; Paiva, N.L. Stress-induced phenylpropanoid metabolism. Plant Cell 1995, 7, 1085–1097. [CrossRef] [PubMed]
5. Liu, Q.Q.; Luo, L.; Zheng, L.Q. Lignins: Biosynthesis and biological functions in plants. Int. J. Mol. Sci. 2018, 19. [CrossRef] [PubMed]
6. Chowdhury, E.M.; Choi, B.S.; Park, S.U.; Lim, H.S.; Bae, H. Transcriptional analysis of hydroxycinnamoyl transferase (HCT) in various tissues of Hibiscus cannabinus in response to abiotic stress conditions. Plant Omics J. 2012, 5, 305–313.
7. Varbanova, M.; Porter, K.; Lu, F.C.; Ralph, J.; Hammerschmidt, R.; Jones, A.D.; Day, B. Molecular and biochemical basis for stress-induced accumulation of free and bound p-coumaraldehyde in cucumber. Plant Physiol. 2011, 157, 1056–1066. [CrossRef] [PubMed]
8. Wang, G.F.; He, Y.J.; Strauch, R.; Olukolu, B.A.; Nielsen, D.; Li, X.; Balint-Kurti, P.J. Maize homologs of hydroxycinnamoyltransferase, a key enzyme in lignin biosynthesis, bind the nucleotide binding leucine-rich repeat Rp1 proteins to modulate the defense response. Plant Physiol. 2015, 169, 2230–2243. [PubMed]
9. Sander, M.; Petersen, M. Distinct substrate specificities and unusual substrate flexibilities of two hydroxycinnamoyltransferases, rosmarinic acid synthase and hydroxycinnamoyl-CoA: Shikimate hydroxycinnamoyl-transferase, from Coleus blumei Benth. Planta 2011, 233, 1157–1171. [CrossRef] [PubMed]

10. Kang, S.; Kang, K.; Chung, G.C.; Choi, D.; Ishihara, A.; Lee, D.S.; Back, K. Functional analysis of the amine substrate specificity domain of pepper tyramine and serotonin N-hydroxycinnamoyltransferases. Plant Physiol. 2006, 140, 704–715. [CrossRef] [PubMed]

11. Eudes, A.; Pereira, J.H.; Yogiswara, S.; Wang, G.; Benites, V.T.; Baidoo, E.E.K.; Lee, T.S.; Adams, P.D.; Keasling, J.D.; Loque, D. Exploiting the substrate promiscuity of hydroxycinnamoyl-CoA: Shikimate hydroxycinnamoyl transferase to reduce lignin. Plant Cell Physiol. 2016, 57, 568–579. [CrossRef] [PubMed]

12. Moglia, A.; Acquadro, A.; Eljounaidi, K.; Milani, A.M.; Cagliero, C.; Rubiolo, P.; Genre, A.; Cankar, K.; Beekwilder, J.; Comino, C. Genome-wide identification of BAHD acyltransferases and in vivo characterization of HQT-like enzymes involved in caffeoylquinic acid synthesis in globe artichoke. Front. Plant Sci. 2016, 7, 1424. [CrossRef] [PubMed]

13. Tuominen, L.K.; Johnson, V.E.; Tsai, C.J. Differential phylogenetic expansions in BAHD acyltransferases across five angiosperm taxa and evidence of divergent expression among Populus paralogues. BMC Genomics 2011, 12, 236. [CrossRef] [PubMed]

14. Bontpart, T.; Cheynier, V.; Ageorges, A.; Terrier, N. BAHD or SCPL acyltransferase? What a dilemma for acylation in the world of plant phenolic compounds. New Phytol. 2015, 208, 695–707. [CrossRef] [PubMed]

15. Sullivan, M.L. A novel red clover hydroxycinnamoyl transferase has enzymatic activities consistent with a role in phaselic acid biosynthesis. Plant Physiol. 2009, 150, 1866–1879. [CrossRef] [PubMed]

16. Shi, R.; Sun, Y.H.; Li, Q.Z.; Heber, S.; Sederoff, R.; Chiang, V.L. Towards a systems approach for lignin biosynthesis in Populus trichocarpa: Transcript abundance and specificity of the monolignol biosynthetic genes. Plant Cell Physiol. 2010, 51, 144–163. [CrossRef] [PubMed]

17. Shinozaki, K.; Yamaguchi-Shinozaki, K.; Seki, M. Regulatory network of gene expression in the drought and cold stress responses. Curr. Opin. Plant Biol. 2003, 6, 410–417. [CrossRef]

18. Ishitani, M.; Xiong, L.M.; Stevenson, B.; Zhu, J.K. Genetic analysis of osmotic and cold stress signal transduction in Arabidopsis: Interactions and convergence of abscisic acid-dependent and abscisic acid-independent pathways. Plant Cell 1997, 9, 1935–1949. [CrossRef] [PubMed]

Chapter 11

THE WRKY TRANSCRIPTION FACTOR GMWRKY12 CONFERS DROUGHT AND SALT TOLERANCE IN SOYBEAN

1. INTRODUCTION

Drought and salinity are the most important abiotic stress factors affecting plants growth and crop yield. On average, 1/3 of cultivable land suffers drought and salinization, which is equivalent to a loss of about 1,500,000 ha of crop land per year [1]. The damage caused by drought and salt are almost the sum of losses caused by other stress factors. Under limited land and water resources, it is necessary to breed new stress-resistant varieties to increase yield and ensure food security. Cultivation of stress-resistant crop varieties is also an important way to ensure high and stable yield of crops. Transgenic technology has become an important way to learn the function of genes in crops [2–4].

Being unable to move, plants encounter numerous biotic and abiotic stresses at different developmental stages which include drought, salinity, temperature changes, nutritional deficiency, pathogen invasion and competition from alien species. To overcome these unfavorable conditions, plants have evolved a complex and efficient signaling network, which can produce a series of responses to external stress signals and induce the expression of stress-related genes to protect the normal activities of the cells [5]. Inducible genes encoding proteins can be divided into three categories based on function: the first is functional genes, which are directly involved in stress response and are located downstream in the signaling network, such as HKT [6,7], SALT [8], NHX [9,10], CAX and CHX [11–13]. Another is transcription factors (TFs) that regulate the expression of functional genes in the middle of the signaling network, like DREB [14,15], MYB [16], WRKY [17,18], NAC [19,20], bZIP [21,22] and ERF [23,24]. The last group includes a variety of protein kinases, which conduct stress signals and are located upstream of the signaling network, such as GST [25], LEA [26] and FNS [27].

Among the three classes of stress-related genes, the TFs form a connecting link between the beginning and end of the signaling network; WRKYs are among the largest family of plant TFs. The WRKY domain is about 60 residues in length and is named by a conserved WRKY domain, containing the WRKYGQK heptapeptide at the N-terminus followed by a zinc-finger motif CX4-5CX22-23HXH or CX7CX23HXC [28,29]. Based on the number of WRKY domains and the structure of zinc finger motifs, WRKY TFs are divided into three groups. Group I includes two WRKY domains and either a CX4-5CX22-23HXH or CX7CX23HXC zinc-finger motif. Group II WRKY proteins contain a single WRKY domain and a CX4-5CX22-23HXH zinc-finger motif; due to differences in the primary amino acid sequence, Group II can be divided into five subgroups IIa-IIe [29,30]. Group III WRKY proteins have a single WRKY domain and a CX7CX23HXC zinc-finger motif.

As one of the members of the plant TF family, WRKY is heavily studied. Researchers have determined that WRKY TFs participate in various physiological and developmental processes [29], such as seed development [31], seed dormancy and germination [32], senescence [33], development [34], plant immune response [35], pathogen defense [18,36] and insect resistance [37,38]. Recent studies have revealed that WRKY proteins are involved in the signal transduction of plant hormones, like abscisic acid (ABA) [39,40], jasmonic acid (JA) [41] and gibberellin (GA) [39]. Numerous studies have demonstrated that WRKY TFs respond to abiotic stresses [42,43], such as salt [4], drought [44], cold [45] and heat [46–48]. There are 74 WRKY TF members in model plant Arabidopsis [49] and 18 WRKYs have been suggested to be induced by exposure to salt stress; overexpression of WRKY25 or WRKY33 was sufficient to increase Arabidopsis NaCl tolerance [50]. Overexpressing TaWRKY2 and TaWRKY19 exhibited salt and drought tolerance in transgenic Arabidopsis [51]. Moreover, researchers found that OsWRKY11 directly bound to the promoter of a drought-responsive gene, RAB21, as well as enhanced heat and drought tolerance in transgenic rice seedlings [52,53]. Ectopic expression of ZmWRKY33 and ZmWRKY58 in Oryza and Arabidopsis improved drought and salt tolerance, respectively, in transgenic plants [54,55]. In addition, there is extensive cross-talk between responses to biotic/abiotic stresses and exogenous hormones, for example drought and salt stress with the plant hormones. Arabidopsis WRKY46, WRKY54 and WRKY70 are involved in Brassinosteroid-mediated drought response and plant growth [43]. Novel cotton WRKY-genes GhWRKY25 and GhWRKY6-like confer tolerance to abiotic and biotic stresses in transgenic Nicotiana and enhanced salt tolerance by activating the ABA signaling pathway and scavenging reactive oxygen species [56]. SA-inducible poplar PtrWRKY73 is also involved in disease resistance in Arabidopsis [37]. All of these studies illustrated that WRKY TFs play a significant role in plant developmental and physiological processes and abiotic and biotic stresses.

Soybean (Glycine max), is an important global cash crop, accounting for 59 percent of the world's oilseed production (http://soystats.com). Currently, due to its high protein content it is often treated as an important source of protein for both human consumption and as fodder. The demand for soybean is thus increasing rapidly and improving soybean yield has become a major research goal. Soybean productivity is greatly affected by growing environment, such as climatic and soil conditions (drought, salt, metallic pollution and fungus infection). Therefore, it is vital to cultivate soybean varieties that are resistant to stressors.

Recently, many studies based on biotechnological and RNA-Seq approaches have been conducted on soybean WRKY TFs. Researchers have identified 188 soybean WRKY genes genome-wide and 66 of the genes have been shown to respond rapidly and transiently to the imposition of salt stress [30]. In the latest version of the soybean genome (Wm82.a2v1), 176 GmWRKY proteins were confirmed and the expression of GmWRKY47 and GmWRKY58 decreased upon dehydration, while GmWRKY92, GmWRKY144 and GmWRKY165 increased under salt treatment [57]. GmWRKY13 may function in plant growth and abiotic stress. GmWRKY21 and GmWRKY54 conferred tolerance to cold stress and salt and drought stress, respectively [58]. Here, based on RNA-Seq and several databases and bioinformatics methods, we identified GmWRKY12, which is associated with abiotic stress tolerance by quantitative RT-PCR. Overexpression of GmWRKY12 could improve tolerance of soybean to drought and salt.

2. RESULTS

2.1 Identification of GmWRKYs Up-Regulated under Drought/Salt Treatment

The GmWRKYs are distributed in different tissues or located upstream of soybean genes to bind the W-box consensus (TTGACY) in the promoters of target genes, initiating functions such as plant development, pathogen defense, insect resistance, response to biotic and abiotic stress and participating in signal transduction mediated by plant hormones [59,60]. In order to identify the function of genes or to explore whether GmWRKY mRNA expression goes up under biotic and abiotic stress, we conducted RNA-Seq (Tables S5 and S6). RNA-Seq data were used to screen GmWRKYs that are responsive to drought and salt. There were 105 GmWRKYs upregulated after drought treatment and fifty-three GmWRKYs were selected based on the rule that log2 (GH_treat/CK1_treat) >1 (Table 1). Nine GmWRKYs were selected from salt treatment RNA-Seq data based on the rule that log2 (NaCl_treat/CK2_treat) >1 (Table 2).

Table 1. Annotation of Glycine max WRKY transcription factors responding to drought stress (up-regulation).

Gene ID [a]	Name [b]	Chr	CDS (bp)	Protein (aa)	Group [c]
GLYMA_14G103100	GmWRKY40	14	849	282	IIb
GLYMA_18G056600	GmWRKY62	18	1689	542	IIb
GLYMA_17G042300	GmWRKY6	17	1173	390	IIe
GLYMA_04G054200	GmWRKY50	4	486	161	IIe
GLYMA_01G222300	GmWRKY22	1	738	245	IIc
GLYMA_02G293400	GmWRKY31	2	1278	425	IIa
GLYMA_04G218700	GmWRKY21	4	591	196	I
GLYMA_06G147100	GmWRKY51	6	591	196	III
GLYMA_01G224800	GmWRKY12	1	714	237	IIc
GLYMA_11G163300	GmWRKY19	11	1647	548	I
GLYMA_06G061900	GmWRKY17	6	885	294	IIb
GLYMA_10G011300	GmWRKY54	10	972	323	IIa
GLYMA_04G223300	GmWRKY58	4	954	317	III
GLYMA_18G213200	GmWRKY57	18	900	299	III
GLYMA_06G125600	GmWRKY53	6	1095	364	IIa
GLYMA_19G217800	GmWRKY23	19	873	290	IId
GLYMA_09G280200	GmWRKY33	9	1632	543	I
GLYMA_03G002300	GmWRKY70	3	747	248	IIc
GLYMA_13G310100	GmWRKY36	13	1845	614	IIc
GLYMA_14G200200	GmWRKY49	14	1728	575	IIc
GLYMA_16G026400	GmWRKY60	16	1122	373	IIc
GLYMA_16G0544001	GmWRKY75	16	588	195	IIb
GLYMA_04G223200	GmWRKY55	4	1020	339	IId
GLYMA_02G232600	GmWRKY39	2	1743	580	III
GLYMA_05G0290001	GmWRKY72	5	1785	594	I
GLYMA_03G220100	GmWRKY41	5	762	253	IIe
GLYMA_08G021900	GmWRKY46	8	1080	356	III
GLYMA_15G003300	GmWRKY27	15	921	306	IIb
GLYMA_17G097900	GmWRKY61	17	1803	600	IIc
GLYMA_01G128100	GmWRKY5	1	1527	508	IId
GLYMA_12G212300	GmWRKY16	12	792	263	IIc
GLYMA_08G082400	GmWRKY28	8	881	293	III
GLYMA_07G227200	GmWRKY3	7	1602	533	IIc
GLYMA_03G256700	GmWRKY43	66	1089	362	IIe

GLYMA_15G168200	GmWRKY42	15	882	293	IIb
GLYMA_13G289400	GmWRKY52	13	798	265	IIc
GLYMA_08G011300	GmWRKY25	8	444	147	IId
GLYMA_09G061900	GmWRKY47	19	1573	296	IIc
GLYMA_17G222300	GmWRKY30	4	555	184	IIa
GLYMA_01G053800	GmWRKY9	1	1368	455	IIc
GLYMA_08G118200	GmWRKY48	7	789	262	IIc
GLYMA_01G056800	GmWRKY32	1	894	297	IId
GLYMA_08G218600	GmWRKY56	8	942	313	III
GLYMA_07G262700	GmWRKY34	7	1554	517	IIb
GLYMA_03G159700	GmWRKY15	1	1017	338	I
GLYMA_11G053100	GmWRKY14	11	963	320	I
GLYMA_05G096500	GmWRKY11	17	1050	334	I
GLYMA_17G222500	GmWRKY63	17	849	278	IIa
GLYMA_08G240800	GmWRKY4	2	1572	523	I
GLYMA_03G176600	GmWRKY29	5	1308	436	IIc
GLYMA_08G325800	GmWRKY35	8	1734	577	IIc
GLYMA_10G138300	GmWRKY1	14	1449	482	IIb
GLYMA_06G077400	GmWRKY37	6	903	300	III

a—The annotated GmWRKYs according to NCBI (https://www.ncbi.nlm.nih.gov/pubmed) and. PlantTFDB (http://planttfdb.cbi.pku.edu.cn/); b—The names of GmWRKYs are given according to SoyDB (http://soykb.org/); c—The grouping is according to [30,61].

Table 2. Annotation of Glycine max WRKY transcription factors responding to salt stress (up-regulation).

Gene ID [a]	Name [b]	Chr	CDS (pb)	Protein (aa)	Group [c]
GLYMA_11G053100	GmWRKY14	9	963	320	I
GLYMA_08G325800	GmWRKY35	8	1734	577	IIc
GLYMA_04G218700	GmWRKY21	10	591	196	I
GLYMA_14G200200	GmWRKY49	18	1728	575	IIc
GLYMA_07G227200	GmWRKY3	18	1602	533	IIc
GLYMA_02G115200	GmWRKY28	8	881	293	III
GLYMA_03G256700	GmWRKY43	16	1089	362	III
GLYMA_06G320700	GmWRKY59	6	2331	776	IIc
GLYMA_01G224800	GmWRKY12	7	714	237	IIc

a—The annotated GmWRKYs according to NCBI (https://www.ncbi.nlm.nih.gov/pubmed) and PlantTFDB (http://planttfdb.cbi.pku.edu.cn/). b—The names of GmWRKYs are given according to SoyDB (http://soykb.org/). c—The grouping is according to [30,61].

2.2 Tissue-Specific Expression Patterns of GmWRKYs

To thoroughly study GmWRKY expression profiles under normal conditions, hierarchical clustering was conducted using expression levels of fifty-three (drought-responsive) and nine (salt-responsive) GmWRKY genes in young leaf, flower, one cm pod, pod shell 10 days after flowering (DAF), pod shell 14 DAF, seed 10 DAF, seed 14 DAF, seed 21 DAF, seed 25 DAF, seed 28 DAF, seed 35 DAF, seed 42 DAF, root and nodule (Figures 1 and 2). Approximately 28% of GmWRKYs from different tissues were expressed at low levels or unexpressed (GmWRKY3, 5, 6, 21, 22, 25, 29, 30, 31, 47, 50, 55, 63, 70 and 72); by contrast, 45% of GmWRKYs were highly expressed in different tissues (GmWRKY4, 9, 11, 14, 16, 17, 19, 28, 32, 34, 35, 36, 37, 39, 41, 42, 46, 48, 49, 52, 56, 57, 60 and 61). Among these GmWRKYs, GmWRKY11 and GmWRKY17 had the highest expression in four different tissues. GmWRKY28, 35, 37, 48 and 57 are highly expressed in nodule, seed 10 DAF, seed 42 DAF, root and flower. Within the nine GmWRKYs related to salt response, GmWRKY3 and GmWRKY21 had low expression and GmWRKY14, 28, 35, 49 and 59 were highly expressed in at least four different tissues. The analysis data are available in Tables S1 and S2.

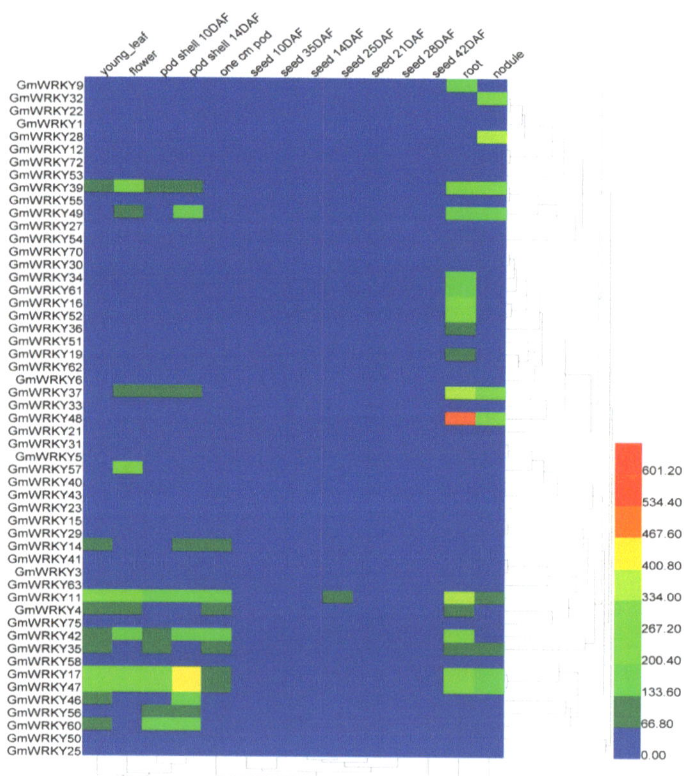

Figure 1. Expression pattern of fifty-three GmWRKYs in six different tissues (young leaf, flower, pod shell, seed, root and nodule). The fifty-three GmWRKYs

were selected from drought treatment RNA-Seq data based on the rule that log2 (GH_treat/CK1_treat) >1. The tissue expression is from SoyDB (http://soykb.org/). The color legend refers to the different expression level under normal condition. "DAF" in the tissue label indicates days after flowering.

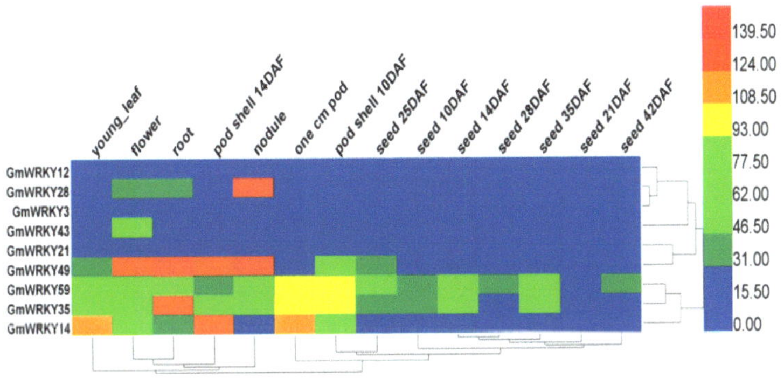

Figure 2. Expression pattern of nine GmWRKYs in six different tissues (young leaf, flower, pod shell, seed, root and nodule). The nine GmWRKYs were selected from salt treatment RNA-Seq data based on the rule that log2 (NaCl_treat/CK2_treat) >1. The tissue expression is from SoyDB (http://soykb.org/). The color legend refers to the different expression level under normal condition. "DAF" in the tissue label indicates days after flowering.

2.3 GmWRKYs Responsive to Both Drought and Salt Treatments

Based on RNA-Seq data and result of Venn method [62], seven GmWRKY genes were found to respond to both drought and salt treatments (GmWRKY3, 12, 14, 21, 35, 43 and 49) (Figure S1A). In order to confirm whether the seven GmWRKY genes are responsive to drought and salt, 10-day-old soybean seedlings were subjected to stress treatments. For drought treatment, soybean seedlings were put on filter paper to stimulate drought and then sampled 0.1 g of leaf on different periods (0, 0.5, 1, 2, 5, 8, 12 and 24 h); for salt treatment, the roots of soybean were soaked in 100 mM NaCl solution then sampled 0.1 g of leaf on different periods (0, 0.5, 1, 2, 5, 8, 12 and 24 h), all samples were submerged immediately in liquid nitrogen and stored at -80 ◦C for RNA extraction then quantitative real-time PCR (qRT-PCR) was conducted. Results confirmed that the seven GmWRKY genes were responsive to both treatments (Figure 3). Under drought treatment, the expression levels of GmWRKY12 and GmWRKY43 were gradually increased at 0, 0.5, 1, 2, 5, 8, 12 and 24 h. GmWRKY12 was highly expressed after 12 h of drought treatment. While GmWRKY14, GmWRKY21 and GmWRKY35 had a tendency to rise first and then decrease, GmWRKY49 was highly expressed at 2 h. Under drought conditions, the expression profiles of five GmWRKY genes were little changed at 0 to 5 h and then increased significantly at 12 h.

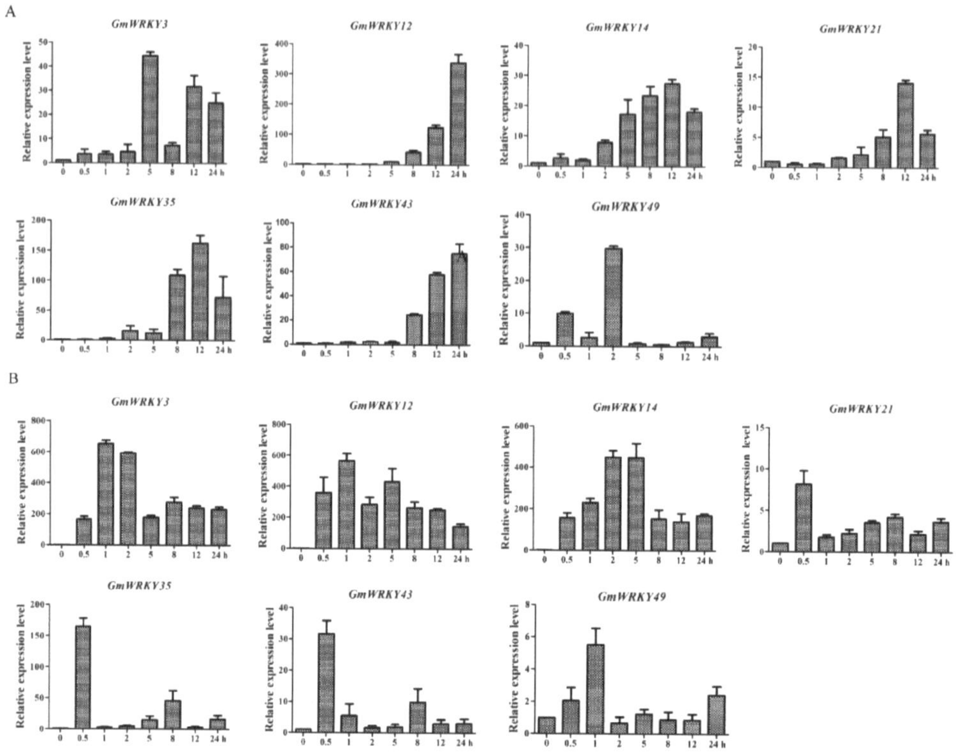

Figure 3. Quantitative RT-PCR of seven GmWRKYs under drought and salt treatment. (A) qRT-PCR of seven GmWRKYs under drought treatment. (B) qRT-PCR of seven GmWRKYs under salt treatment. The expression level of GmActin as a loading control. The data represent means ± SD of three biological replications.

Under salt treatment, the expression profile increased first and then decreased, meanwhile, there was a notable change at 0 to 0.5 h and GmWRKY3, 12, 14 and 35 were highly expressed. GmWRKY12, which was 714 bp in length, encoded 237 amino acids and had low expression in different tissues under normal conditions but was highly expressed under drought and salt treatments was selected for further investigation (Figure S1B).

2.4 Multiple Sequence Alignment and Phylogenetic Analysis of GmWRKY12

Although WRKYGQK sequence is a conservative motif of WRKY proteins, WRKY variant domains, such as WRKYGEK, WRKYGKK, WQKYGQK, WSKYGQK and WRKYGM have been found in the genomes of Arabidopsis [28], rice [63], grape [64] and tomato [65]. This difference may be a variation of WRKY TFs developed over long-term evolution. The domain of these variations is unique and may represent a new type. Therefore, to identify conservation of GmWRKY12, WRKY12 from 20 different species were selected for multiple sequence alignment (Figure 4A). Results showed that 20 species only harbored one WRKY variant WRKYGQK, with amino acid sequence similarity of 75%, which illustrated that GmWRKY12 was highly conserved. To further evaluate the evolutionary relationship between GmWRKY12 and WRKY12 of 32 different species, a phylogenetic tree was constructed with the neighbor-joining method [66]. Phylogenetic results showed that the relationship between GmWRKY12 and VrWRKY12 (XP_014515898.1) was the closest (Figure 4B).

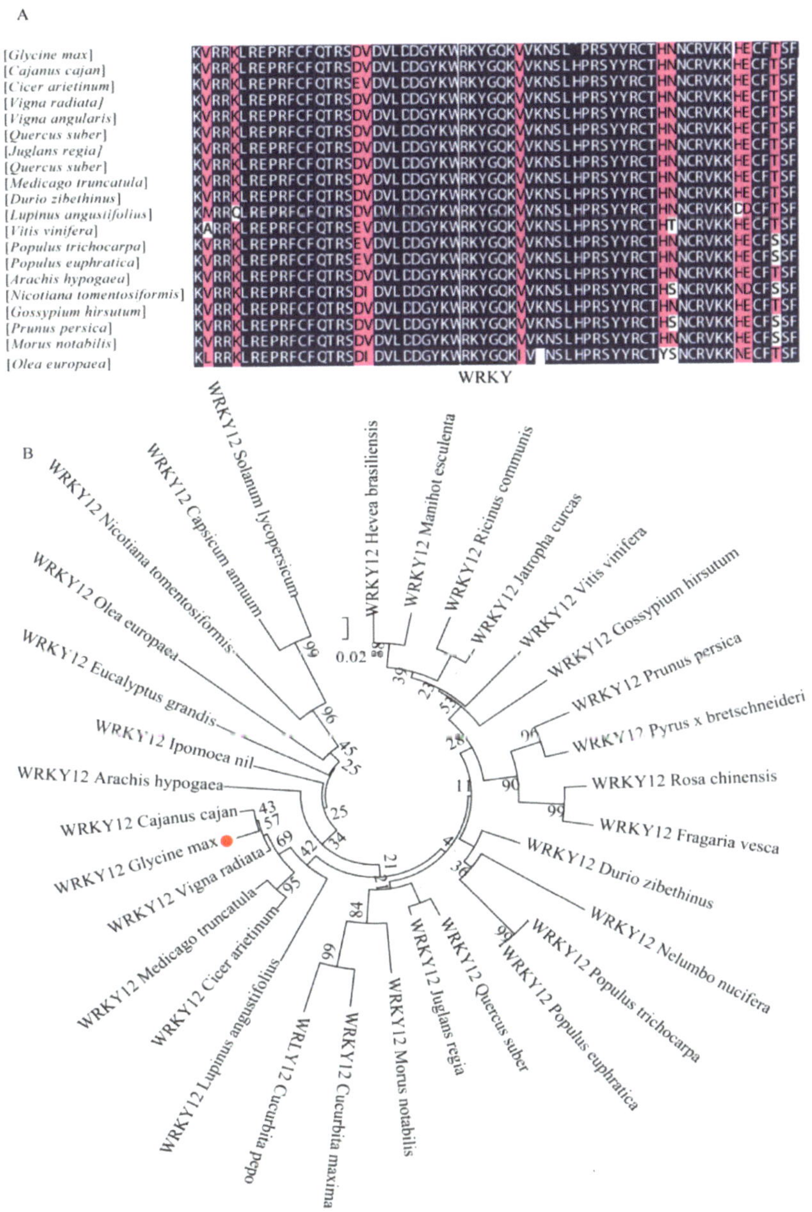

Figure 4. Multiple alignment and phylogenetic relationship of GmWRKY12 with different species.

(A) Multiple alignment of GmWRKY12 with other WRKY12 proteins from other species (B) Phylogenetic relationship of GmWRKY12 in different species. The red dot in (B) means GmWRKY12. The number of nodes is the bootstrap value and the number on the branch is the evolutionary distance. Bootstrap replications are 1000.

2.5 Expression Patterns of GmWRKY12 under Different Treatments

GmWRKY12 was responsive to drought and salt treatments (Figure 3). WRKY proteins are reported to be involved in signal transductions of plant hormones [39]. In order to identify whether GmWRKY12 was responsive to other abiotic stresses, expression patterns were identified using qRT-PCR. Results indicated that GmWRKY12 not only participated in drought and salt response but was also responsive to ABA and SA. Under low concentrations of SA, the expression profile of GmWRKY12 was increased about 50-fold (Figure 5).

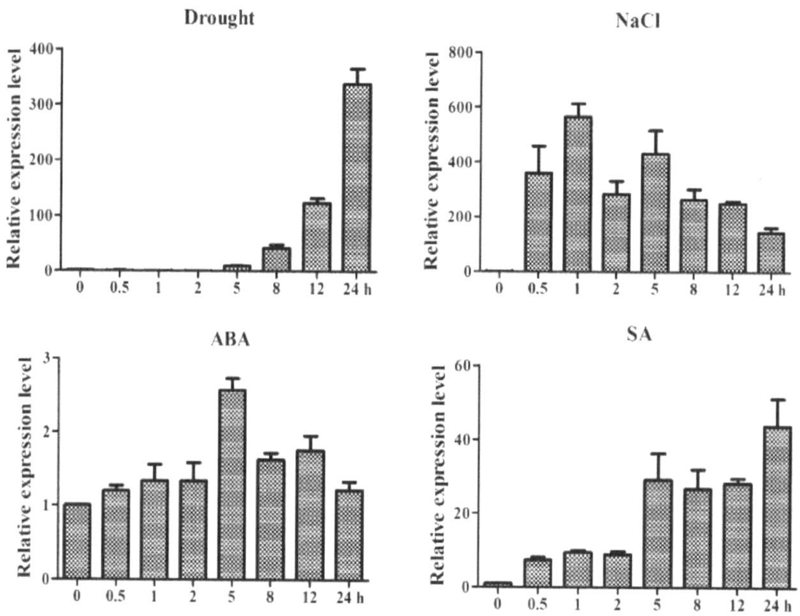

Figure 5. Expression patterns of GmWRKY12 under drought, NaCl, exogenous ABA and SA. The ordinates are the relative expression level (fold) of GmWRKY12 compared to non-stressed control. The horizontal ordinate is treatment time for 0, 0.5, 1, 2, 5, 8, 12 and 24 h. The expression level of GmActin as a loading control. All experiments were repeated three times. Error bars represent standard deviations (SDs). All data represent the means ± SDs of three independent biological replicates.

2.6 Cis-Acting Elements in Promoter

To further understand the regulatory mechanism of GmWRKY12, we isolated its promoter region. Cis-elements correlated to stress were present in the promoter region, including the ABA and wound responsive elements ABER4 and MYC, drought responsive element MYB, salt stress responsive element GT-1 and wound responsive element W-box. In addition, there was another element that participated in heat and GA response in the promoter region of GmWRKY12 (Table 3). This analysis suggested that GmWRKY12 may function in abiotic stress response.

Table 3. Cis-elements analysis of GmWRKY12 promotor.

Cis-Elements	Numbers	Target Sequences	Functions
MYC	32	CANNTG	ABA and wound responsive element
W-box	21	TTGAC/TTTGACY/TGACY	SA responsive element
ABER4	18	ACGT	ABA responsive element
MYB	14	C/TAACNA/G	Drought responsive element
CCAATB	10	CCAAT	Heat-responsive element
GT-1	7	GAAAAA	Salt stress responsive element
DPBF	6	ACACNNG	Dehydration-responsive element
GARE	2	TAACAAR	GA-responsive element
"Numbers" corresponds to the number of cis-elements of each type present in the promoter.			

2.7 GmWRKY12 was Located in the Nucleus

To investigate GmWRKY12 subcellular localization, GmWRKY12 were fused to the N-terminus of the humanized green fluorescent protein (hGFP) and co-transformed into wheat mesophyll protoplasts with the nucleus marker AT2G03340 (AtWRKY3)-mCherry [67,68]. The 35S::GFP vector was transformed as the control. Fluorescence of GmWRKY12 was specifically detected in the nucleus, whereas GFP fluorescence was distributed throughout the cell (Figure 6).

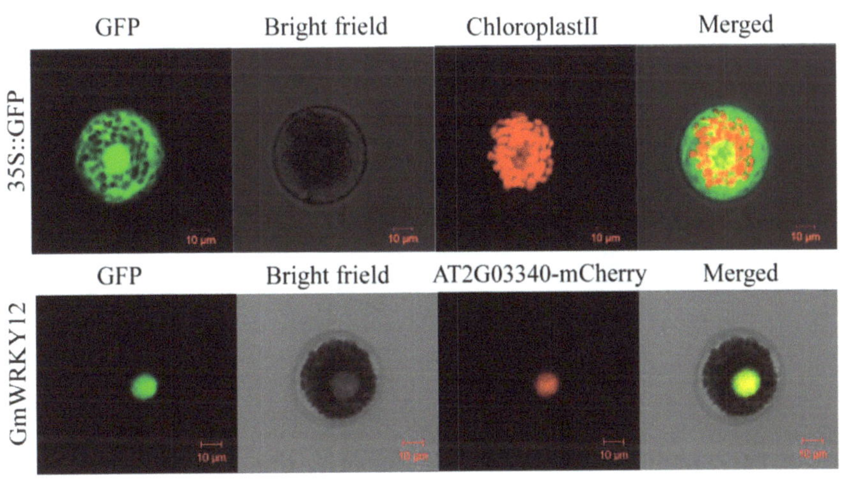

Figure 6. Co-localization of GmWRKY12. The recombinant plasmid of GmWRKY12-GFP and At2G03340-mCherry were co-transformed into wheat mesophyll

protoplasts under the control of the CaMV 35S promoter. GmWRKY12 was localized in the nucleus of wheat mesophyll protoplasts protoplasts. Results were observed by a confocal laser scanning microscope (LSM700; CarlZeiss, Oberkochen Germany) after incubating in darkness at 22 ◦C for 18–20 h. Scale bars = 10 μm.

2.8 GmWRKY12 Improved Drought and Salt Tolerance of Soybean

We further used transgenic hairy root assays to investigate the roles of GmWRKY12 in abiotic stress responses. Amplified cDNA sequence of GmWRKY12 was constructed into pCAMBIA3301 to create an overexpression transgenic line and the control was pCAMBIA3301 plant expression vector with CaMV35S promoter. Two constructs were transferred into Agrobacterium rhizogenes strain K599 (NCPPB2659) [69] then transformed into soybean hairy roots as previously described [70,71]. After drought treatment for 20 days, both control and over-expression soybean seedlings had leaf shedding to different degrees, especially the old leaves of the plants (Figure 7A). However, compared with transgenic soybean seedlings, the control seedlings were severely wilted and almost 99% of the leaves had serious dehydration and drying. By contrast, there was slight shedding of the old leaves of transgenic soybean seedlings but the new leaves were still growing vigorously. Results of proline and malondialdehyde (MDA) content determination showed that overexpression of GmWRKY12 increased proline content in transgenic lines, while the MDA content was decreased due to drought stress (Figure 7B,C). Fresh weight and main length of transgenic soybean hair roots under drought treatment were measured (Figure 8E,F), results showed overexpressed GmWRKY12 in soybean roots enhanced drought tolerance of soybean by increasing the length of transgenic hair roots and the number of transgenic hair roots.

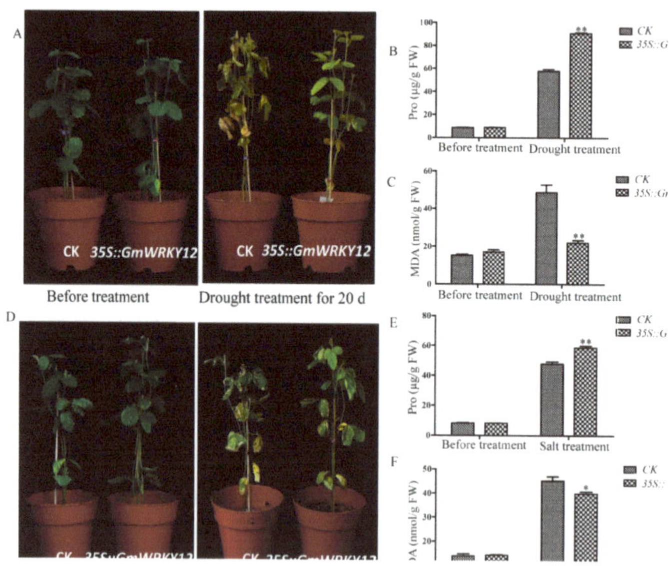

Figure 7. Phenotype identification of GmWRKY12 under drought and salt treatments. (A) Images of drought stress resistance phenotypes of CK and

35S::GmWRKY12 soybean seedlings after drought treatment for 20 days. (B) Proline contents in CK and 35S::GmWRKY12 soybean seedlings under normal growth conditions and drought treatment. (C) MDA contents in in CK and 35S::GmWRKY12 soybean seedlings under normal growth conditions and drought treatment. (D) Images of salt stress resistance phenotypes of CK and 35S::GmWRKY12 soybean seedlings after 200 mM NaCl treatment for 7 days. (E) Proline contents in CK and 35S::GmWRKY12 soybean seedlings under normal growth conditions and salt treatment. (F) MDA contents in CK and 35S::GmWRKY12 soybean seedlings under normal growth conditions and salt treatment. All data represent the means ± SDs of three independent biological replicates. ANOVA tests demonstrated that there were significant differences (* $p < 0.05$, $p < 0.01$).**

Meanwhile, under NaCl (200 mM) treatment, control and overexpression soybean seedlings had different degrees of leaf shedding (Figure 7D). Compared with the control, transgenic soybean seedlings were slightly wilted and slowly drying out, while the control seedlings were almost dry due to the osmotic stress. Results of Pro and MDA content in transgenic lines (Figure 7E,F) fresh weight and main length of transgenic soybean hair roots (Figure 8H,I) also showed that GmWRKY12 improved salt tolerance of soybean. These results demonstrated that GmWRKY12 confers stress tolerance in transgenic hairy roots.

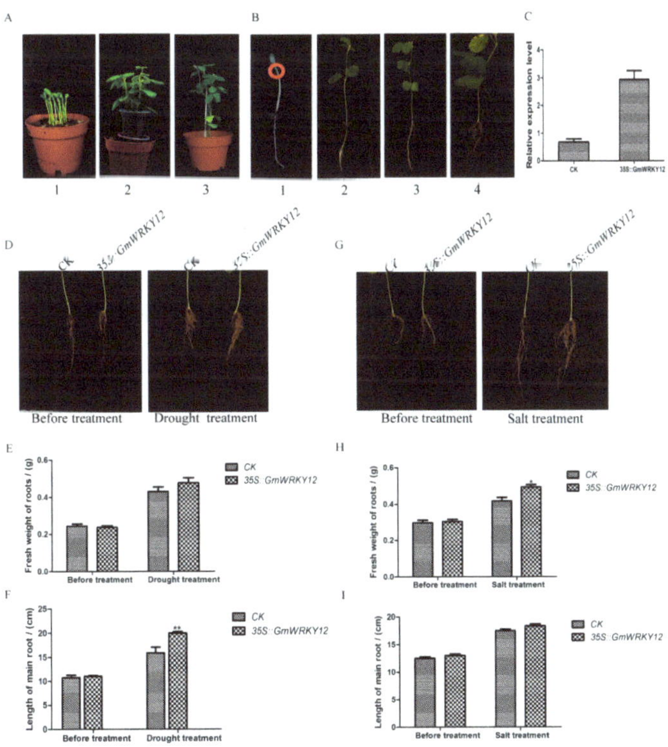

Figure 8. Different growth stage of transgenic soybean seedlings and phenotypes of

transgenic soybean hair roots. (A) Images of different growth stage of transgenic soybean seedlings cultivated in flowerpot before any treatment. (A1) Soybean seedlings of 5-days-old without injected A. rhizogenes carrying GmWRKY12. (A2) Soybean seedlings which have injected A. rhizogenes carrying GmWRKY12 for 7 days. (A3) Soybean seedlings which have injected A. rhizogenes carrying GmWRKY12 for 14 days (The original main roots were removed by cutting from 1 cm below the infection site and the hairy roots of the seedlings were cultivated in nutritious soil with full water and grown with 16 h light (100 μM photons m−2 ·s−1)/8 h dark at 25 °C). (B) Images of different growth stage of signal transgenic soybean seedling before any treatment. (B1) Soybean seedling of 5-days-old without injected A. rhizogenes carrying GmWRKY12 and the red circle shows the inject site of A. rhizogenes. (B2) Soybean seedling which have injected A. rhizogenes carrying GmWRKY12 for 7 days and new hair roots have generated. (B3) Soybean seedling which have injected A. rhizogenes carrying GmWRKY12 for 14 days. (B4) Soybean seedling that have salt treatment for 7days. (C) Relative expression of CK and 35S::GmWRKY12 transgenic soybean hair roots under normal growth conditions. (D) Images of drought stress resistance phenotypes of CK and 35S::GmWRKY12 transgenic soybean hair roots after drought treatment for 20 days. (E) Fresh weight in CK and 35S::GmWRKY12 transgenic soybean hair roots under normal growth conditions and drought treatment. (F) Length in CK and 35S::GmWRKY12 transgenic soybean hair roots under normal growth conditions and drought treatment. (G) Images of salt stress resistance phenotypes of CK and 35S::GmWRKY12 transgenic soybean hair roots after 200 mM NaCl treatment for 7 days. (H) Fresh weight in CK and 35S::GmWRKY12 transgenic soybean hair roots under normal growth conditions and salt treatment. (I) Length in CK and 35S::GmWRKY12 transgenic soybean hair roots under normal growth condition and salt treatment. All data represent the means ± SDs of three independent biological replicates. ANOVA tests demonstrated that there were significant differences (* $p < 0.05$, ** $p < 0.01$).

3. DISCUSSION

The WRKY transcription factor superfamily, as a recently described member of the TF family, has been studied by many researchers due to its numerous and diverse biological functions. Since the first reports of WRKY TFs [72], research conducted in different species [4,52,57,73,74] has shown that WRKY TFs play significant roles in plant development and stress responses. Recently, many studies of GmWRKY TFs have been based on biotechnological and RNA-Seq approaches [30,57]. However, these studies mainly reported genome-wide annotation of the WRKYs and structure analysis of some genes involved in response to abiotic and biotic stresses. Although these genes have been identified through biochemistry and bioinformatics approaches, knowledge about soybean stress tolerance was limited. In this study, based on qRT-PCR and RNA-Seq data, GmWRKY12 was selected for investigation of stress tolerance in soybean (Figure S1).

According to classifications in the WRKY family [18,28,75], WRKY12 belongs to Group IIc and contains a single WRKY domain and a CX4-5CX22-23HXH zinc-finger motif. Recent studies have shown that the WRKYGQK heptapeptide, which can specifically recognize and bind to the W-box consensus sequence (TTGACY) in the promoters of target genes, can be replaced by WRKYGKK, WRKYGEK, WKKYEDK, or WKKYCEDK; variations of the WRKYGQK motif might change the DNA binding specificities to downstream target genes [75]. However, multiple sequence alignment results showed that WRKY12 in different species only harbor the same WRKYGQK heptapeptide, demonstrating that WRKY12 protein is evolutionarily conserved and can recognize and bind to downstream target genes (Figure 4A). The result was consistent with the results observed in other species [54,57,65,76,77]. Structural conservation determines functional specificity: in rice, OsWRKY12 was related to normal plant growth and expression of OsWRKY12 was low at the seedling stage but increased gradually with growth [78]; similar results were found in specific tissues in our study. GmWRKY12 has low expression in young leaf, flower, one cm pod, pod shell 10 DAF, seed 10 DAF, seed 14 DAF, seed 21 DAF, seed 25 DAF, seed 28 DAF, seed 35 DAF, seed 42 DAF and root under normal conditions. At the pod shell 14 DAF and nodule stages, the expression levels gradually increase (Table S2), which may be because genes are differentially expressed at different growth stages, or may perform different activities, such as metabolism, nutrient absorption or material transformation. For example, at the nodule stage, plants are primarily vegetative, while at seed 42 DAF, plants are accumulating nutrients [57]. In addition, WRKY12 was related to plant flowering time: Arabidopsis plants overexpressing MlWRKY12 showed early flowering phenotype [79]. WRKY12 and WRKY13 have opposite effects on flowering time in the action of GA [80]. Overexpression of three Triticum genes, TaWRKY12, TaWRKY18 and TaZFP2 induced the expression of some genes related to Pi absorption and transportation, enhancing the abilities of Pi uptake and Pi use efficiency in plants under low-Pi stress conditions [81]. Thus, GmWRKY12, like other WRKYs, is involved in plant growth and development.

There are many cis-acting elements in the GmWRKY12 promoter region, such as MYC (ABA and wound responsive element), W-box (SA responsive element), ABER4 (ABA responsive element), MYB (drought responsive element), CCAATB (heat-responsive element), GT-1 (salt stress responsive element), DPBF (dehydration-responsive element) and GARE (GA-responsive element) (Table 3). The presence of these elements indicates that GmWRKY12 may take part in various biotic and abiotic responses except for growth and development of plants. Research of tobacco transcription factors NtWRKY12 and TGA2.2 found that NtWRKY12 alone was able to induce PR-1a::GUS expression to high levels, the PR-1a gene was salicylic acid-inducible to activate the expression of SA-inducible genes [82]. SA is an important endogenous molecule that activates plant hypersensitive response and systemic acquired resistance, which are often involved in disease resistance of plants [83]. As the closest orthologue of AtWRKY12, BrWRKY12 from Chinese cabbage conferred enhanced resistance to Pectobacterium carotovorum ssp. carotovorum (Pcc) through transcriptional activation of defense-related genes [84]. Furthermore, LrWRKY12 were responsive to SA and methyl jasmonate (MeJA) treatments and conferred more resistance to B. cinerea than in wild-type plants [85]. These results show that WRKY12 plays an important role in disease defense of plants, mainly because WRKYGQK specifically binds to the W-box to induce expression of downstream target genes.

In addition to the significant roles of WRKY12 identified in development and disease defense of plants, WRKY12 also functions in plant stress responses. Under treatment with NaCl and PEG, the expression level of THWRKY12 in Tamarix tissues was increased, the expression pattern of THWRKY12 after ABA treatment was approximately the same as the expression level changes under NaCl and PEG treatment, showing that the gene may participate in regulating salt and drought tolerance through the signaling pathway regulated by ABA [86]. In our study, GmWRKY12 was first screened following both drought and salt treatment using RNA-Seq. In order to confirm whether it was responsive to salt and drought stress, qRT-PCR was conducted and further showed that GmWRKY12 was highly expressed under drought and salt treatment, which indicated that the gene was related to drought and salt tolerance (Figure 3). Cis-acting elements and expression pattern analysis of GmWRKY12 also showed that it may participate in the ABA signaling pathway (Table 3 and Figure 5). However, compared to the high expression level under drought and salt treatment, on the condition of ABA, GmWRKY12 had low expression. Resistance identification of GmWRKY12 using a soybean hairy root assay further showed that GmWRKY12 may be involved in regulating salt and drought tolerance by promoting the combination of cis-acting elements with drought and salt-related genes, thereby enhancing plant resistance (Figure 7). Similar results were also found in other studies [87–89].

4. MATERIALS AND METHODS

4.1 Identification and Annotation of GmWRKYs Response to Drought/Salt Stress

Identification of the response of GmWRKYs to drought/salt stress was based on

RNA-seq data collected from a set of drought and salt stress experiments (Tables S5 and S6). Seeds of Williams 82 were cultivated in a 10 × 10 cm flowerpot (vermiculite: nutritious soil is 1:3), fresh leaf of 10-day-old soybean seedlings were used for RNA-Seq.CK1_treat-Expression represented two independent replicates of plants sampled before any treatment; GH_treat-Expression related to drought treatment for 5 h (put on the filter paper to simulate drought) of soybean plants at room temperature; CK2_treat-Expression without NaCl treatment; and NaCl_treat-Expression salt treatment that soaking soybean roots with 100 mM NaCl solution for 1 h and then sampled for RNA-seq [57,68]. Both log2 (GH_treat/CK1_treat) >1, log2 (NaCl_treat/CK2_treat) >1 and up-regulated were treated as the rule to select GmWRKYs responding to drought/salt stress. Several databases: NCBI (https://www.ncbi.nlm.nih.gov/pubmed), PlantTFDB (http://planttfdb.cbi.pku.edu.cn/), Phytozome (https://phytozome.jgi.doe.gov/pz/portal.html) and SoyDB (http://soykb.org/), were used to annotate Gene ID, Name, Chromosomal localization, CDS, Protein and Group.

4.2 Tissue-Specific Expression Patterns of GmWRKYs

Data of six different tissues (young leaf, flower, pod shell, seed, root and nodule) from different growth periods was available from SoyBase (https://www.soybase.org/soyseq/). Heml1.0 software (http://www.patrick-wied.at/static/heatmapjs/) was used to perform hierarchical clustering of fifty-three and nine GmWRKYs under normal conditions. The analysis data are available in Tables S1 and S2.

4.3 RNA Extraction and qRT-PCR

Seeds of Williams 82 was cultivated in a 10 × 10 cm flowerpot (vermiculite: nutritious soil is 1:3), fresh leaf tissue of 10-day-old soybean seedlings were used for RNA extraction of different stress treatment. For drought treatment, soybean seedlings were dried on filter paper then sampled 0.1 g of leaf on different periods (0, 0.5, 1, 2, 5, 8, 12 and 24 h), for salt, ABA and SA treatment, the roots of soybean seedlings were soaked in 100 mM NaCl, 100 μmol·L−1 ABA and 100 μmol·L−1 SA solution, respectively [68]. Then sampled 0.1 g of leaf on different periods (0, 0.5, 1, 2, 5, 8, 12 and 24 h), all samples were submerged immediately in liquid nitrogen and stored at -80 °C for RNA extraction using RNA prep plant kit (TIANGEN, Beijing, China); cDNA was synthesized using a Prime Script First-Strand cDNA Synthesis Kit (TransGen, Beijing, China) following the manufacturer's instructions. cDNA of treatment for 0 h was used for screen one highly expressed gene from seven GmWRKYs that response to both drought and salt treatment (Figure S1B). qRT-PCR was performed with Super Real PreMix Plus (TransGen, Beijing, China) on an ABI Prism 7500 system (Applied Biosystems, Foster City, CA, USA). Specific primers of GmWRKY3, 12, 14, 21, 28, 35, 43, 49 and soybean actin primers are listed in Table

S4. Three biological replicates were used for qRT-PCR analysis. The 2-ΔΔCt method was used for quantification.

4.4 Gene Isolation and Phylogenetic Analysis of GmWRKY12

Venn2.0 (http://bioinfogp.cnb.csic.es/tools/venny/index.html) was used to screen GmWRKYs that respond to both drought and salt treatment, then qRT-PCR was used to find genes highly expressed under stresses. Full-length GmWRKY12 was amplified by PCR with specific primers from soybean cDNA (Williams 82); primers of GmWRKY12 are available in Table S4. PCR products were cloned into pLB vector (TIANGEN, Beijing, China) and sequenced for further study. The amino acid sequence of WRKY12 in different species were searched for in the NCBI database on account of the amino acid similarity between GmWRKY12 and WRKY12 in different species is more than 50%. DNAMAN was applied for multiple sequence alignment on the basis of the amino acid similarity between GmWRKY12 and WRKY12 in different species is more than 60%. Phylogenetic trees were constructed using MEGA 6.0 with the neighbor-joining method [66] and 1000 bootstrap replications. Information of WRKY12 in different species is listed in Table S3.

4.5 Co-Localization of GmWRKY

Seeds of Kenong199 were cultivated in a 10 × 10 cm flowerpot (vermiculite: nutritious soil is 1:3), fresh leaf tissue of 7-day-old wheat seedlings were used for preparation of wheat protoplasts. Amplified cDNA sequence of GmWRKY12 was cloned into the N-terminus hGFP protein driven by the CaMV35S promoter. The cDNA coding sequences of AT2G03340 (AtWRKY3) which located in the nucleus [67] were fused to the N-terminus of the mCherry protein (WRKY25-RFP) under the control of the CaMV 35S promoter [68]. The recombinant plasmid of GmWRKY12-GFP and AtWRKY3-mCherry were co-transformed into wheat mesophyll protoplasts via the PEG4000-mediated method. The 35S::GFP vector was transformed as the control. Fluorescence was observed using a confocal laser scanning microscope (LSM700; CarlZeiss, Oberkochen, Germany) after incubating in darkness at 22 ◦C for 18–20 h. Primers are available in Table S4.

4.6 Cis-acting Elements in Promoter

The 2.0 kb promoter region upstream of the ATG start codon in the promoter of GmWRKY12 was obtained from soybean genomic DNA in the Ensembl Plants website, cis-acting elements were analyzed by PLACE (http://www.dna.affrc.go.jp/PLACE/).

4.7 A. rhizogenes-mediated Drought and Salt Stress Assays

To generate a transgenic line of soybean the amplified cDNA sequence of GmWRKY12 was constructed into pCAMBIA3301 for an overexpression transgenic line (35S::GmWRLY12) and the control was pCAMBIA3301 plant vector with CaMV35S promoter (CK) and two constructs transferred into A. rhizogenes strain K599 (NCPPB2659) [69]. Primers are available in Table S4. Williams 82 was cultivated in a 10 × 10 cm flowerpot (vermiculite: nutritious soil is 1:3) for stress experiments (Figure 8A1), soybean seeds were grown under a photoperiod of 16 h light (100 μM photons m−2·s−1)/8 h dark at 25 ◦C. When plants displayed two cotyledons (Figure 8A1), A. rhizogenes strain K599 harboring pCAMBIA3301 (CK) and K599 harboring 35S::GmWRLY12 were injected at the cotyledonary node and/or hypocotyl (Figure 8B1). A plastic cup was used to surround the inoculated soybean seedlings to provide high humidity conditions. After 3 days, nutritious soil was prepared and used to fill the gaps in the plastic cup so that soybean seedlings could grow new roots (Figure 8A2); plants typically need two weeks to generate new roots (2–10 cm) at the infection site (Figure 8B2,B3). The original main roots were removed by cutting from 1 cm below the infection site and the hairy roots of the seedlings were cultivated in nutritious soil with full water and grown with 16 h light (100 μM photons m−2·s−1)/8 h dark at 25 ◦C for 5 days [70,71]. Each flowerpot cultivated 5 transgenic soybean seedlings and 5 replications of each stress treatment (Figure 8A3). Afterward, the transgenic soybean seedlings were subjected to natural dehydration and 200 mM NaCl for drought and salt stress assays [19,68]. For drought stress assay, both CK and transgenic soybean seedlings were grown without water for 20 days. For salt stress assay, CK and transgenic soybean seedlings were treated with 200 mM NaCl solution for 7 days. There are some Supplement Materials need to prepare for culturing A. rhizogenes strain K599 that harbored (35S::GmWRLY12) and the control (CK)., eg: Solidified LB medium with streptomycin sulfate (100 mg/L) and Kanamycin solution(100 mg/L) (10 g tryptone, 5 g yeast extract, 10 g NaCl, 15 g agar per liter), Liquid LB medium containing streptomycin sulfate (100 mg/L) and Kanamycin solution (100 mg/L) [69].

4.8 Measurements of Proline and MDA Contents

Both proline and MDA content were measured with the Pro and MDA assay kit (Comin, Beijing, China) based on the manufacturer's protocols; all measurements were from three biological replicates.

4.9 Measurements of Fresh Weight and Main Length

Transgenic soybean hair roots were used to measure the fresh weight and main length. All data represent the means ± SDs of three independent biological replicates.

5. CONCISE SUMMARY

This study utilized RNA-Seq to identify 62 GmWRKY genes in the soybean genome that exhibited differential expression across six distinct tissues under normal conditions. Seven GmWRKYs exhibited a response to both drought and salinity treatment. According to the qRT-PCR analysis, GmWRKY12, a nuclear protein of 237 amino acids, was identified as part of WRKY Group II. It exhibited responsiveness to salinity, aridity, and the exogenous hormones abscisic acid and salicylic acid. The results of the Agrobacterium rhizogenes-mediated hairy root assay indicated that the overexpression of GmWRKY12 may enhance drought and salt tolerance in soybean. These findings offered novel insights into the functions of soybean WRKY genes in responses to abiotic stress.

REFERENCES

1. Peng, X.; Ma, X.; Fan, W.; Man, S.; Cheng, L.; Alam, I.; Lee, B.H.; Qi, D.; Shen, S.; Liu, G. Improved drought and salt tolerance of Arabidopsis thaliana by transgenic expression of a novel DREB gene from Leymus chinensis. Plant Cell Rep. 2011, 30, 1493–1502.
2. Hong, C.; Cheng, D.; Zhang, G.; Zhu, D.; Chen, Y.; Tan, M. The role of ZmWRKY4 in regulating maize antioxidant defense under cadmium stress. Biochem. Bioph. Res. Commun. 2016, 482, 1504–1510. [CrossRef]
3. Fu, J.; Liu, Q.; Wang, C.; Liang, J.; Liu, L.; Wang, Q. ZmWRKY79 positively regulates maize phytoalexin biosynthetic gene expression and is involved in stress response. J. Exp. Bot. 2017, 69, 497–510. [CrossRef]
4. Liu, Q.L.; Xu, K.D.; Pan, Y.Z.; Jiang, B.B.; Liu, G.L.; Jia, Y.; Zhang, H.Q. Functional Analysis of a Novel Chrysanthemum WRKY Transcription Factor Gene Involved in Salt Tolerance. Plant Mol. Biol. Rep. 2014, 32, 282–289. [CrossRef]
5. Zhu, J.K. Cell signaling under salt, water and cold stresses. Curr. Opin. Plant Biol. 2001, 4, 401–406. [CrossRef]
6. Horie, T.; Hauser, F.; Schroeder, J.I. HKT transporter-mediated salinity resistance mechanisms in Arabidopsis and monocot crop plants. Trends Plant Sci. 2009, 14, 660–668. [CrossRef]
7. Takahashi, R.; Liu, S.; Takano, T. Cloning and functional comparison of a high-affinity K+ transporter gene PhaHKT1 of salt-tolerant and salt-sensitive reed plants. J. Exp. Bot. 2007, 58, 4387–4395. [CrossRef] [PubMed]
8. Guan, R.; Qu, Y.; Guo, Y.; Yu, L.; Liu, Y.; Jiang, J.; Chen, J.; Ren, Y.; Liu, G.; Tian, L. Salinity tolerance in soybean is modulated by natural variation in GmSALT3. Plant J. 2015, 80, 937–950. [CrossRef]
9. Yokoi, S.; Quintero, F.J.; Cubero, B.; Ruiz, M.T.; Bressan, R.A.; Hasegawa, P.M.; Pardo, J.M. Differential expression and function of Arabidopsis thaliana NHX Na+/H+ antiporters in the salt stress response. Plant J. 2010, 30, 529–539. [CrossRef]

10. Rodríguez-Rosales, M.P.; Gálvez, F.J.; Huertas, R.; Aranda, M.N.; Baghour, M.; Cagnac, O.; Venema, K. Plant NHX cation/proton antiporters. Plant Signal. Behav. 2009, 4, 265–276. [CrossRef]

11. Shigaki, T.; Hirschi, K. Characterization of CAX-like genes in plants: Implications for functional diversity. Gene 2000, 257, 291–298. [CrossRef]

12. Shigaki, T.; Hirschi, K.D. Diverse functions and molecular properties emerging for CAX cation/H+ exchangers in plants. Plant Biol. 2006, 8, 419–429. [CrossRef] [PubMed]

13. Qi, X.; Li, M.W.; Xie, M.; Liu, X.; Ni, M.; Shao, G.; Song, C.; Kay-Yuen, Y.A.; Tao, Y.; Wong, F.L. Identification of a novel salt tolerance gene in wild soybean by whole-genome sequencing. Nat. Commun. 2014, 5. [CrossRef] [PubMed]

14. Wang, J.W.; Yang, F.P.; Chen, X.Q.; Liang, R.Q.; Zhang, L.Q.; Geng, D.M.; Zhang, X.D.; Song, Y.Z.; Zhang, G.S. Induced expression of DREB transcriptional factor and study on its physiological effects of drought tolerance in transgenic wheat. Acta Genetica Sinica 2006, 33, 468–476. [CrossRef]

15. Morran, S.; Eini, O.; Pyvovarenko, T.; Parent, B.; Singh, R.; Ismagul, A.; Eliby, S.; Shirley, N.; Langridge, P.; Lopato, S. Improvement of stress tolerance of wheat and barley by modulation of expression of DREB/CBF factors. Plant Biotechnol. J. 2011, 9, 230–249. [CrossRef] [PubMed]

16. Dubos, C.; Stracke, R.; Grotewold, E.; Weisshaar, B.; Martin, C.; Lepiniec, L. MYB transcription factors in Arabidopsis. Trends Plant Sci. 2010, 15, 573–581. [CrossRef] [PubMed]

17. Eulgem, T.; Rushton, P.J.; Robatzek, S.; Somssich, I.E. The WRKY superfamily of plant transcription factors. Trends Plant Sci. 2000, 5, 199–206. [CrossRef]

18. Eulgem, T.; Somssich, I.E. Networks of WRKY transcription factors in defense signaling. Curr. Opin. Plant Biol. 2007, 10, 366–371. [CrossRef] [PubMed]

19. Hao, Y.J.; Wei, W.; Song, Q.X.; Chen, H.W.; Zhang, Y.Q.; Wang, F.; Zou, H.F.; Lei, G.; Tian, A.G.; Zhang, W.K. Soybean NAC transcription factors promote abiotic stress tolerance and lateral root formation in transgenic plants. Plant J. 2011, 68, 302–313. [CrossRef]

20. Jin, H.X.; Huang, F.; Cheng, H.; Song, H.N.; Yu, D.Y. Overexpression of the GmNAC2 Gene, an NAC Transcription Factor, Reduces Abiotic Stress Tolerance in Tobacco. Plant Mol. Biol. Rep. 2013, 31, 435–442. [CrossRef]

21. Guerinot, M.L. The ZIP family of metal transporters. BBA Biomembr. 2000, 1465, 190–198. [CrossRef]

22. Liao, Y.; Zou, H.F.; Wei, W.; Hao, Y.J.; Tian, A.G.; Huang, J.; Liu, Y.F.; Zhang,

J.S.; Chen, S.Y. Soybean GmbZIP44, GmbZIP62 and GmbZIP78 genes function as negative regulator of ABA signaling and confer salt and freezing tolerance in transgenic Arabidopsis. Planta 2008, 228, 225–240. [CrossRef]

23. Xu, Z.S.; Chen, M.; Ma, L.C.; Ma, Y.Z. Functions and application of the AP2/ERF franscription factor family in crop improvement. Bull Bot. 2011, 61, 570–585.
24. Xu, Z.S.; Chen, M.; Li, L.C.; Ma, Y.Z. Functions of the ERF transcription factor family in plants. Botany 2008, 86, 969–977. [CrossRef]
25. Fukuda, A.; Okada, Y.; Suzui, N.; Fujiwara, T.; Yoneyama, T.; Hayashi, H. Cloning and characterization of the gene for a phloem-specific glutathione S -transferase from rice leaves. Physiol. Plant. 2010, 120, 595–602. [CrossRef] [PubMed]
26. Hundertmark, M.; Hincha, D.K. LEA (Late Embryogenesis Abundant) proteins and their encoding genes in Arabidopsis thaliana. BMC Genomics 2008, 9. [CrossRef]
27. Yan, J.; Wang, B.; Jiang, Y.; Cheng, L.; Wu, T. GmFNSII-Controlled Soybean Flavone Metabolism Responds to Abiotic Stresses and Regulates Plant Salt Tolerance. Plant Cell Physiol. 2014, 55, 74–86. [CrossRef]
28. Fei, C.; Yue, H.; Vannozzi, A.; Wu, K.; Cai, H.; Yuan, Q. The WRKY transcription factor family in model plants and crops. Crit. Rev. Plant Sci. 2018, 36, 1–25.
29. Rushton, P.J.; Somssich, I.E.; Ringler, P.; Shen, Q.J. WRKY transcription factors. Plant Signal. Behav. 2010, 15, 247–258. [CrossRef]
30. Yu, Y.; Wang, N.; Hu, R.; Xiang, F. Genome-wide identification of soybean WRKY transcription factors in response to salt stress. Springerplus 2016, 5. [CrossRef]
31. Lagacã, M.; Matton, D.P. Characterization of a WRKY transcription factor expressed in late torpedo-stage embryos of Solanum chacoense. Planta 2004, 219, 185–189. [CrossRef] [PubMed]
32. Zentella, R.; Zhang, Z.; Park, M.; Thomas, S.; Endo, A.; Murase, K.; Fleet, C.; Jikumaru, Y.; Nambara, E.; Kamiya, Y. Global analysis of della direct targets in early gibberellin signaling in Arabidopsis. Plant Cell 2007, 19, 3037–3057. [CrossRef] [PubMed]
33. Silke, R.; Imre, E.S. Targets of AtWRKY6 regulation during plant senescence and pathogen defense. Genes Dev. 2002, 16, 1139–1149.
34. Johnson, C.S.; Kolevski, B.; Smyth, D.R. Transparent TESTA glabra2, a trichome and seed coat development gene of Arabidopsis, encodes a WRKY transcription factor. Plant Cell 2002, 14, 1359–1375. [CrossRef] [PubMed]

35. Pandey, S.P.; Somssich, I.E. The role of WRKY transcription factors in plant immunity. Plant Physiol. 2009, 150, 1648–1655. [CrossRef] [PubMed]

36. Birkenbihl, R.P.; Diezel, C.; Somssich, I.E. Arabidopsis WRKY33 is a key transcriptional regulator of hormonal and metabolic responses toward Botrytis cinerea infection. Plant Physiol. 2012, 159, 266–285. [CrossRef] [PubMed]

37. Duan, Y.; Jiang, Y.; Ye, S.; Karim, A.; Ling, Z.; He, Y.; Yang, S.; Luo, K. PtrWRKY73, a salicylic acid-inducible poplar WRKY transcription factor, is involved in disease resistance in Arabidopsis thaliana. Plant Cell Rep. 2015, 34, 831–841. [CrossRef]

38. Qiu, Y.P.; Yu, D.Q. Over-expression of the stress-induced OsWRKY45 enhances disease resistance and drought tolerance in Arabidopsis. Environ. Exp. Bot. 2009, 65, 35–47. [CrossRef]

39. Zhang, L.; Gu, L.; Ringler, P.; Smith, S.; Rushton, P.J.; Shen, Q.J. Three WRKY transcription factors additively repress abscisic acid and gibberellin signaling in aleurone cells. Int. J. Exp. Plant Biol. 2015, 236, 214–222. [CrossRef]

40. Zou, X.; Seemann, J.R.; Neuman, D.; Shen, Q.J. A WRKY gene from creosote bush encodes an activator of the abscisic acid signaling pathway. J. Biol. Chem. 2004, 279, 55770–55779. [CrossRef]

41. Shimono, M.; Sugano, S.; Nakayama, A.; Jiang, C.J.; Ono, K.; Toki, S.; Takatsuji, H. Rice WRKY45 plays a crucial role in benzothiadiazole-inducible blast resistance. Plant Cell 2007, 19, 2064 2076. [CrossRef] [PubMed]

42. Chen, L.; Song, Y.; Li, S.; Zhang, L.; Zou, C.; Yu, D. The role of WRKY transcription factors in plant abiotic stresses. BBA Gene Regul. Mech. 2012, 1819, 120–128. [CrossRef] [PubMed]

43. Chen, J.; Nolan, T.; Ye, H.; Zhang, M.; Tong, H.; Xin, P.; Chu, J.; Chu, C.; Li, Z.; Yin, Y. Arabidopsis WRKY46, WRKY54 and WRKY70 transcription factors are involved in brassinosteroid-regulated plant growth and drought response. Plant Cell 2017, 29, 1425–1439. [CrossRef] [PubMed]

44. Erpen, L.; Devi, H.S.; Grosser, J.W.; Dutt, M. Potential use of the DREB/ERF, MYB, NAC and WRKY transcription factors to improve abiotic and biotic stress in transgenic plants. Plant Cell Tissue Organ 2018, 132, 1–25. [CrossRef]

45. Taji, T.; Ohsumi, C.; Iuchi, S.; Seki, M.; Kasuga, M.; Kobayashi, M.; YamaguchiShinozaki, K.; Shinozaki, K. Important roles of drought- and cold-inducible genes for galactinol synthase in stress tolerance in Arabidopsis thaliana. Plant J. 2010, 29, 417–426. [CrossRef]

46. Wang, C.T.; Ru, J.N.; Liu, Y.W.; Li, M.; Zhao, D.; Yang, J.F.; Fu, J.D.; Xu, Z.S. Maize WRKY Transcription Factor ZmWRKY106 Confers Drought and Heat Tolerance in Transgenic Plants. Int. J. Mol. Sci. 2018, 19. [CrossRef] [PubMed]

47. He, G.H.; Xu, J.Y.; Wang, Y.X.; Liu, J.M.; Li, P.S.; Ming, C.; Ma, Y.Z.; Xu, Z.S. Drought-responsive WRKY transcription factor genes TaWRKY1 and TaWRKY33 from wheat confer drought and/or heat resistance in Arabidopsis. BMC Plant Biol. 2016, 16. [CrossRef]

48. Wang, C.T.; Ru, J.N.; Liu, Y.W.; Yang, J.F.; Li, M.; Xu, Z.S.; Fu, J.D. The Maize WRKY transcription factor ZmWRKY40 confers drought resistance in transgenic Arabidopsis. Int. J. Mol. Sci. 2018, 19. [CrossRef]

49. Ulker, B.; Somssich, I.E. WRKY transcription factors: From DNA binding towards biological function. Curr. Opin. Plant Biol. 2004, 7, 491–498. [CrossRef]

50. Jiang, Y.; Qiu, Y.; Hu, Y.; Yu, D. Heterologous expression of AtWRKY57 confers drought tolerance in Oryza sativa. Front. Plant Sci. 2016, 7. [CrossRef]

51. Niu, C.F.; Wei, W.; Zhou, Q.Y.; Tian, A.G.; Hao, Y.J.; Zhang, W.K.; Ma, B.; Lin, Q.; Zhang, Z.B.; Zhang, J.S. Wheat WRKY genes TaWRKY2 and TaWRKY19 regulate abiotic stress tolerance in transgenic Arabidopsis plants. Plant Cell Environ. 2012, 35, 1156–1170. [CrossRef] [PubMed]

52. Wu, X.; Shiroto, Y.; Kishitani, S.; Ito, Y.; Toriyama, K. Enhanced heat and drought tolerance in transgenic rice seedlings overexpressing OsWRKY11 under the control of HSP101 promoter. Plant Cell Rep. 2009, 28, 21–30. [CrossRef] [PubMed]

53. Lee, H.; Cha, J.; Choi, C.; Choi, N.; Ji, H.S.; Park, S.R.; Lee, S.; Hwang, D.J. Rice WRKY11 plays a role in pathogen defense and drought tolerance. Rice 2018, 11. [CrossRef] [PubMed]

54. Cai, R.; Zhao, Y.; Wang, Y.; Lin, Y.; Peng, X.; Li, Q.; Chang, Y.; Jiang, H.; Xiang, Y.; Cheng, B. Overexpression of a maize WRKY58 gene enhances drought and salt tolerance in transgenic rice. Plant Cell Tissue Organ 2014, 119, 565–577. [CrossRef]

55. Li, H.; Gao, Y.; Xu, H.; Dai, Y.; Deng, D.; Chen, J. ZmWRKY33, a WRKY maize transcription factor conferring enhanced salt stress tolerances in Arabidopsis. Plant Growth Regul. 2013, 70, 207–216. [CrossRef]

56. Ullah, A.; Sun, H.; Hakim; Yang, X.; Zhang, X. A novel cotton WRKY-gene, GhWRKY6-like, improves salt tolerance by activating the ABA signalling pathway and scavenging of reactive oxygen species. Physiol. Plant 2017, 162, 439–454. [CrossRef] [PubMed]

57. Song, H.; Wang, P.; Hou, L.; Zhao, S.; Zhao, C.; Xia, H.; Li, P.; Zhang, Y.; Bian, X.; Wang, X. Global analysis of WRKY genes and their response to dehydration and salt stress in soybean. Front. Plant Sci. 2016, 7. [CrossRef]

58. Zhou, Q.Y.; Tian, A.G.; Zou, H.F.; Xie, Z.M.; Lei, G.; Huang, J.; Wang, C.M.; Wang, H.W.; Zhang, J.S.; Chen, S.Y. Soybean WRKY-type transcription factor genes, GmWRKY13, GmWRKY21 and GmWRKY54, confer differential tolerance to abiotic stresses in transgenic Arabidopsis plants. Plant Biotechnol. J. 2010, 6, 486–503. [CrossRef]

59. Li, J.; Wang, J.; Wang, N.; Guo, X.; Gao, Z. GhWRKY44, a WRKY transcription factor of cotton, mediates defense responses to pathogen infection in transgenic Nicotiana benthamiana. Plant Cell Tissue Organ 2015, 121, 127–140. [CrossRef]

60. Liu, X.; Song, Y.; Xing, F.; Wang, N.; Wen, F.; Zhu, C. GhWRKY25, a group I WRKY gene from cotton, confers differential tolerance to abiotic and biotic stresses in transgenic Nicotiana benthamiana. Protoplasma 2015, 253, 1–17. [CrossRef]

61. Wang, Y.; Jiang, L.; Chen, J.; Tao, L.; An, Y.; Cai, H. Overexpression of the alfalfa WRKY11 gene enhances salt tolerance in soybean. PLoS ONE 2018, 13, e0192382. [CrossRef]

62. Mehdi, P.; Vijayaraj, N.; Youping, D. GeneVenn—A web application for comparing gene lists using Venn diagrams. Bioinformation 2007, 1, 420–422.

63. Berri, S.; Abbruscato, P.; FaivreRampant, O.; Brasileiro, A.C.; Fumasoni, I.; Satoh, K.; Kikuchi, S.; Mizzi, L.; Morandini, P.; Pè, M.E. Characterization of WRKY co-regulatory networks in rice and Arabidopsis. BMC Plant Biol. 2009, 9, 1–22. [CrossRef] [PubMed]

64. Guo, C.; Guo, R.; Xu, X.; Gao, M.; Li, X.; Song, J.; Zheng, Y.; Wang, X. Evolution and expression analysis of the grape (Vitis vinifera L.) WRKY gene family. J. Exp. Bot. 2014, 65, 1513–1528. [CrossRef] [PubMed]

65. Huang, S.; Gao, Y.; Liu, J.; Peng, X.; Niu, X.; Fei, Z.; Cao, S.; Liu, Y. Genome-wide analysis of WRKY transcription factors in Solanum lycopersicum. Mol. Genet. Genomics 2012, 287, 495–513. [CrossRef] [PubMed]

66. Tamura, K.; Stecher, G.; Peterson, D.; Filipski, A.; Kumar, S. MEGA6: Molecular evolutionary genetics analysis version 6.0. Mol. Biol. Evol. 2013, 30, 2725–2729. [CrossRef] [PubMed]

67. Miller, M.J.; Barrett-Wilt, G.A.; Zhihua, H.; Vierstra, R.D. Proteomic analyses identify a diverse array of nuclear processes affected by small ubiquitin-like modifier conjugation in Arabidopsis. Proc. Natl. Acad. Sci. USA 2011, 107, 16512–16517. [CrossRef]

68. Du, Y.T.; Zhao, M.J.; Wang, C.T.; Gao, Y.; Wang, Y.X.; Liu, Y.W.; Chen, M.; Chen, J.; Zhou, Y.B.; Xu, Z.S. Identification and characterization of GmMYB118 responses to drought and salt stress. BMC Plant Biol. 2018, 18. [CrossRef]

69. Kereszt, A.; Li, D.; Indrasumunar, A.; Nguyen, C.D.; Nontachaiyapoom, S.; Kinkema, M.; Gresshoff, P.M. Agrobacterium rhizogenes-mediated transformation of soybean to study root biology. Nat. Protoc. 2007, 2, 948–952. [CrossRef]

70. Cao, D.; Hou, W.S.; Song, S.K.; Sun, H.B.; Wu, C.X.; Gao, Y.S.; Han, T.F. Assessment of conditions affecting Agrobacterium rhizogenes-mediated transformation of soybean. Plant Cell Tissue Organ 2009, 96, 45–52. [CrossRef]

71. Wang, F.; Chen, H.W.; Li, Q.T.; Wei, W.; Li, W.; Zhang, W.K.; Ma, B.; Bi, Y.D.; Lai, Y.C.; Liu, X.L. GmWRKY27 interacts with GmMYB174 to reduce expression of GmNAC29 for stress tolerance in soybean plants. Plant J. 2015, 83, 224–236. [CrossRef] [PubMed]

72. Ishiguro, S.; Nakamura, K. Characterization of a cDNA encoding a novel DNA-binding protein, SPF1, that recognizes SP8 sequences in the 5t upstream regions of genes coding for sporamin and β -amylase from sweet potato. Mol. Genet. Genomics 1994, 244, 563–571. [CrossRef]

73. Wei, K.F.; Chen, J.; Chen, Y.F.; Wu, L.J.; Xie, D.X. Molecular phylogenetic and expression analysis of the complete WRKY transcription factor family in Maize. DNA Res. 2012, 19, 153–164. [CrossRef] [PubMed]

74. Wu, K.L. The WRKY family of transcription factors in rice and Arabidopsis and their origins. DNA Res. 2005, 12, 9–26. [CrossRef]

75. Jue, D.; Sang, X.; Liu, L.; Shu, B.; Wang, Y.; Liu, C.; Xie, J.; Shi, S. Identification of WRKY Gene Family from Dimocarpus longan and its expression analysis during flower induction and abiotic stress responses. Int. J. Mol. Sci. 2018, 19. [CrossRef] [PubMed]

76. Kim, C.Y.; Vo, K.T.X.; Cong, D.N.; Jeong, D.H.; Lee, S.K.; Kumar, M.; Kim, S.R.; Park, S.H.; Kim, J.K.; Jeon, J.S. Functional analysis of a cold-responsive rice WRKY gene, OsWRKY71. Plant Biotechnol. Rep. 2016, 10, 13–23. [CrossRef]

77. Ding, M.; Chen, J.; Jiang, Y.; Lin, L.; Cao, Y.; Wang, M.; Zhang, Y.; Rong, J.; Ye, W. Genome-wide investigation and transcriptome analysis of the WRKY gene family in Gossypium. Mol. Genet. Genomics 2015, 290, 151–171. [CrossRef]

78. Shi, J.N.; Li-Yun, L.I.; Wen-Jing, X.U.; Guan, M.L.; Xue-Jiao, L.I.; Niu, D.D.; Lan, J.P.; Dou, S.J.; Liu, L.J.; Liu, G.Z. Expression analysis of eight WRKY transcription factors in rice leaf growth and disease resistance response. Acta Phytopathol.

Sin. 2014, 44, 54–64.

79. Yu, Y.; Hu, R.; Wang, H.; Cao, Y.; He, G.; Fu, C.; Zhou, G. MlWRKY12, a novel Miscanthus transcription factor, participates in pith secondary cell wall formation and promotes flowering. Int. J. Exp. Plant Biol. 2013, 212, 1–9. [CrossRef]

80. Li, W.; Wang, H.; Yu, D. Arabidopsis WRKY Transcription Factors WRKY12 and WRKY13 oppositely regulate flowering under short-day conditions. Mol. Plant 2016, 9, 1492–1503. [CrossRef]

81. Li, X.J.; Guo, C.J.; Lu, W.J.; Duan, W.W.; Zhao, M.; Ma, C.Y.; Gu, J.T.; Xiao, K. expression pattern analysis of Zinc finger protein genes in wheat (Triticum aestivum L.) under phosphorus deprivation. J. Integr. Agric. 2014, 13, 1621–1633. [CrossRef]

82. Van Verk, M.C.; Neeleman, L.; Bol, J.F.; Linthorst, H.J. Tobacco transcription factor NtWRKY12 interacts with TGA2.2 in vitro and in vivo. Front. Plant Sci. 2011, 2, 1085–1091. [CrossRef] [PubMed]

83. Thurow, C.; Schiermeyer, A.; Krawczyk, S.; Butterbrodt, T.; Nickolov, K.; Gatz, C. Tobacco bZIP transcription factor TGA2.2 and related factor TGA2.1 have distinct roles in plant defense responses and plant development. Plant J. 2010, 44, 100–113. [CrossRef] [PubMed]

84. Kim, H.S.; Park, Y.H.; Nam, H.; Lee, Y.M.; Song, K.; Choi, C.; Ahn, I.; Park, S.R.; Lee, Y.H.; Hwang, D.J. Overexpression of the Brassica rapa transcription factor WRKY12 results in reduced soft rot symptoms caused by Pectobacterium carotovorum in Arabidopsis and Chinese cabbage. Plant Biol. 2015, 16, 973–981. [CrossRef] [PubMed]

85. Cui, Q.; Yan, X.; Gao, X.; Zhang, D.M.; He, H.B.; Jia, G.X. Analysis of WRKY transcription factors and characterization of two Botrytis cinerea-responsive LrWRKY genes from Lilium regale. Plant Physiol. Biochem. 2018, 525–536. [CrossRef] [PubMed]

86. Wang, J.X.; Tang, Y.K.; Liu, Q.; Liang, C.Q.; Wang, Y.C. Cloning and expression analysis of THWRKY12 gene from Tamarix hispida. Chin. J. Agric. Biotechol. 2013, 5, 55–57.

87. Baranwal, V.K.; Negi, N.; Khurana, P. Genome-wide identification and structural, functional and evolutionary analysis of WRKY components of Mulberry. Sci. Rep. 2016, 6. [CrossRef]

88. Tao, X.; Chen, C.; Li, C.; Liu, J.; Liu, C.; He, Y. Genome-wide investigation of WRKY gene family in pineapple: Evolution and expression profiles during development and stress. BMC Genomics 2018, 19. [CrossRef]

89. Tan, C.K.; Carey, A.J.; Cui, X.; Webb, R.I.; Ipe, D.; Crowley, M.; Cripps, A.W.; Benjamin, B.W., Jr.; Ulett, K.B.; Schembri, M.A. Genome-wide mapping of cystitis due to Streptococcus agalactiae and Escherichia coli in mice identifies a unique bladder transcriptome that signifies pathogen-specific antimicrobial defense against urinary tract infection. Infect. Immun. 2012, 80, 3145–3160. [CrossRef]

12

Chapter

COMPREHENSIVE GENOME CHARACTERIZATION AND IDENTIFICATION OF TRIHELIX TRANSCRIPTION FACTORS AND EXPRESSION PROFILING IN RESPONSE TO ABIOTIC STRESSES IN RICE (ORYZA SATIVA L.).

1. INTRODUCTION: AN OVERVIEW

Transcription factors are prevalent in plants. They are integral to numerous growth and development processes and react to abiotic stressors [1]. Prior research indicated the existence of over 60 families of transcription factors in plants [2,3]. Nonetheless, limited information exists for some significant families of transcription factors. Trihelix transcription factors are exclusive to plants. They were initially identified and isolated from pea (Pisum sativum) in the 1990s. They attach to the core sequence of 5-G-Pu-(T/A)-A-A-(T/A)-3' inside the promoter region of the rbcS-3A gene to modulate light-dependent expression [4]. They were originally designated as GT factors due to their affinity for light-responsive GT elements. The DNA-binding domain of the GT factors exhibits a characteristic tandem trihelix (helix-loop-helix-loop-helix) configuration, thereafter designated as the trihelix transcription factor. Subsequent research revealed that the trihelix structure of the GT factors resembles the solution structure of the Myb/SANT-LIKE DNA-binding domain [5]. GT factors evolved from Myb/SANT-LIKE proteins in plants. Gaps between helix pairs created different recognition sequences between GT factors and Myb/SANT-LIKE proteins [5,6]. According to databases like Pfam, the Myb/SANT-LIKE domain represents the trihelix conserved domain. Trihelix is a family of transcription factors that have only recently received attention. However, the trihelix genes have been systematically studied mainly in dicotyledonous plants such as Arabidopsis, tomato and chrysanthemum, while almost no research has been carefully carried out in a monocotyledonous plant. In Arabidopsis, 30 GT family members were identified and divided into the GT-1, GT-2, GT γ, SH4, and SIP1 subfamilies named after their founding members [7]. The 96 trihelix proteins of tomato (Solanum lycopersicum) were classified into six subfamilies (clades GT-1, GT-2, SH4, SIP1, GT γ, and GT δ). The GT δ subfamily is apparently missing in Arabidopsis [8]. Most of the trihelix gene subfamily structures vary substantially, especially at the C-terminus. The exceptions are GT1 and GT2. Earlier studies identified the trihelix family genes as a class of light regulators. Nevertheless, the roles of GT factors in light regulation must be systematically established. In Arabidopsis, the GT1 subfamily genes may participate in salt stress and pathogen responses and their expression was induced by light in 3-d seedlings [9]. In contrast, the rice GT-1 gene RML1 (OsMSL21 in the present study) was repressed by light in etiolated seedlings [10]. The trihelix transcription factors in soybean, GmGT-2A and GmGT-2B, were induced by ABA (abscisic acid), drought, high salt levels, and cold in soybean seedlings [11]. Loss-of-function analysis of GTL1 revealed that gtl1 mutants had fewer stomata than wild type plants. In this way, the former had comparatively lower water loss and higher drought tolerance than the latter [12]. The expression of the rice GT γ clade gene OsGT γ -1 increased 2.5 to 10 times in response to salt stress and was also upregulated by ABA treatment [13]. On the other hand, the expression of several trihelix genes in Chrysanthemum was downregulated by ABA [14]. Trihelix transcription factors are also associated with plant morphogenesis. The trihelix transcription factor PETAL LOSS (PTL) determines the number of petals per flower and sepal fusion in Arabidopsis. The rice SH4 clade gene (SH4) promotes the abscission layer development and function in mature seed peduncles [15]. However, the function of the SH4 clade has not yet been investigated. The Arabidopsis SIP1 genes ASIL1 and ASIL2 downregulated the LEA (Late Embryogenesis Abundant) genes in Arabidopsis seedlings [7]. The trihelix genes also have multiple functions throughout plant development. The molecular mechanisms of their stress responses and their involvement in the signaling pathway require elucidation.

Rice (Oryza sativa L.) is a significant global cereal crop and a vital resource in plant research. This study found 41 rice trihelix genes utilizing HMM-search in silico based on the Myb/SANT-LIKE domain. We examined their chromosomal locations, gene synteny, evolutionary relationships, gene architectures, motif compositions, cis-regulatory elements, and expression patterns across various tissues, developmental stages, and environmental stress responses. This work aimed to evaluate the structure and function of rice trihelix genes and the evolutionary relationships between rice trihelix proteins and those of other species, including dicotyledonous and monocotyledonous plants. To ascertain the function of trihelix genes in response to stress, we assessed their reactions to abiotic stressors, including drought and elevated salinity, as well as to stress signaling chemicals, such as abscisic acid and hydrogen peroxide. Our findings establish a theoretical foundation for the functional investigation of rice trihelix family genes, particularly concerning abiotic stress responses.

2. RESULTS

2.1 Identification of Trihelix Genes in Rice

The HMM (Hidden Markov Model) for the Myb/SANT-LIKE domain identified 117 gene candidates and a rice-specific Myb/SANT-LIKE domain was built using them. The HMM profile search was performed on the whole rice genome with the rice-specific Myb/SANT-LIKE domain and 79 new candidate genes were found. Only genes with E-value < 0.01 were classified in the trihelix family. Putative genes were verified in the Pfam and InterPro databases to confirm the existence of the complete Myb/SANT-LIKE domains. Finally, 41 trihelix genes were identified.

All trihelix genes mapped onto the rice chromosomes, they were named OsMSL01-OsMSL41 according to the gene distribution order on the chromosomes. OsMSL25 and OsMSL34 have two alternative splicing. "MSL" stands for "Myb/SANT-LIKE". The characteristics of OsMSLs including the gene MSU_Locus ID, the chromosomes locations, the lengths of the CDS (coding sequence) and amino acid sequences, the number of exons, the protein sizes, and the isoelectric points are summarized in Table 1. OsMSL19 was the smallest protein with 266 amino acids, whereas OsMSL12 was the largest with 882 amino acids. The protein MW (Molecular Weight) ranged from 28.62 kDa to 97.37 kDa. Their predicted isoelectric points varied from 4.45 (OsMSL09) to 11.38 (OsMSL17). Twenty-nine of the trihelix transcription factors were localized in the nucleus, ten in the chloroplast, and two in the peroxisome (OsMSL04 and OsMSL29).

Table 1. Detailed information of all trihelix family genes identified in the rice genome.

Gene Name	Gene Locus	Chr	ORF (bp)	Exon No.	Length (aa)	MW (kD)	pI	Localization
OsMSL01	LOC_Os01g11200.1	Chr1	828	3	275	31.8	8.57	Nucleus
OsMSL02	LOC_Os01g27590.1	Chr1	1131	2	376	41.08	8.6	Nucleus
OsMSL03	LOC_Os01g34400.1	Chr1	1227	3	408	46.06	9.22	Chloroplast

OsMSL04	LOC_ Os01g36850.1	Chr1	1170	2	389	57.61	9.46	Peroxisome
OsMSL05	LOC_ Os01g48320.1	Chr1	999	1	332	35.92	9.62	Nucleus
OsMSL06	LOC_ Os01g52090.1	Chr1	969	1	322	35.76	5.54	Nucleus
OsMSL07	LOC_ Os01g62410.1	Chr1	1764	7	587	64.1	8.5	Nucleus
OsMSL08	LOC_ Os02g01380.1	Chr2	1113	1	370	40.5	5.33	Nucleus
OsMSL09	LOC_ Os02g07800.1	Chr2	1308	2	435	46.24	4.45	Chloroplast
OsMSL10	LOC_ Os02g08450.1	Chr2	1968	4	655	74.35	7.09	Chloroplast
OsMSL11	LOC_ Os02g31160.1	Chr2	1125	2	374	39.55	5.17	Nucleus
OsMSL12	LOC_ Os02g33610.1	Chr2	2649	18	882	97.37	8.97	Chloroplast
OsMSL13	LOC_ Os02g33770.1	Chr2	1233	1	410	46.67	6.28	Nucleus
OsMSL14	LOC_ Os02g35690.1	Chr2	1260	1	419	44.44	6.11	Nucleus
OsMSL15	LOC_ Os02g43300.1	Chr2	1887	3	628	67.74	4.87	Nucleus
OsMSL16	LOC_ Os03g44130.1	Chr3	1104	3	367	41.74	6.61	Chloroplast
OsMSL17	LOC_ Os03g46350.1	Chr3	1038	2	345	42.49	11.38	Nucleus
OsMSL18	LOC_ Os04g21860.1	Chr4	1254	1	417	47.15	9.1	Nucleus
OsMSL19	LOC_ Os04g30890.1	Chr4	801	1	266	28.62	9.64	Nucleus
OsMSL20	LOC_ Os04g36790.1	Chr4	1254	1	417	43.93	6.53	Nucleus
OsMSL21	LOC_ Os04g40930.1	Chr4	1158	5	385	41.93	5.82	Nucleus
OsMSL22	LOC_ Os04g45750.1	Chr4	1587	2	528	57.46	5.74	Nucleus
OsMSL23	LOC_ Os04g57530.1	Chr4	1173	2	390	41.44	9.01	Nucleus
OsMSL24	LOC_ Os05g03740.1	Chr5	1002	1	333	36.95	5.85	Nucleus
OsMSL25	LOC_ Os05g48690.1	Chr5	1041	1	346	37.44	9.92	Nucleus

OsMSL26	LOC_ Os06g32944.1	Chr6	876	3	291	32.82	8.22	Nucleus
OsMSL27	LOC_ Os07g02500.1	Chr7	1104	3	367	41.74	6.61	Chloroplast
OsMSL28	LOC_ Os07g10950.1	Chr7	1437	8	478	54.23	9.06	Chloroplast
OsMSL29	LOC_ Os08g08130.1	Chr8	1170	2	389	44.28	6.2	Peroxisome
OsMSL30	LOC_ Os08g12950.1	Chr8	1254	1	417	47.15	8.92	Nucleus
OsMSL31	LOC_ Os08g37810.1	Chr8	948	1	315	35.06	7.09	Nucleus
OsMSL32	LOC_ Os08g44690.1	Chr8	912	1	303	34.83	9.29	Nucleus
OsMSL33	LOC_ Os09g03570.1	Chr9	1932	7	643	73.57	9.06	Chloroplast
OsMSL34	LOC_ Os09g38570.1	Chr9	1011	1	336	36.34	6.58	Nucleus
OsMSL35	LOC_ Os10g33030.1	Chr10	1104	3	367	41.7	6.61	Chloroplast
OsMSL36	LOC_ Os10g41460.1	Chr10	1011	1	336	35.64	8.9	Nucleus
OsMSL37	LOC_ Os11g06410.1	Chr11	1492	2	483	55.06	6.24	Nucleus
OsMSL38	LOC_ Os11g17954.1	Chr11	1545	4	514	58.1	8.3	Nucleus
OsMSL39	LOC_ Os11g38660.1	Chr11	1938	5	645	72.97	7.34	Chloroplast
OsMSL40	LOC_ Os12g06640.1	Chr12	1299	1	432	48.77	6.13	Nucleus
OsMSL41	LOC_ Os12g10550.1	Chr12	888	2	295	33.5	7.74	Nucleus

2.2 Chromosomal Distributions and Synteny Analysis of Rice Trihelix Genes

The extraction of the chromosomal information of the OsMSLs identified their chromosomal locations. As shown in Figure 1, all OsMSLs had precise positions in the chromosomes. Each rice chromosome contains ≥1 OsMSL. The OsMSLs are unevenly and non-randomly distributed on 12 chromosomes. Chr2 (chromosomal 2) contains the largest number of OsMSLs (eight) whereas Chr6 contains only one. The first four chromosomes contain 23 trihelix genes while chromosomes 5–12 have on average only 2–3 genes per chromosome. Therefore, OsMSLs are distributed mainly on the first four rice chromosomes. Although Chr2 is relatively short, it contains the most OsMSLs. Chr1 is the longest in rice and also contains numerous

OsMSLs. Chr10, the shortest chromosome, contains two OsMSLs. In contrast, Chr6 is longer than Chr10 but contains only one OsMSL. There is no apparent correlation between the chromosome length and OsMSL gene distribution. Moreover, only OsMSL12 and OsMSL13 form gene clusters on Chr2.

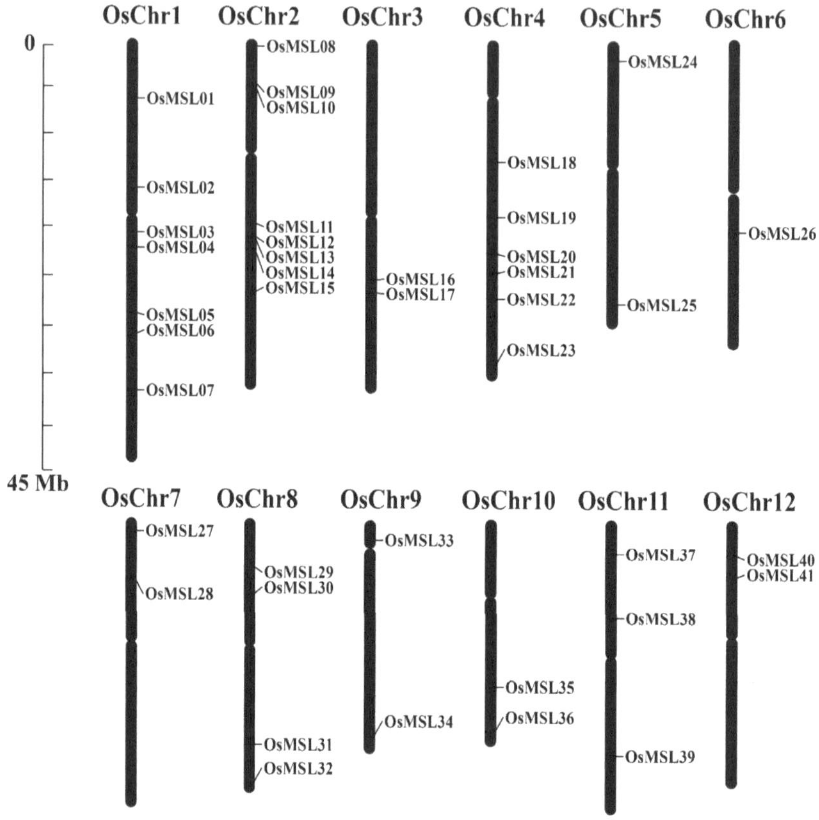

Figure 1. Chromosomal locations of rice trihelix genes. Black bars represent the chromosomes. Chromosome numbers are shown at the tops of the bar. Trihelix genes are labeled at the right of the chromosomes. Scale bar on the left indicates the chromosome lengths (Mb).

Synteny was also used to analyze rice trihelix gene duplication. Chromosomal region within 200 kb containing two or more genes is defined as a tandem duplication event [16]. As shown in Figure 1, four rice trihelix genes (OsMSL09/10 and OsMSL12/13) were clustered into two tandem duplication event regions on rice chromosomal 2. Besides the tandem duplication events, segmental duplications were also investigated by BLASTP and MCScanX methods [17]. Four segmental duplication events with eight rice trihelix genes were also identified, which are located on duplicated segments on chromosomes 1, 2, 4, 5, and 11 (Figure 2). This finding is consistent with the highly divergent, non-conservative evolution of OsMSLs.

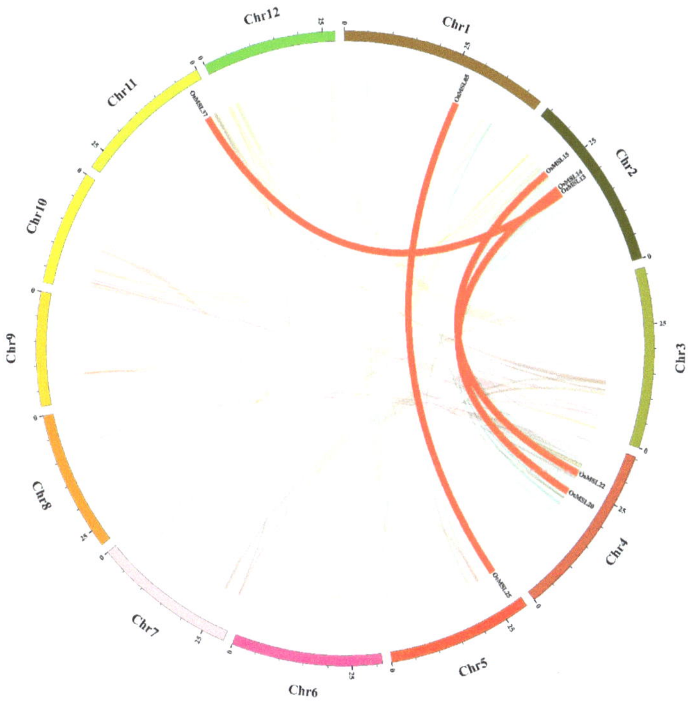

Figure 2. Schematic representations of segmental duplications of rice trihelix genes. Different color lines indicate all synteny blocks in rice genome between each chromosome, and the thick red lines indicate duplicated trihelix gene pairs. The chromosome number is indicated at the bottom of each chromosome. Scale bar marked on the chromosome indicating chromosome lengths (Mb).

To further understand the gene duplication mechanisms of the rice trihelix family, we constructed four comparative syntenic maps of rice associated with four representative species, including one dicots (Arabidopsis) (Figure 3A) and three monocots (Brachypodium distachyon, wheat and maize) (Figure 3B). A total of 23 rice trihelix genes showed a syntenic relationship with those in maize, followed by wheat (21), Brachypodium distachyon (19) and Arabidopsis (2), indicating that in comparison with monocotyledonous plants, rice trihelix genes show a high evolution divergence with dicotyledonous plants. Congruously, previous research reported that 14 pairs of orthologous trihelix genes were found between tomato and Arabidopsis [8]. Some OsMSLs were found to be associated with at least three syntenic gene pairs, such as OsMSL14, OsMSL17, and OsMSL21. These genes may have played a crucial role in the trihelix gene family during evolution. To better understand the evolutionary constraints acting on the trihelix gene family, the Ka/Ks ratios of the trihelix gene pairs were calculated (Tables S1–S5). All segmental and tandem duplicated OsMSL gene pairs, and the majority of orthologous trihelix gene pairs had Ka/Ks < 1, suggesting that the rice trihelix gene family might have experienced a strong purifying selective pressure during evolution.

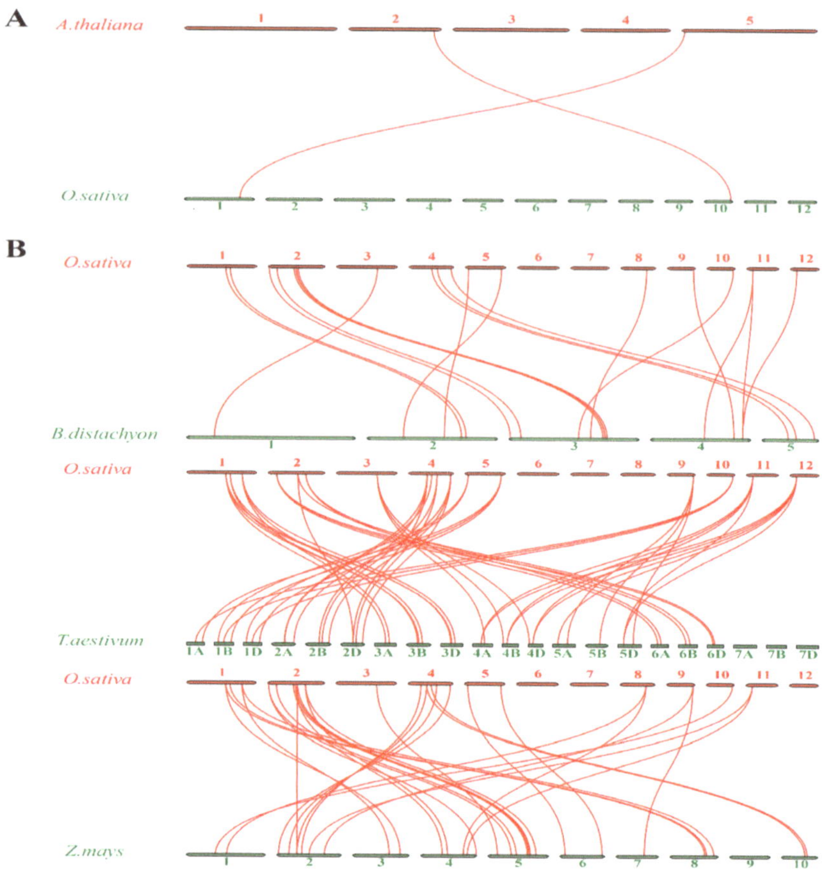

Figure 3. Synteny analysis of trihelix genes between rice and (A) dicotyledonous plant Arabidopsi thaliana, (B) monocotyledonous plant Brachypodium distachyon, wheat and maize. Gray lines in the background indicate the collinear blocks within rice and other plant genomes, while the red lines highlight the syntenic trihelix gene pairs. The species names with the prefixes 'A. thaliana', 'B. distachyon', 'T. aestivum', 'Z. mays' and 'O. sativa' indicate Arabidopsi thaliana, Brachypodium distachyon, Triticum aestivum, Zea mays and Oryza sativa, respectively. Red or green bars represent the chromosomes. The chromosome number is labeled at the top or bottom of each chromosome.

2.3 Phylogenetic Analysis, Gene Structure, and Motif Composition of Trihelix Genes

To better understand the phylogenetic relationships of trihelix genes, a maximum likelihood phylogenetic tree was built based on the multiple sequence alignment of Myb/SANT-LIKE domains among rice and other species which include dicotyledonous plants such as Arabidopsis, soybean, tomato, chrysanthemum and monocotyledonous plant such as maize, wheat, wild rice, Brachypodium distachyon. As shown in Figure 4, OsMSLs were divided into five subfamilies named

SIP1, GT γ, GT, SH4, and GT δ according to the characteristics of their trihelix DNA binding domains. Some genes that have been classified previously such as SIGT-4/7/12/18/36 in tomato [8], CmTH2/6/12/17/19/20 in chrysanthemum [14], GmGT-2A and GmGT-2B in soybean [11] was as a classified marker. The GT clade was the largest subfamily, containing 28 trihelix genes, whereas the SH4 clade was the smallest, consisting of 13 members, indicating that trihelix genes were distributed unevenly in the different clades. All clades consisted of genes both from dicot and monocot species. There is a similar classification in rice which was previously named GT δ in tomato and two tomato trihelix genes SIGT-4 and SIGT-12 have been found in this subfamily. To demonstrate the evolutionary relationships among OsMSLs, we constructed an unrooted phylogenetic tree using the full-length amino acid sequences of the OsMSLs. Of the 43 transcripts of the 41 rice trihelix genes, nine belonged to SIP1, 10 belonged to GT γ, 11 belonged to GT, five belonged to SH4, and eight belonged to GT δ (Figure 5A). Most of the duplicated genes were present in the GT δ classification. The phylogenetic tree of the all MSLs between rice and Arabidopsis was constructed and is shown in Figure S1. However, we found that the GT δ subfamily does not contain Arabidopsis trihelix genes.

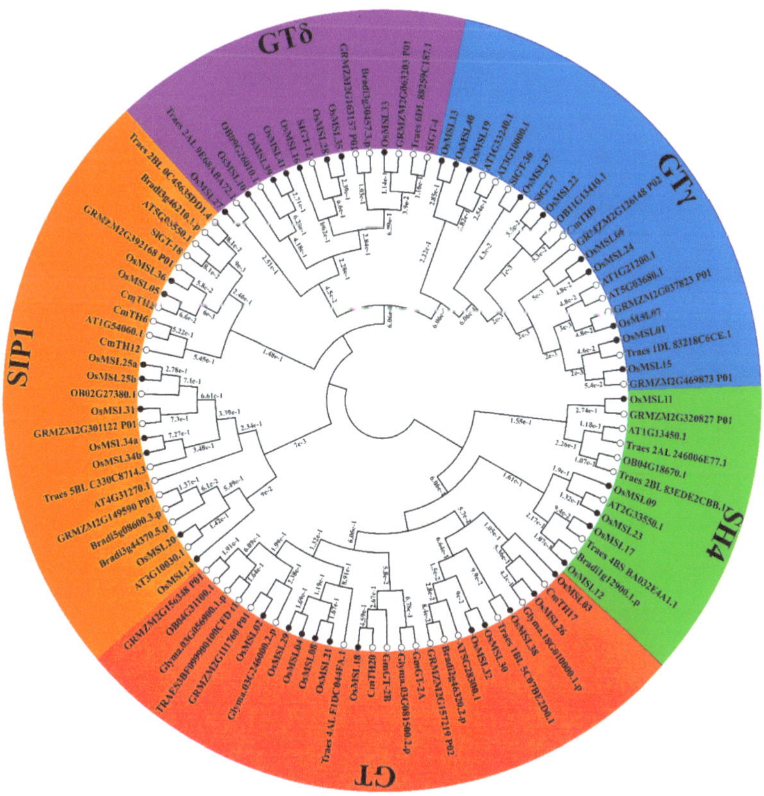

Figure 4. Phylogenetic relationships among 105 trihelix proteins in rice, Arabidopsis, soybean, maize, tomato, wheat, chrysanthemum, wild rice and

Brachypodium distachyon. The maximum likelihood tree was created using MEGA v. 7.0 (bootstrap value = 1000) and the bootstrap value of each branch is displayed. Forty-three OsMSL proteins are marked with black circles and other species are marked with white circles. The phylogenetic tree was clustered into SIP1, GT γ, GT, SH4, and GT δ.

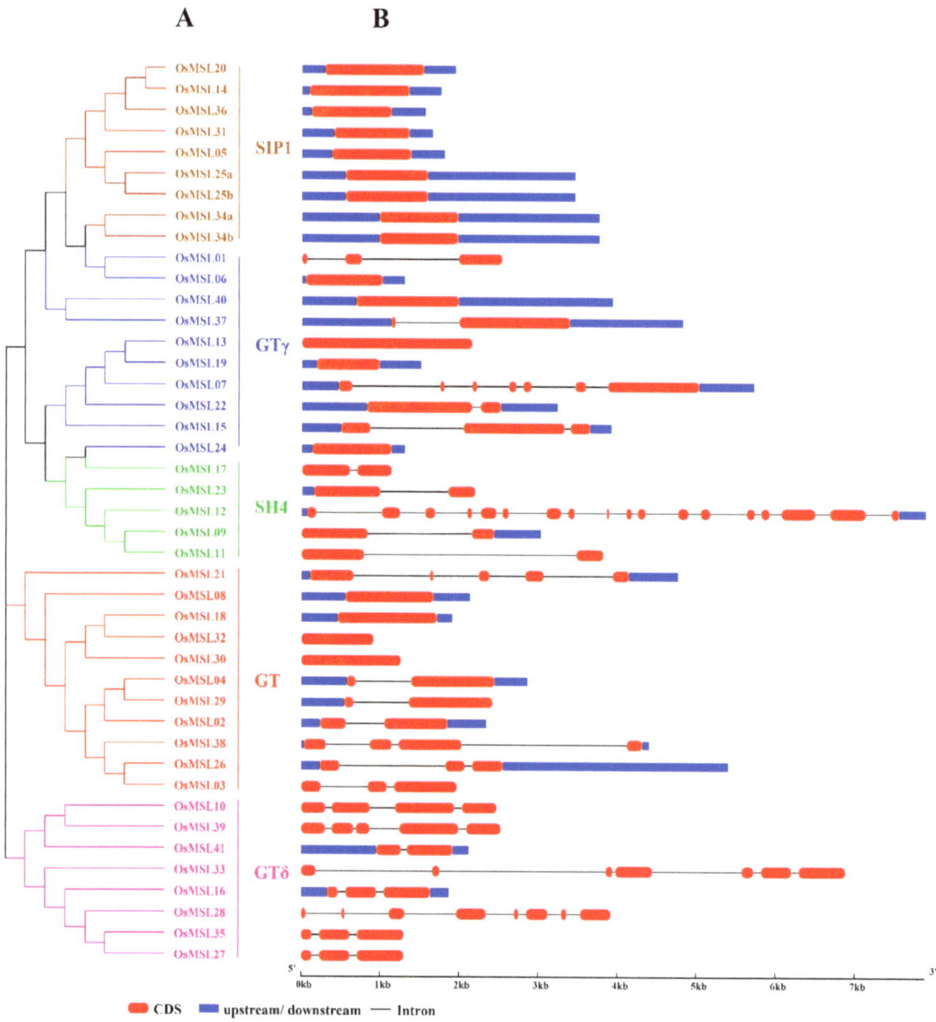

Figure 5. Phylogenetic analysis and gene structure of the rice trihelix family. (A) Phylogenetic analysis of the rice trihelix family. The phylogenetic tree was constructed based on the full-length amino acid sequences of the rice trihelix proteins by using MEGA v. 7.0 with the maximum-likelihood method. Bootstrap = 1,000. SIP1, GT γ, GT, SH4, and GT δ are marked with different colors. (B) Gene structures of the rice trihelix family. These were analyzed by the Gene Structure Display Server (GSDSv. 2.0). Exons, introns, and untranslated regions are marked by round red rectangles, black lines, and blue rectangles, respectively. The scale bar at the bottom estimates the lengths of the exons, introns, and untranslated regions.

To identify the differences between the rice trihelix family genes, we analyzed the OsMSL gene structure by comparing each coding sequence with its corresponding genomic sequence. As shown in Figure 5B, the number of OsMSL exons is discontinuously distributed from 1 through 18. Combining the gene structure with the phylogenetic tree, we found that the OsMSL exon-intron distribution is related to its classification. Closely related genes usually have homologs. Therefore, their gene structures are similar. For example, the OsMSL genome sequences in the SIP1 subfamily have no introns and only one exon. Therefore, the evolution of this gene subfamily is relatively conservative. The genes in the GT δ subfamily have no UTR region and only exons and introns except for OsMSL16 and OsMSL41. In contrast, the structures of the various genes in the GT γ, GT, and SH4 subfamilies are relatively different. These results indicate that although the OsMSLs are subdivided into five families, their genes are relatively conservative.

To determine the functions of the trihelix family genes, the OsMSL motif composition was analyzed by amino acid sequence in the MEME program. Ten motifs with E < 1.8×10−45 were identified. These resemble the MSLs in chrysanthemum. The genes for each subfamily were classified [14]. As shown in Figure 6, except for OsMSL01 and OsMSL09, most trihelix family genes contain motif 1 (Myb-type DNA-binding domain) located at the N-terminus of the amino acid sequence. Motifs 2, 6, and 8 are various trihelix DNA binding domains (WWW, WWF, and WWI). These determine OsMSL classification, structure, and function [18]. As the gene structure analysis indicated, the gene motifs and distribution patterns are closely related to their subfamilies. SIP1 contains motif 8, GT γ contains motif 6, and only OsMSL06 contains an extra motif 8. Both GT and SH4 contain motif 2 but that in SH4 is longer than that in GT. OsMSL09 and OsMSL12 in SH4 also contain an additional motif 8. OsMSL02, OsMSL03, OsMSL04, OsMSL26, OsMSL29, and OsMSL38 in the GT subfamily also contain motif 6. Motif 2 with other functional domains and conservative sequences are contained in the rice-specific GT δ subfamily. Although their functions have yet to be elucidated, they may indicate that the GT δ gene in rice has multiple functions.

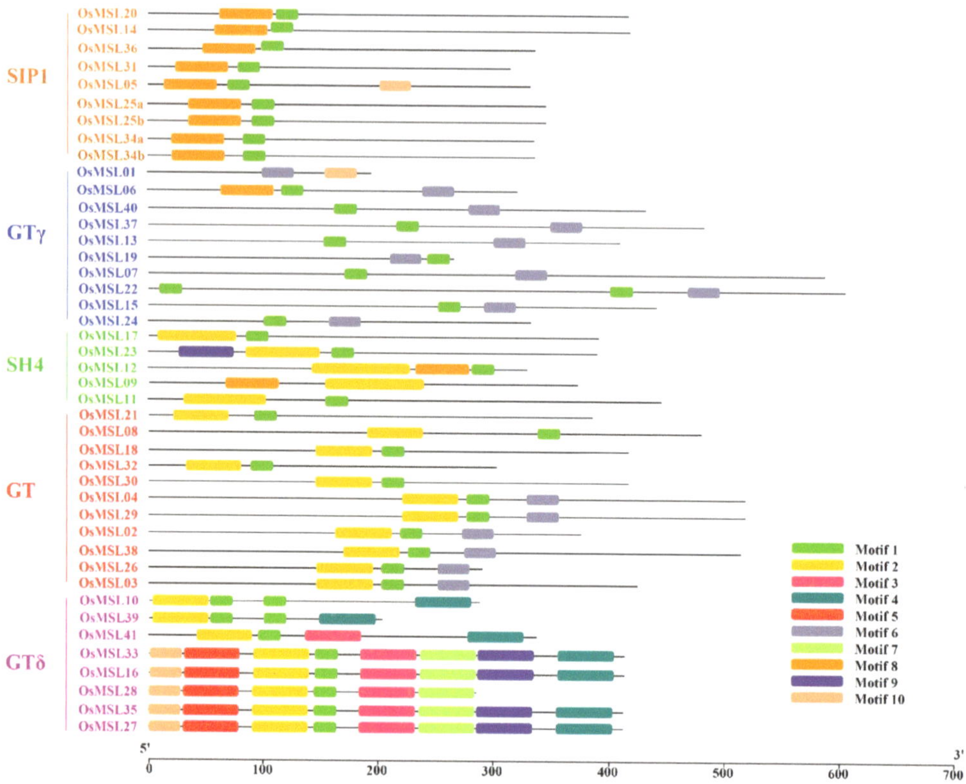

Figure 6. Motif composition of rice trihelix proteins. Motif analysis was performed using the MEME program as described in the methods section. The trihelix proteins are listed on the left. Boxes of different colors represent the various motifs. Their location in each sequence is marked. Motif sequences are shown in Figure S2. The scale bar at the bottom indicates the lengths of the trihelix protein sequences.

2.4 Cis-Element Analysis of Rice Trihelix Genes

To understand the genetic functions, metabolic networks, and regulatory mechanisms of rice trihelix genes, the shared cis-elements in the promoter regions of the OsMSLs were analyzed. The 1500-bp upstream OsMSL sequence was obtained and identified as a hypothetical promoter. The potential shared OsMSL cis-element was scanned and screened out and its distribution and function were analyzed. Two dehydration-responsive- and three light-responsive cis-elements common to all OsMSLs were identified and labeled by different colors in the promoter sequence (Figure 7).

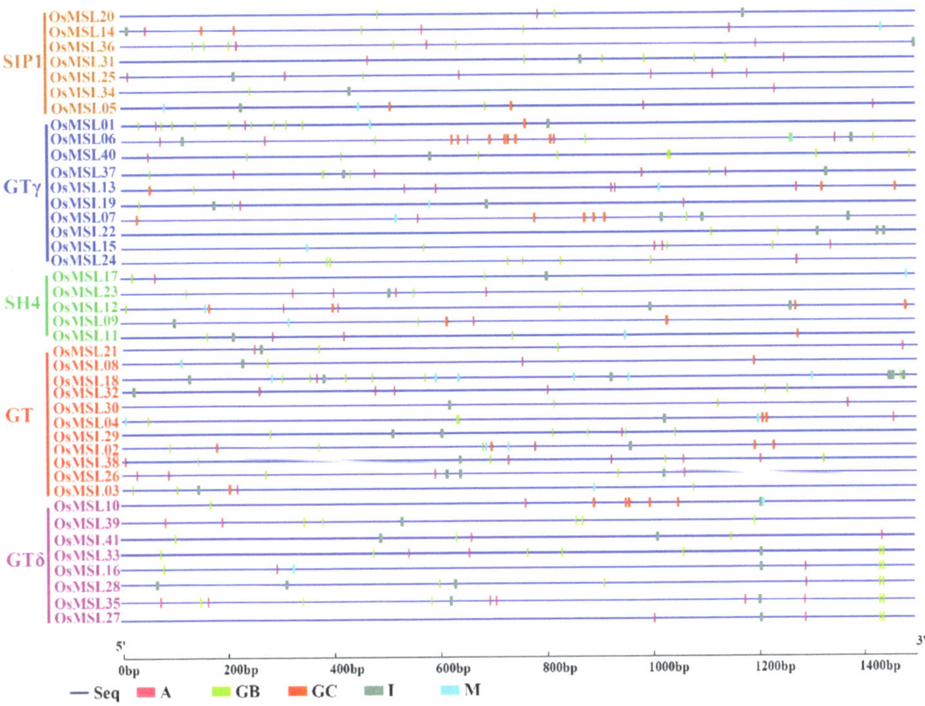

Figure 7. Predicted cis-elements in the promoter regions of the rice trihelix genes. All promoter sequences (-1500 bp) were analyzed. The trihelix genes are shown on the left side of the figure. The scale bar at the bottom indicates the length of promoter sequence. Green bar (GB): GATABOX element; purple bar (A): ACGTATERD1 element; red bar (GC): GT1CONSENSUS element; gray bar (I): INRNTPSADB element; blue bar (M): GT1GMSCAM4 element.

As shown in Figure 7, A (ACGTATERD1) and M (GT1GMSCAM4) element are two dehydration-responsive elements and M is a core element. Therefore, OsMSLs probably participate in dehydration (including drought and salt) stress responses. GB (GATABOX element), GC (GT1CONSENSUS), and I (INRNTPSADB) are three light-responsive elements. They indicate that the OsMSLs family potentially consists of light-inducible/repressible genes. Light responsiveness is typical of the GT factor (now known as the trihelix family gene) and was confirmed in our cis-element study. To verify whether OsMSLs are regulated by light under both normal- and stress conditions, a dark treatment was added to the OsMSL expression analysis.

2.5 Expression Profiles of Trihelix Genes in Rice Tissues and Developmental Stages

The expression profiles of the various rice tissues including the root, stem, leaf, and sheath of four-leaf rice seedlings were investigated (Figure 8). As shown in Figure 8A, all 43 transcripts expressed in all tissues but their expression levels varied greatly in each tissue. Particularly, there are more highly expressing trihelix genes in the leaves and sheaths than the other organs. There are almost no genes with low expression levels and most of the genes remained at high expression levels in the leaves. In contrast, comparatively fewer genes with extremely high expression exist in the stems and none of them express at an extremely high level in the roots.

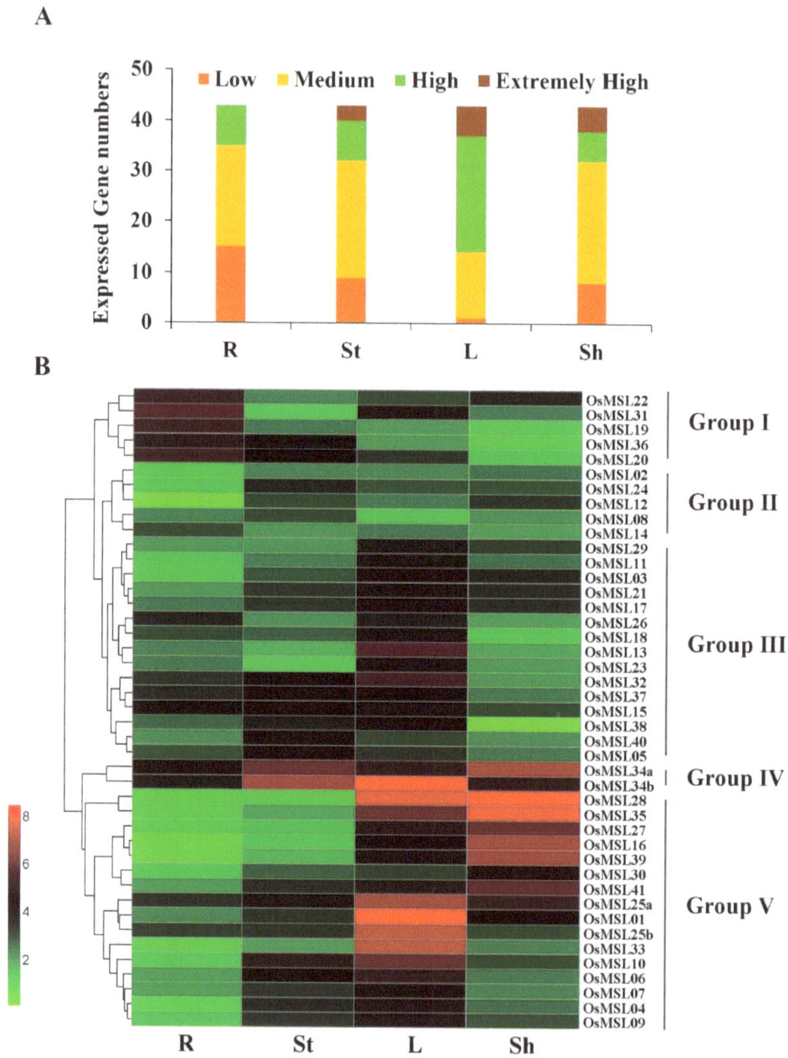

Figure 8. Expression of the rice trihelix gene family in various tissues (R: Root; St: Stem; L: Leaf; Sh: Sheath). (A) Numbers of expressed genes in each tissue. Expression data of the rice trihelix gene family were retrieved from the Expression Atlas database. Extremely high: Expression value > 6, high: 6 ≥ expression value > 4, medium: 4 ≥ expression value > 2, low: 2 ≥ expression value > 0; (B) Expression patterns of the trihelix genes in various rice tissues. Heatmaps were created in HemI v.1.0 and based on the expression data. Expression levels are depicted by different colors on the scale. Green and red represent low and high expression levels, respectively.

The OsMSLs were clustered into groups I to V according to their expression characteristics in different tissues (Figure 8B). OsMSLs in groups I and II expressed at lower levels in all tissues, but members in group I displayed relatively higher expression in root tissue. For instance, although OsMSL19 maintained very low expression levels in most tissues, its expression displayed a distinct enrichment in root. In contrast, the genes in group II, especially OsMSL02, OsMSL08, and OsMSL12, remained low in the various tissues. On the other hand, in Arabidopsis, the At5g63420 gene, the orthology of OsMSL12, is highly expressed specifically in the seeds, suggesting that also OsMSL12 could show a dominant expression level in seeds rather than in any of these four tissues [19]. Contrary to the group I, most of the genes in group III expressed at higher levels in the stems or leaves and at lower levels in the roots. However, the transcription levels of OsMSL34a and OsMSL34b in group IV were high in all four tissues. Group V members also displayed tissue specific expression, but expressed at higher levels than group I and III. Besides, the genes in group V, especially OsMSL16, OsMSL27, OsMSL28, OsMSL35, and OsMSL39, expressed extremely highly in the leaves and sheaths, whereas their expression levels in the roots and stems were lower than those of genes in group I and III, respectively. In conclusion, OsMSLs displayed tissue expression specificity, indicating their potential roles in different mechanisms.

We also investigated the OsMSL expression profiles at different rice developmental stages including 7 (S1), 20 (S2), 40 (S3), 80 (S4), 100(S5), and 140(S6) days after sowing. As shown in Figure 9A, the numbers of genes with high- and extremely high expression levels are greater in S2 and S3 than in the other stages. In contrast, no genes expressed at high levels during S4, S5, or S6. The genes were subdivided into five groups according to their expression characteristics at different stages (Figure 9B). OsMSLs in group I, were at comparatively higher transcription levels during the S2 and S3 stages except for OsMSL28, which displayed high expression levels at S1 stage as well. In general, the expression levels of the genes in groups II and III were similar to those in group I, but their expression levels in S2 and S3 were lower than in group I. Notably, the group IV genes had lower expression levels than the other four clusters at almost all developmental stages. Contrary to groups II and III, high gene expression levels were observed mainly at S1 in group V. Besides, with the exception of OsMSL35 and OsMSL06, all other genes in group V expressed at significantly lower levels by S2. In conclusion, OsMSLs play a potential role at early developmental stages.

2.6 Quantitative Real-Time PCR Analysis of Rice Trihelix Genes in Responses to Different Treatments under Normal and Dark Conditions

A prediction of the cis-elements of the OsMSLs suggested that they may participate in rice dehydration stress tolerance and light-mediated signaling pathways. To verify this hypothesis, we subjected rice seedlings to ABA, hydrogen peroxide, drought, and high salt then carefully selected 12 genes expressing positively in the leaves (Figures 10 and 11). On the whole, although trihelix genes were induced by multiple treatments, their expression levels under one treatment were much higher than other treatments. For instance, six OsMSLs (OsMSL25a, OsMSL25b, OsMSL28, OsMSL34a, OsMSL34b, and OsMSL35) were induced by multiple treatments, but the transcript levels significantly increased under hydrogen peroxide treatment compared to other treatments. Several genes were induced after being repressed, such as OsMSL41, which remained downregulated up until 12 h of ABA treatment. Some trihelix genes showed high transcript levels under multiple treatments. For example, OsMSL39 simultaneously responded to all treatments. OsMSL28 was significantly induced by three tested treatments except ABA treatment. In conclusion, all 12 genes are induced by various abiotic stress or stress signaling molecules, but the expression levels are different and there was no significant correlation between gene expression and its classification.

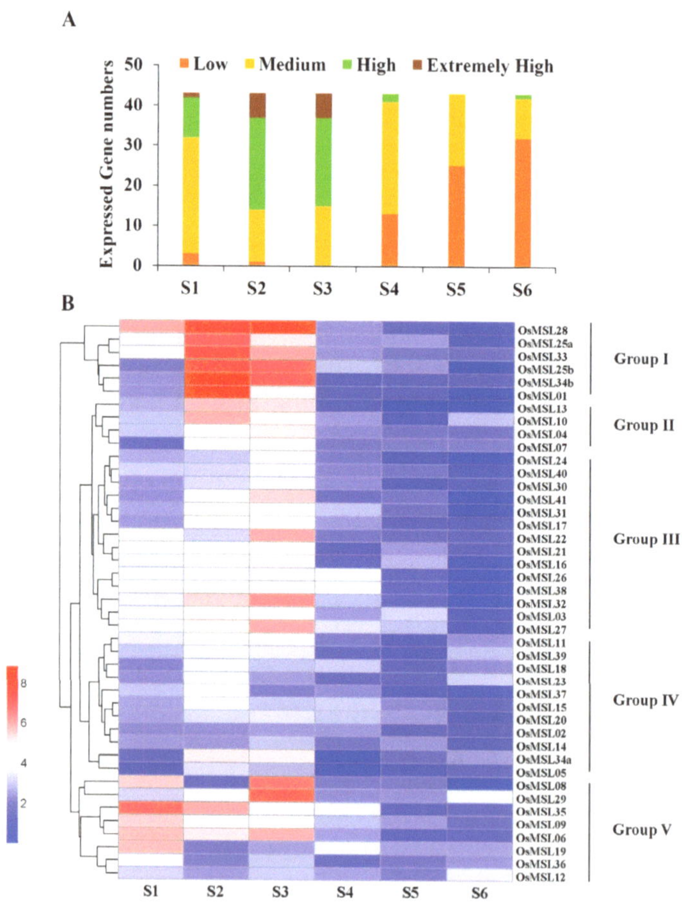

Figure 9. Expression of the rice trihelix gene family at different developmental stages (S1: 7 days after sowing; S2: 20 days after sowing; S3: 40 days after sowing;

S4: 80 days after sowing; S5: 100 days after sowing; S6: 140 days after sowing). (A) Numbers of expressed genes in various developmental stages. Expression data of the rice trihelix gene family genes were retrieved from the Expression Atlas database. Extremely high: Expression value > 6, high: 6 ≥ expression value > 4, medium: 4 ≥ expression value > 2, low: 2 ≥ expression value > 0; (B) Expression patterns of trihelix genes in various rice developmental stages. Heatmaps were created in HemI v.1.0 and based on the expression data. Expression levels are depicted by different colors on the scale. Blue and red represent low and high expression levels, respectively.

Figure 10. Expression profiles of 12 selected OsMSLs in response to (A) ABA (abscisic acid), (B) hydrogen peroxide treatments. Data were normalized to the

β-actin gene. Vertical bars indicate standard deviations. Asterisks indicate corresponding genes significantly upregulated or downregulated compared with the control. (* $p < 0.05$; ** $p < 0.01$; Student's t-test).

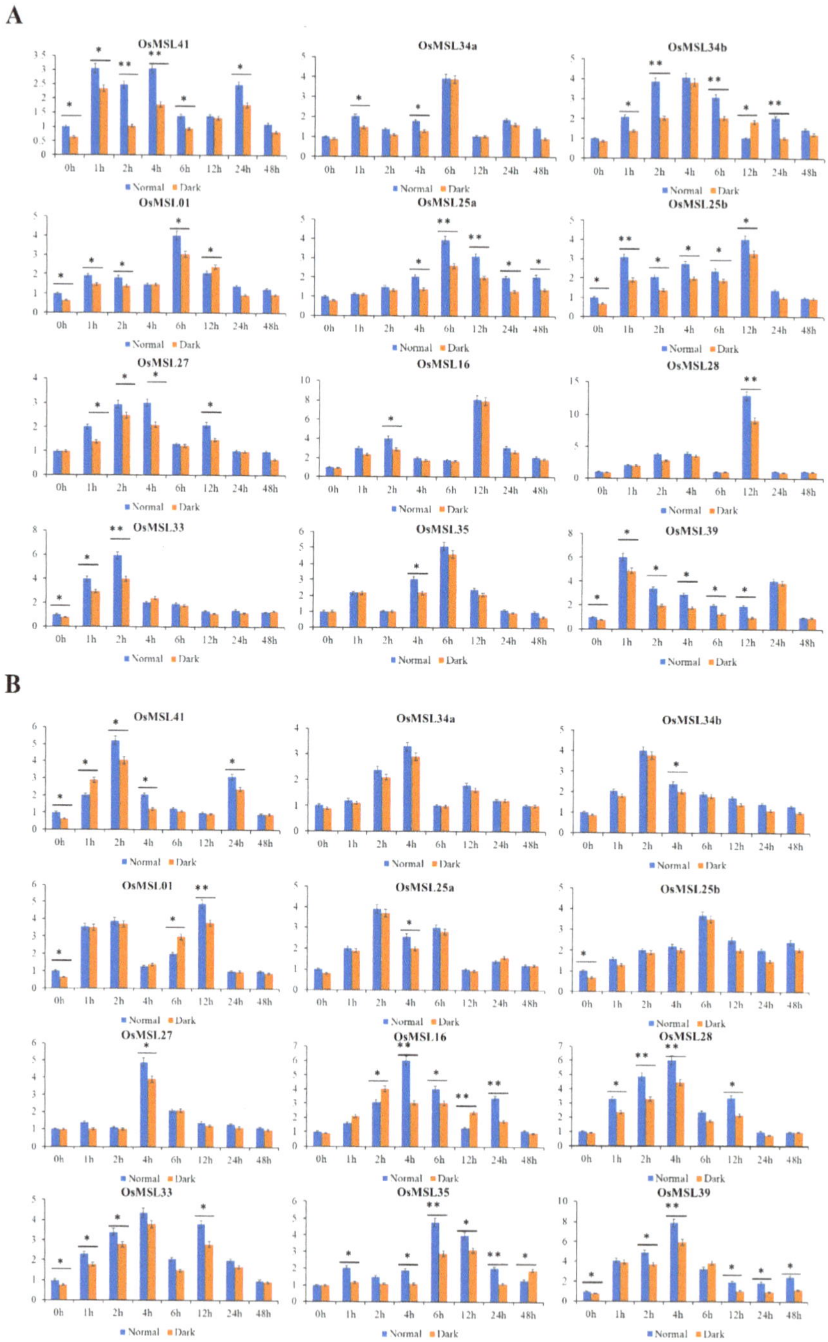

Figure 11. Expression profiles of 12 selected OsMSLs in response to (A) drought, (B) high salt stress, treatments. Data were normalized to the β-actin gene.

Vertical bars indicate standard deviations. Asterisks indicate corresponding genes significantly upregulated or downregulated compared with the control. (* $p < 0.05$; ** $p < 0.01$; Student's t-test).

To determine whether light influences trihelix genes expression, the expression levels of 12 OsMSLs from each treatment group under light and dark conditions were investigated (Figures 10 and 11). Overall, some OsMSLs were induced under normal conditions. For example, at 0 h, OsMSL01, OsMSL25b, OsMSL33, OsMSL39, and OsMSL41 expression levels significantly differed between the two conditions. Several genes began to be light-regulated expressed under different treatments. For instance, after ABA exposure, the expression levels of OsMSL16, OsMSL25a, OsMSL25b, OsMSL28, OsMSL34a, and OsMSL39 at different time points under normal conditions were significantly higher than those at the same intervals under dark conditions. In contrast, some genes were repressed by light at different points, such as the expression levels of OsMSL41 at 2 h under ABA treatment, OsMSL28 at 6 h under hydrogen peroxide treatment, OsMSL34b at 12 h under drought stress and OsMSL16 at 12 h under high salty stress were higher in the dark than under normal conditions. Therefore, the relationship between trihelix genes and light is complex and trihelix gene expression is not directly regulated by light but may be controlled by multiple regulatory mechanisms.

3. DISCUSSION

The trihelix DNA binding domain is usually associated with that of Myb/SANT LIKE. There are three strongly conserved, regularly spaced tryptophan (W) residues in each repeating Myb α-helix. The residues of the Myb α-helix regions are also strongly conserved between the GT factors and the Myb/SANT-LIKE proteins. Individual helices are longer, so the trihelix domains of the GT factors and the Myb/SANT LIKE proteins are related [20]. In contrast, other amino acids at this location have longer Myb repeat sequences than the helix-turn-helix structure formed by the Myb-type DNA-binding domain. Therefore, the trihelix family genes have functions and target gene sequences differing from those of the MYB transcription factor family [20]. In the present study, 41 rice trihelix genes were identified using the Myb-type (Myb/SANT-LIKE) DNA-binding domain. However, in a previous study, only 31 rice trihelix genes were identified [21]. Because the repeated search method was performed in the present study, the trihelix gene could be identified in the rice genome more comprehensively. Based on the wide ranges of OsMSL protein MW, isoelectric point, and subcellular localization, we speculated that OsMSLs are not conservatively evolved. Besides, in Arabidopsis, the chimeric trihelix gene At4g17060 was misannotated in the genome [22]. Its ortholog LOC_Os10g41460 (OsMSL36), was identified in this study.

In general, gene families expand by tandem and segmental duplications [23]. Evolutionary conservatism increased with the number of duplicated genes in a gene family [16]. Chromosomal distribution and gene duplication analyses indicated that there are six pairs of duplicated genes in rice among total 41 trihelix gene, including two pairs of tandem genes and four pairs of segmental genes. However, in Arabidopsis, fifteen pairs of duplicated trihelix genes were previously detected among 34 trihelix genes [24]. These results suggested that OsMSLs are less conserved, and most genes may not originate from the same ancestor. On the other hand, these results demonstrate that the rice trihelix family has a high degree of evolutionary divergence and is non-conservative. These properties may account for the substantial differences among the rice trihelix proteins.

We conducted a phylogenetic analysis to elucidate the evolutionary relationships within the rice and other species trihelix gene family. In a previous study, no GT δ was designated for rice or Arabidopsis [18] because the Arabidopsis family Myb/SANT-LIKE DNA-binding domain or protein sequence was aligned with the rice genome in the attempt to identify the potential rice trihelix gene. However, there is a relatively long evolutionary distance between rice and Arabidopsis. Therefore, this method may overlook the specific trihelix genes in rice. Misclassification may have resulted in deviations because of the influence of the Arabidopsis trihelix genes. In the present study, a class of rice trihelix genes and some tomato trihelix genes previously assigned to this subfamily have been found in the GT δ clade. Subsequent investigation revealed that this subfamily has a high structure similarity. For this reason, the evolutionary relationships of its members are more conservative than those in other subgroups. In previous studies, the GT clade was divided into the GT1 and GT2 subfamilies cited [13,14,24]. According to our evolutionary analysis, there are two GT subfamily clusters (Figure 3A). The difference between GT1 and GT2 is smaller than those among SIP1, SH4, GT γ, and GT δ. The genes in GT1 each have one trihelix DNA binding domain with three conserved tryptophans. Those in GT2 each have an additional trihelix DNA binding domain with two tryptophans and one phenylalanine [25]. Consequently, we classified both the GT1 and GT2 clades in the GT subfamily.

Although the evolution of the trihelix family was not conservative, our gene structure and conservative functional domain analyses indicated that the genes within the same subfamilies (especially SIP1 and GT δ) were still relatively conserved. In fact, most duplicated genes occurred in GT δ. The MEME analysis revealed that the functional domain distribution of each OsMSL was related to its classification (Figure 4). Therefore, these conserved functional domains play central roles in group-specific functions. In contrast, the gene structures among the various groups differed greatly from the conserved functional domains. For this reason, they may have different downstream regulatory genes and participate in different signaling pathways.

The distribution and type of cis-elements on the gene promoter may determine OsMSL functions. In this study, we identified five cis-elements shared by all genes out of a large number among 41 OsMSLs. The results showed that OsMSLs were mainly involved in abiotic stress and light-induced responses. To date, little functional analysis has been conducted on the trihelix transcription factors in plants. Previous studies showed that they participate in responses to pathogens and abiotic stress, light induction, and nitrogen metabolism [18]. The light-induced process is a major feature of the trihelix genes. Light induces massive reprogramming of the plant transcriptome and upregulates or downregulates gene expression and its corresponding signaling pathway [26]. Light signaling coordinates the induction or repression of specific downstream genes like bHLH [27], bZIP [28], R2R3-MYB [29], FAR1 [30], and FHY3 [31]. Studies on light regulatory mechanisms in plants focused on the long-term effects of light exposure. However, little attention has been paid to the transient light-responsive processes of transcription factors in plant stress reactions. In the present study, the OsMSL expression profiles disclosed that their responses to light are transient and change with the processing time. Therefore, OsMSLs may be regulated by light in response to abiotic stress in the same way that phototropism, chloroplast movement, and stomatal opening participate in rapid light-responsive processes and are not under extensive transcriptional regulation. This mechanism substantially differs from that observed in relation to gene expression changes in response to the long-term effects of light on photoperiod.

Gene expression specificity in plant tissues and developmental stages may indicate possible gene functions. Previous studies showed that certain trihelix genes in tomato and chrysanthemum exhibited stable expression levels in all tissues [8,14]. However, most OsMSLs do not maintain stable expression levels in different tissues. OsMSL expression levels may vary substantially among tissues. For example, the expression levels of OsMSL16, OsMSL27, OsMSL28, OsMSL35, and OsMSL39 were extremely high in the leaves and sheaths but comparatively low in the roots and stems. Subcellular localization revealed that these genes are expressed in the chloroplasts, and these are present only in the rice leaves and sheaths. The expression patterns of OsMSL16, OsMSL27, OsMSL28, OsMSL35, and OsMSL39 in GT δ were similar. Therefore, the different OsMSLs within the same group explain the parallel functions of the rice trihelix family. OsMSL02, OsMSL08, and OsMSL14 in Group II were expressed at low levels in all four tissues. Either they are inducible or they are only upregulated under special conditions [32].

High OsMSL expression was observed mainly in 7–40 days seedlings (from germination to early tillering). Therefore, they indicate that OsMSLs contribute primarily to the early stages of rice growth, as do many other genes. For example, ZFP182 overexpression enhanced salt, drought, and cold tolerance in transgenic rice seedlings [33]. Loss of the ABA transporter OsPM1 in 35 days rice seedlings conferred greater drought sensitivity than that seen in the WT [34]. The expression patterns of OsMSL34a and OsMSL34b disclosed that they were, in fact, different variable splicing forms of the same gene with very different expression levels. Since OsMSL34b might be the primary variable splicing form of the gene, OsMSL34a was downregulated to some extent.

In the previous study on rice trihelix genes, their responses to various plant hormones were highlighted [21], and the present study focused on abiotic stress and stress signaling molecules. Other than plant growth and development, MSLs participate in stress responses [18]. Although no cis-elements related to the ABA signaling pathway were found in the OsMSL promoter region, ABA nonetheless, induced the rice trihelix genes. ABA accumulates when plants are subjected to a water deficit. It regulates the expression of drought stress-related genes and modulates the molecular, cellular, and physiological mechanisms for adaptation to environmental stress [35]. In chrysanthemum, however, ABA downregulated the trihelix genes but others were upregulated after prolonged ABA exposure [14]. In contrast, GmGT-2A and GmGT-2B in soybean were upregulated by ABA [11]. In the present study, twelve OsMSLs from all subfamilies have been induced by ABA. These results suggest that the signaling mechanisms of the trihelix family genes vary with species. Whether the ABA presence has a negative regulation on the trihelix gene requires further experimental verification.

ROS (reactive oxygen species) are produced in response to most environmental stress. Excessive ROS accumulation may irreversibly damage cells [36,37]. In previous studies, trihelix family genes had at least one response during plant osmotic stress defense [18]. It follows that trihelix family genes may also participate in ROS scavenging and enhance plant tolerance to various stresses. However, little is known about the mechanism of the peroxide reaction mediated by the trihelix family. We performed a quantitative PCR analysis on 12 OsMSLs subjected to hydrogen peroxide. All 12 OsMSLs responded to hydrogen peroxide stress. Therefore, they may help improve the permeation tolerance by increasing the ROS scavenging capacity in rice.

The trihelix transcription factors bind to GT elements on the light-regulating genes [25]. In darkness, photo regulatory genes are repressed and their associated trihelix family genes are also affected. Nevertheless, certain trihelix family genes are downregulated in response to light exposure, apparently because they must be repressed to be able to downregulate target whose expression is light-dependent. Certain constitutively expressed trihelix genes occur in Arabidopsis. Their expression is ubiquitous and indifferent to the light regime. They are concentrated mainly in GT1 and GT2 [38]. In the present study, however, the expression of the 12 rice trihelix genes changed direction at least once after light or dark treatment. Therefore, they may be inducible rather than constitutive. It remains to be determined whether OsMSLs are regulated by one wavelength or a wide light spectrum. Further investigation of the light path and its components is necessary. The results of our study have helped initiate the research of the rice trihelix transcription factors. In future research, the relationships among the OsMSLs, ABA-mediated dehydration stress tolerance, and ROS scavenging ability under light regulation should be explored.

4. MATERIALS AND METHODS

4.1 Identification and Sequence Analysis of Trihelix Transcription Factor Family in Rice

The rice trihelix transcription factors were identified according to a previously described method with minor changes [39]. The Hidden Markov Model of Myb/SANT-LIKE domain (PF13837) was downloaded from the Pfam database (http://pfam.xfam.org/) [40]. The entire rice amino acid, genome, and CDS sequence assembly and corresponding annotation were downloaded from the EnsemblPlants database (http://plants.ensembl.org/index.html) [41]. The candidate proteins were sought by the HMMSEARCH program (https://www.ebi.ac.uk/Tools/hmmer/search/hmmsearch) base on the Bio-Linux system (Dr Tracey Timms-Wilson, Centre for Ecology & Hydrology (CEH), Oxfordshire, UK). The domain sequences of these candidate proteins were extracted and used to build a rice-specific Hidden Markov Model. All rice proteins were detected by the rice-specific Hidden Markov Model. Those with E-value < 0.01 were selected. The trihelix proteins were verified using the Pfam and InterPro databases (http://www.ebi.ac.uk/interpro/) [42]. Proteins obtained by the domain and database screening confirmation were considered trihelix family members. The corresponding CDS and gene sequences were extracted according to their protein identifications.

The MEME program (http://meme-suite.org/) identified conserved motifs of the trihelix family proteins with the following parameters: Any number of repetitions; minimum seven motifs; maximum 49 motifs; optimum 10–200 amino acids; expected E-value < 1 × 10-48. The trihelix family gene structures were displayed by comparing the coding and genomic sequences with the Gene Structure Display Server tools (http://gsds.cbi.pku.edu.cn/) [43]. The chromosomal locations of the trihelix family genes were mapped onto the rice linkage map with an online tool according to their TIGR numbers [44]. The isoelectric points and molecular weights of trihelix family proteins were estimated with ExPASy (http://expasy.org/) [45].

4.2 Phylogenetic Analysis

A multiple alignment was performed with the full-length amino acid sequences of the rice trihelix family proteins using MEGA v. 7.0 (https://www.megasoftware.net/) [46]. Unrooted trees were constructed by the maximum-likelihood (ML) method with the following parameters: Poisson correction; pairwise deletion; 1000 bootstrap replicates.

4.3 Gene Duplication and Ka/Ks Analysis

Synteny blocks of the rice genome were downloaded from the Plant Genome Duplication Database (PGDD, http://chibba.agtec.uga.edu/duplication/) [47]. Duplicated OsMSLs pairs were connected by solid lines.

4.4 Cis-Element Analysis of Trihelix Transcription Factor Family

Promoters of the trihelix family genes were downloaded from the Phytozome database (https://phytozome.jgi.doe.gov/pz/portal.html#) [48]. The PLACE database (https://sogo.dna.affrc.go.jp/) was used to analyze the cis-regulatory elements of the trihelix family gene promoters [49].

4.5 Plant Growth Conditions and Treatments

Nipponbare rice seeds (O. sativa L. ssp. japonica) were surface-sterilized with 10% sodium hypochlorite solution for 30 min then sown on 1/2 MS (Murashige & Skoog) solid medium and cultured in a light incubator. After 2 weeks, the seedlings were at the two true leaf stage. They were transplanted into Hoagland's nutrient solution and cultured in an artificial climate chamber under controlled conditions (14 h light at 28 ◦C/10 h dark at 22 ◦C; relative humidity 70%). The rice seedlings were subjected to various stresses at the three-leaf stage (4 weeks) [50].

For the drought, salt, and hydrogen peroxide stress treatments, the rice seedlings were transferred to Hoagland's nutrient solution containing 20% polyethylene glycol (PEG)-6000 (w/v), 150 mM NaCl, or 2% hydrogen peroxide (v/v), respectively. For the ABA treatment, the rice seedlings were cultured on 1/2 MS solid medium containing 10 μM ABA. The control group was maintained on normal nutrient solution or medium. All other culture conditions were the same as described above. Treated rice tissues were harvested at 0 h, 1 h, 2 h, 4 h, 6 h, 12 h, 24 h, and 48 h. The samples were immediately placed in liquid nitrogen and stored at -80 ◦C until use. Untreated material was used as a control. The experimental procedure was repeated at least three times.

4.6 Expression Analysis of Trihelix Transcription Factor Family

Total RNA was extracted from rice tissues by the TRIzol method (Thermo Fisher Scientific, Waltham, MA, USA) and treated with DNase to eliminate any DNA contamination. RNA quality was assessed by electrophoresis and stored at -80 ◦C until use. First-strand cDNA (10 μL) was synthesized according to the instructions for the PrimeScript™ RT Master Mix (Takara Biomedical Technology (Beijing) Co., Ltd., Beijing, China). Primers were designed with Primer Premier v. 5.0 (PREMIER Biosoft International, Palo Alto, CA, USA) and were based on the trihelix gene family transcript sequences. Gene specific primers for quantitative real-time PCR are listed in Table S6. Primer amplification specificity was verified in the rice genome database using Blast from NCBI (https://www.ncbi.nlm.nih.gov/) [51]. Rice _-actin was the internal reference gene. Quantitative real-time PCR was performed in the ABI 7300 Real Time PCR System (Applied Biosystems, Foster City, CA, USA) using SYBR Green chemistry and reaction mix consists of 10 μL SYBR qPCR Master Mix (Vazyme Biotech Co.,Ltd., Nanjing, China), 0.4 μL upstream and downstream primers respectively, 0.4 μL ROX, 2 μL cDNA (10 times dilution) and 6.8 μL ddH2O to 20 μL. The PCR reaction protocol was 95 _C for 5 min; 95 _C for 10 s; 60 _C for 20 s; 72 _C for 20 s; 45 cycles. Gene expression levels were calculated by the 2-ΔΔCT method: ΔΔCT = (CTtarget - CTactin) at time x - (CTtarget - CTactin) at time 0 [52]. The test was repeated three times. Expression data for the rice trihelix family genes were retrieved from the Expression Atlas database (https://www.ebi.ac.uk/gxa/home) [53]. Heatmaps were created in HemI v.1.0 (The CUCKOOWorkgroup, Hubei, China) and based on the expression data [54].

5. CONCISE SUMMARY

Trihelix transcription factors are involved in numerous plant biological processes, however they have not been comprehensively examined. Bioinformatics analysis found 41 rice trihelix transcription factors. Gene synteny analysis indicated that OsMSLs exhibit less conservation, suggesting that the majority of these genes may not derive from a common ancestor. Phylogenetic studies classified them into five subfamilies. The gene architectures and conserved functional domains of rice trihelix transcription factors exhibited significant variation; nonetheless, they together possessed five cis-elements regulating responses to light and dehydration stress. Analysis of expression patterns indicated that trihelix transcription factors had the highest expression levels during the early stages of rice growth, with the majority of significantly elevated genes concentrated in the leaves and sheaths. Real-time quantitative PCR study of the trihelix family genes exposed to various stressors or ABA demonstrated their induction in response to drought, high salinity, hydrogen peroxide, and ABA. Nevertheless, these reactions were also modulated by light, and the expressions of individual OsMSL varied depending on the presence or absence of light. Our study clarified the biological roles of trihelix transcription factors in rice.

REFERENCES

1. Lindemose, S.; O'Shea, C.; Jensen, M.K.; Skriver, K. Structure, Function and Networks of Transcription Factors Involved in Abiotic Stress Responses. Int. J. Mol. Sci. 2013, 14, 5842–5878. [CrossRef] [PubMed]
2. Jin, J.P.; Tian, F.; Yang, D.C.; Meng, Y.Q.; Kong, L.; Luo, J.C.; Gao, G. PlantTFDB 4.0: Toward a central hub for transcription factors and regulatory interactions in plants. Nucleic. Acids Res. 2017, 45, D1040–D1045. [CrossRef] [PubMed]
3. Riechmann, J.L.; Heard, J.; Martin, G.; Reuber, L.; Jiang, C.; Keddie, J.; Adam, L.; Pineda, O.; Ratcliffe, O.J.; Samaha, R.R. Arabidopsis transcription factors: Genome-wide comparative analysis among eukaryotes. Science 2000, 290, 2105. [CrossRef] [PubMed]
4. Green, P.J.; Kay, S.A.; Chua, N.H. Sequence-specific interactions of a pea nuclear factor with light-responsive elements upstream of the rbcS-3A gene. EMBO J. 1987, 6, 2543–2549. [CrossRef] [PubMed]
5. Nagano, Y. Several Features of the GT-Factor Trihelix Domain Resemble Those of the Myb DNA-Binding Domain. Plant Physiol. 2000, 124, 491–493. [CrossRef] [PubMed]
6. Qin, Y.; Ma, X.; Yu, G.H.; Wang, Q.; Wang, L.; Kong, L.R.; Kim, W.; Wang, H.W. Evolutionary History of Trihelix Family and Their Functional Diversification. DNA Res. 2014, 21, 499–510. [CrossRef]
7. Gao, M.J.; Lydiate, D.J.; Li, X.; Lui, H.; Gjetvaj, B.; Hegedus, D.D.; Rozwadowski, K. Repression of Seed Maturation Genes by a Trihelix Transcriptional Repressor in Arabidopsis Seedlings. Plant Cell 2009, 21, 54–71. [CrossRef] [PubMed]
8. Yu, C.Y.; Cal, X.F.; Ye, Z.B.; Li, H.X. Genome-wide identification and expression profiling analysis of trihelix gene family in tomato. Biochem. Biophys. Res. Commun. 2015, 468, 653–659. [CrossRef] [PubMed]
9. Murata, J.; Takase, H.; Hiratsuka, K. Characterization of a Novel GT-box Binding Protein from Arabidopsis. Plant Biotechnol. 2002, 19, 103–112. [CrossRef]
10. Wang, R.H.G.; Han, B. Transcript abundance of rml1, encoding a putative GT1-like factor in rice, is up-regulated by Magnaporthe grisea and down-regulated by light. Gene 2004, 324, 105–115. [CrossRef] [PubMed]
11. Xie, Z.M.; Zou, H.F.; Lei, G.; Wei, W.; Zhou, Q.Y.; Niu, C.F.; Liao, Y.; Tian, A.G.; Ma, B.; Zhang, W.K. Soybean Trihelix Transcription Factors GmGT-2A and GmGT-2B Improve Plant Tolerance to Abiotic Stresses in Transgenic Arabidopsis. PLoS ONE 2009, 4, e6898. [CrossRef]
12. Chan, Y.Y.; Pence, H.E.; Jing, B.J.; Miura, K.; Gosney, M.J.; Hasegawa, P.M.; Mickelbart, M.V. The Arabidopsis GTL1 Transcription Factor Regulates Water

Use Efficiency and Drought Tolerance by Modulating Stomatal Density via Transrepression of SDD1. Plant Cell 2010, 22, 4128.

13. Fang, Y.; Xie, K.; Xin, H.; Hu, H.; Xiong, L. Systematic analysis of GT factor family of rice reveals a novel subfamily involved in stress responses. Mol. Genet. Genom. 2010, 283, 157–169. [CrossRef] [PubMed]
14. Song, A.P.; Wu, D.; Fan, Q.Q.; Tian, C.; Chen, S.M.; Guan, Z.Y.; Xin, J.J.; Zhao, K.K.; Chen, F.D. Transcriptome-Wide Identification and Expression Profiling Analysis of Chrysanthemum Trihelix Transcription Factors. Int. J. Mol. Sci. 2016, 17, 198. [CrossRef] [PubMed]
15. Li, C.; Zhou, A.; Sang, T. Rice Domestication by Reducing Shattering. Science 2006, 311, 1936–1939. [CrossRef] [PubMed]
16. Holub, E.B. The arms race is ancient history in Arabidopsis, the wildflower. Nat. Rev. Genet. 2001, 2, 516–527. [CrossRef]
17. Wang, Y.; Tang, H.; DeBarry, J.D.; Tan, X.; Li, J.; Wang, X.; Lee, T.-H.; Jin, H.; Marler, B.; Guo, H. MCScanX: A toolkit for detection and evolutionary analysis of gene synteny and collinearity. Nucleic Acids Res. 2012, 40, e49. [CrossRef] [PubMed]
18. KaplanLevy, R.N.; Brewer, P.B.; Quon, T.; Smyth, D.R. The trihelix family of transcription factors—Light, stress and development. Trends Plant Sci. 2012, 17, 163–171. [CrossRef] [PubMed]
19. Markus, S.; Davison, T.S.; Henz, S.R.; Pape, U.J.; Monika, D.; Martin, V.; Bernhard, S.L.; Detlef, W.; Lohmann, J.U. A gene expression map of Arabidopsis thaliana development. Nat. Genet. 2005, 37, 501–506.
20. Nagata, T.; Niyada, E.; Fujimoto, N.; Nagasaki, Y.; Noto, K.; Miyamoto, Y.; Murata, J.; Hiratsuka, K.; Katahira, M. Solution structures of the trihelix DNA-binding domains of the wild-type and a phosphomimetic mutant of Arabidopsis GT-1: Mechanism for an increase in DNA-binding affinity through phosphorylation. Proteins 2010, 78, 3033–3047. [CrossRef] [PubMed]
21. Jianhui, J.; Yingjun, Z.; Hehe, W.; Liming, Y. Genome-wide analysis and functional prediction of the Trihelix transcription factor family in rice. Hereditas 2015, 37, 1228. (In Chinese)
22. Geraldo, N.; Baurle, I.; Kidou, S.; Hu, X.; Dean, C. FRIGIDA Delays Flowering in Arabidopsis via a Cotranscriptional Mechanism Involving Direct Interaction with the Nuclear Cap-Binding Complex. Plant Physiol. 2009, 150, 1611. [CrossRef]
23. Cannon, S.B.; Mitra, A.; Baumgarten, A.; Young, N.D.; May, G. The roles of segmental and tandem gene duplication in the evolution of large gene families in Arabidopsis thaliana. BMC Plant Biol. 2004, 4, 10. [CrossRef] [PubMed]

24. Ali, M.A.; Yasmeen, E.; Riaz, M.; Azeem, F.; Sultan, S.; Abbas, A.; Riaz, K. Genome-wide analysis of trihelix Transcription factor gene family in Arabidopsis thaliana. Pak. J. Agric. Sci. 2016, 53, 439–448.

25. Lam, E. Domain analysis of the plant DNA-binding protein GT1a: Requirement of four putative alpha-helices for DNA binding and identification of a novel oligomerization region. Mol. Cell. Biol. 1995, 15, 1014–1020. [CrossRef] [PubMed]

26. Jiao, Y.; Lau, O.S.; Deng, X.W. Light-regulated transcriptional networks in higher plants. Nat. Rev. Genet. 2007, 8, 217–230. [CrossRef] [PubMed]

27. Toledoortiz, G.; Huq, E.; Quail, P.H. The Arabidopsis Basic/Helix-Loop-Helix Transcription Factor Family. Plant Cell 2003, 15, 1749–1770. [CrossRef]

28. Jakoby, M.; Weisshaar, B.; Dröge-Laser, W.; Vicente-Carbajosa, J.; Tiedemann, J.; Kroj, T.; Parcy, F. bZIP transcription factors in Arabidopsis. Trends Plant Sci. 2002, 7, 106–111. [CrossRef]

29. Ballesteros, M.L.; Bolle, C.; Lois, L.M.; Moore, J.M.; Viellecalzada, J.P.; Grossniklaus, U.; Chua, N.H. LAF1, a MYB transcription activator for phytochrome A signaling. Gene Dev. 2001, 15, 2613–2625. [CrossRef]

30. Hudson, M.; Ringli, C.; Boylan, M.T.; Quail, P.H. The FAR1 locus encodes a novel nuclear protein specific to phytochrome A signaling. Gene Dev. 1999, 13, 2017–2027. [CrossRef]

31. Wang, H.; Deng, X.W. Arabidopsis FHY3 defines a key phytochrome A signaling component directly interacting with its homologous partner FAR1. EMBO J. 2002, 21, 1339–1349. [CrossRef]

32. Tang, Y.; Qin, S.; Guo, Y.; Chen, Y.; Wu, P.; Chen, Y.; Li, M.; Jiang, H.; Wu, G. Genome-Wide Analysis of the AP2/ERF Gene Family in Physic Nut and Overexpression of the JcERF011 Gene in Rice Increased Its Sensitivity to Salinity Stress. PLoS ONE 2016, 11, e0150879. [CrossRef] [PubMed]

33. Huang, J.; Sun, S.; Xu, D.; Lan, H.; Sun, H.; Wang, Z.; Bao, Y.; Wang, J.; Tang, H.; Zhang, H. A TFIIIA-type zinc finger protein confers multiple abiotic stress tolerances in transgenic rice (Oryza sativa L.). Plant Mol. Biol. 2012, 80, 337. [CrossRef] [PubMed]

34. Yao, L.; Cheng, X.; Gu, Z.; Huang, W.; Li, S.; Wang, L.; Wang, Y.F.; Xu, P.; Ma, H.; Ge, X. The AWPM-19 Family Protein OsPM1 Mediates Abscisic Acid Influx and Drought Response in Rice. Plant Cell 2018, 30, 1258–1276. [CrossRef]

35. Yamaguchi-Shinozaki, K.; Shinozaki, K. Transcriptional regulatory networks in cellular responses and tolerance to dehydration and cold stresses. Annu. Rev. Plant Biol. 2006, 57, 781–803. [CrossRef] [PubMed]

36. Apel, K.; Hirt, H. Reactive oxygen species: Metabolism, oxidative stress, and signal transduction. Annu. Rev. Plant Biol. 2004, 55, 373–399. [CrossRef] [PubMed]

37. Miller, G.; Suzuki, N.; Ciftci-Yilmaz, S.; Mittler, R. Reactive oxygen species homeostasis and signalling during drought and salinity stresses. Plant Cell Environ. 2010, 33, 453–467. [CrossRef] [PubMed]

38. Gilmartin, P.M.; Memelink, J.; Hiratsuka, K.; Kay, S.A.; Chua, N.H. Characterization of a gene encoding a DNA binding protein with specificity for a light-responsive element. Plant Cell 1992, 4, 839–849. [CrossRef]

39. Lozano, R.; Hamblin, M.T.; Prochnik, S.; Jannink, J.L. Identification and distribution of the NBS-LRR gene family in the Cassava genome. BMC Genom. 2015, 16, 360. [CrossRef] [PubMed]

40. El-Gebali, S.; Mistry, J.; Bateman, A.; Eddy, S.R.; Luciani, A.; Potter, S.C.; Qureshi, M.; Richardson, L.J.; Salazar, G.A.; Smart, A. The Pfam protein families database in 2019. Nucleic Acids Res. 2018, 28, 263–266. [CrossRef]

41. Kersey, P.J.; Allen, J.E.; Allot, A.; Barba, M.; Boddu, S.; Bolt, B.J.; Carvalho-Silva, D.; Christensen, M.; Davis, P.; Grabmueller, C. Ensembl Genomes 2018: An integrated omics infrastructure for non-vertebrate species. Nucleic Acids Res. 2018, 46, D802–D808. [CrossRef] [PubMed]

42. Mitchell, A.L.; Attwood, T.K.; Babbitt, P.C.; Blum, M.; Bork, P.; Bridge, A.; Brown, S.D.; Chang, H.-Y.; El-Gebali, S.; Fraser, M.I. InterPro in 2019: Improving coverage, classification and access to protein sequence annotations. Nucleic Acids Res. 2018. [CrossRef] [PubMed]

43. Hu, B.; Jin, J.; Guo, A.Y.; Zhang, H.; Luo, J.; Gao, G. GSDS 2.0: An upgraded gene feature visualization server. Bioinformatics 2014, 31, 1296. [CrossRef] [PubMed]

44. Kurata, N.; Yamazaki, Y. Oryzabase. An integrated biological and genome information database for rice. Plant Physiol. 2006, 140, 12–17. [CrossRef] [PubMed]

45. Artimo, P.; Jonnalagedda, M.; Arnold, K.; Baratin, D.; Csardi, G.; De Castro, E.; Duvaud, S.; Flegel, V.; Fortier, A.; Gasteiger, E. ExPASy: SIB bioinformatics resource portal. Nucleic Acids Res. 2012, 40, W597–W603. [CrossRef] [PubMed]

46. Kumar, S.; Stecher, G.; Tamura, K. MEGA7: Molecular evolutionary genetics analysis version 7.0 for bigger datasets. Mol. Biol. Evol. 2016, 33, 1870–1874. [CrossRef] [PubMed]

47. Lee, T.H.; Tang, H.; Wang, X.; Paterson, A.H. PGDD: A database of gene and genome duplication in plants. Nucleic Acids Res. 2013, 41, 1152–1158. [CrossRef] [PubMed]

48. Goodstein, D.M.; Shu, S.; Howson, R.; Neupane, R.; Hayes, R.D.; Fazo, J.; Mitros, T.; Dirks, W.; Hellsten, U.; Putnam, N. Phytozome: A comparative platform for green plant genomics. Nucleic Acids Res. 2011, 40, D1178–D1186. [CrossRef]
49. Higo, K.; Ugawa, Y.; Iwamoto, M.; Korenaga, T. Plant cis-acting regulatory DNA elements (PLACE) database: 1999. Nucleic Acids Res. 1999, 27, 297–300. [CrossRef] [PubMed]
50. Ren, J.; Gao, F.; Wu, X.; Lu, X.; Zeng, L.; Lv, J.; Su, X.; Luo, H.; Ren, G. Bph32, a novel gene encoding an unknown SCR domain-containing protein, confers resistance against the brown planthopper in rice. Sci. Rep. 2016, 6, 37645. [CrossRef] [PubMed]
51. Coordinators, N.R. Database resources of the national center for biotechnology information. Nucleic Acids Res. 2017, 45, D12.
52. Livak, K.J.; Schmittgen, T.D. Analysis of relative gene expression data using real-time quantitative PCR and the 2(-Delta Delta C(T)) Method. Methods 2001, 25, 402–408. [CrossRef] [PubMed]
53. Papatheodorou, I.; Fonseca, N.A.; Keays, M.; Tang, Y.A.; Barrera, E.; Bazant, W.; Burke, M.; Füllgrabe, A.; Fuentes, A.M.-P.; George, N. Expression Atlas: Gene and protein expression across multiple studies and organisms. Nucleic Acids Res. 2017, 46, D246–D251. [CrossRef] [PubMed]
54. Deng, W.; Wang, Y.; Liu, Z.; Cheng, H.; Xue, Y. HemI: A Toolkit for Illustrating Heatmaps. PLoS ONE 2014, 9, e111988. [CrossRef] [PubMed]

13 Chapter

IDENTIFICATION, CHARACTERIZATION, AND EXPRESSION PATTERNS OF TCP GENES AND MICRORNA319 IN COTTON

1. INTRODUCTION: AN OVERVIEW

Transcription factors are essential for the control of gene expression. Gene expression can be regulated by transcription factors that either activate or repress transcription, so they are vital for many cell biological process [1]. The TEOSINTE BRANCHED 1, CYCLOIDEA, and PROLIFERATING CELL FACTORS (TCP) gene family is a small group of transcription factors exclusive to higher plants [2]. This class of transcription factors Transcription factors are crucial for regulating gene expression. Transcription factors regulate gene expression by either activating or repressing transcription, making them essential for numerous cellular biological processes [1]. The TEOSINTE BRANCHED 1, CYCLOIDEA, and PROLIFERATING CELL FACTORS (TCP) gene family comprises a limited collection of transcription factors exclusive to higher plants. This category of transcription factors performs numerous roles in regulating various stages of plant growth and development by modulating cell proliferation [3]. They possess a highly conserved 59-amino-acid basic helix–loop–helix (bHLH) motif at the N-terminus, referred to as the TCP domain [4]. This domain facilitates DNA binding, nuclear localization, and participates in protein–protein interactions [5]. TCP family members can be categorized into two classes based on variations in the TCP domain: Class I (the PCF or TCP-P class) and Class II (the TCP-C class) [6,7]. Class II is further categorized into the CINCINNATA (CIN) and CYC/TB1 subgroups. Moreover, numerous class II members within the TCP domain exhibit an arginine-rich motif of 18 to 20 residues [8]. The R domain is anticipated to adopt a hydrophilic α-helix or coiled-coil configuration that facilitates protein–protein interactions. has many functions in regulating diverse plant growth and development processes by controlling cell proliferation [3]. They are characterized by a highly conserved 59-amino-acid basic helix–loop–helix (bHLH) motif at the N-terminus designated as the TCP domain [4]. This domain is responsible for DNA binding, nuclear targeting, and is involved in protein–protein interactions [5]. Based on variation in the TCP domain, TCP family members can be classified into two classes: Class I (also known as the PCF or TCP-P class) and class II (also known as the TCP-C class) [6,7]. Class II is further subdivided into the CINCINNATA (CIN) and CYC/TB1 subgroups. In addition to the TCP domain, several class II members possess an 18–20-residue arginine-rich motif [8]. This so-called R domain was predicted to form a hydrophilic α-helix or a coiled-coil structure that mediates protein–protein interactions [2].

It has been reported that many TCP transcription factors participate in the regulation of diverse physiological and biological processes, such as phytohormone biosynthesis and signal transduction, branching, leaf morphogenesis flower development and senescence, pollen development, and regulation of the circadian clock in various plants [9–11]. In Arabidopsis thaliana seeds, TCP14 was expressed in the vascular tissues of embryos. It promotes germination through antagonism of abscisic acid signaling [12]. TCP1 is expressed in restricted areas of the flower meristem, leaf vasculature, and at the junctions of roots and hypocotyls. It mediates the expression of a key brassinosteroid (BR) biosynthetic gene by directly associating with the two GGNCCC motifs in the promoter region of DWARF4 (DWF4) [13]. DWF4 encodes a 22-hydroxylase and is responsible for multiple 22-hydroxylation steps during BR biosynthesis [14]. The expression levels of DWF4 were positively correlated with TCP1 abundance in planta. In Arabidopsis flowers, the gynoecium and silique development was modulated by TCP15 through partly regulating auxin biosynthesis. The ectopic expression of Arabidopsis TCP15 represses style and stigma development, thus producing gynoecia with decreased stigmatic tissue and/or carpel fusion defects in apical parts [15]. TCP17 and its two closely related homologs, TCP5 and TCP13, play an important role in mediating shade-induced hypocotyl elongation by up-regulating auxin biosynthesis via a PHYTOCHROME INTERACTING FACTORS (PIF)-dependent and a PIF-independent pathway [16]. In rice (Oryza sativa), OsTCP19 was upregulated under salt and water-deficit stress. Overexpression of OsTCP19 in Arabidopsis caused upregulation of INDOLE-3-ACETIC ACID3(IAA3), ABSCISIC ACID INSENSITIVE 3(ABI3), and ABI4, and downregulation of LIPOXYGENASE2 (LOX2), thus leading to developmental abnormalities, such as less lateral roots [17]. MicroRNAs (miRNAs) are a class of smallnon-coding RNAs generated from single-strand hairpin RNA precursors. They regulate gene expression by binding to complementary sequences within target mRNAs [18]. Considerable progress has been made in identifying the targets of plant miRNAs. In Arabidopsis, five CIN-like TCP genes (TCP2, TCP3, TCP4, TCP10, and TCP24) were targeted by miR319 and have been implicated in regulating leaf morphogenesis [19]. Knockdown of a subset of Class II TCP transcription factors by overexpression of miR319 increases tolerance to dehydration and salinity stress in bentgrass (Agrostis stolonifera) [20]. Accumulated functional characterization of TCPs indicated their diverse function in a developmental-, tissue-, and signal-dependent context. In addition to their importance as transcriptional regulators of cell-cycle genes, TCPs have other functions with comparable impact on plant development. Characterization of TCPs and their signaling pathway will be beneficial to unravel their exact role in the control of plant development and evolution.

Allotetraploid upland cotton (Gossypium hirsutum L.) constitutes over 90% of global cultivated cotton, serves as the primary source of renewable textile fibers, and is additionally planted for oilseed production. Sequencing has proven challenging due to its intricate allotetraploid (AtDt) genome [21]. Recently, the entire genome was sequenced by combining whole-genome shotgun reads, bacterial artificial chromosome-end sequences, and genotype-by-sequencing genetic maps [22,23]. Repeated sequences constitute 67.2% of the AtDt genome, with transposable elements derived from Dt exhibiting greater activity than those from At [23]. The availability of genomic information presents a significant opportunity to find and describe TCP genes in this plant species for the first time. This study discovered and characterized 73 unique TCP transcription factors in the G. hirsutum genome. Comprehensive details concerning their genomic architectures, chromosomal positions, and a phylogenetic tree were also supplied. We used RNA-seq data to examine transcript patterns across various tissues, encompassing distinct developmental phases of ovule and fiber, in addition to their responses to heat, drought, and salt stress. Additionally, the TCP genes targeted by miR319 were described.

2. RESULTS

2.1 Identification and Characterization of TCP Proteins in G. hirsutum

To identify the TCP genes in the G. hirsutum genome, protein sequences of Arabidopsis and rice TCPs serve as BLAST search queries, and multiple-alignment was performed. A total of 73 TCP genes were identified. All candidate TCP genes were confirmed to encode the conserved TCP domain using the InterProScan database and NCBI's CDD, the Conserved Domain Database [24]. Seven GhTCP genes were found to possess the R domain. These characteristic features suggested that they were members of the TCP gene family. Detailed characteristics of the TCP transcription factors in G. hirsutum are offered in Table S1. The GhTCP proteins are different in their length, molecular weight (Mw), and theoretical isoelectric point (pI). The mean length and Mw of these proteins was 347 amino acids and 37.58 kDa, respectively. The pI varied from pH 5.80 (GhTCP9) to 10.07 (GhTCP38) with an average of pH 8.09. All the GhTCP proteins were predicted to localize in the nucleus. Proteins were localized at their appropriate subcellular compartment to perform their desired function [3,25].

Unrooted phylogenetic trees were constructed based on the multiple sequence alignment of 73 GhTCP protein sequences and their Arabidopsis and rice homologs. The TCP transcription factors from the three species were distributed in almost all clades, indicating that the TCP family diversified before divergence of these plants. The phylogenetic tree placed the GhTCPs into two classes (Figure 1), as was also found for all species so far. Class I was named the TCP-P or PCF class, and class II was named the TCP-C class. The class II genes were further divided into two groups: CYC/TB1 and CIN. In G. hirsutum, CYC/TB1 and CIN were a larger family: For CYC/TB1, approximately twice the size of those of Arabidopsis and rice; and for CIN, approximately five times the size. Seven GhTCP genes belonged to the CYC/TB1 group–in Arabidopsis and rice, three of 24 AtTCPs and three of 21 OsTCPs were grouped into this subfamily. Fifty GhTCPs belonged to the PCF group, and13 AtTCPs and10 OsTCPs were also grouped into this subfamily. CYC/TB-type proteins were divided into two subgroups. One group contained four G. hirsutum TCPs, but only one Arabidopsis TCP and none from rice, which indicated that this group was either acquired after the divergence of monocots and dicots or was lost in rice. In G. hirsutum, the number of TCP genes was significantly higher than those in tomato, Citrullus lanatus, Arabidopsis, rice, and Prunus mume (Figure 2).

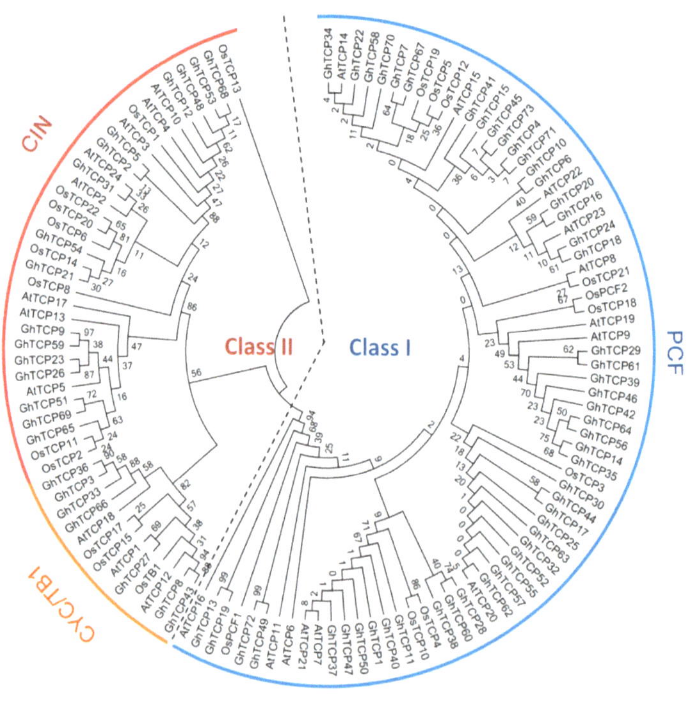

Figure 1. Phylogenetic analysis of TCP proteins from G. hirsutum, Arabidopsis, and rice. The deduced full-length amino acid sequences were aligned using ClustalX

2.0 and the phylogenetic tree was constructed using MEGA 6.0 by the Neighbor-Joining (NJ) method with 1000 bootstrap replicates. The three subclasses are indicated with different colors.

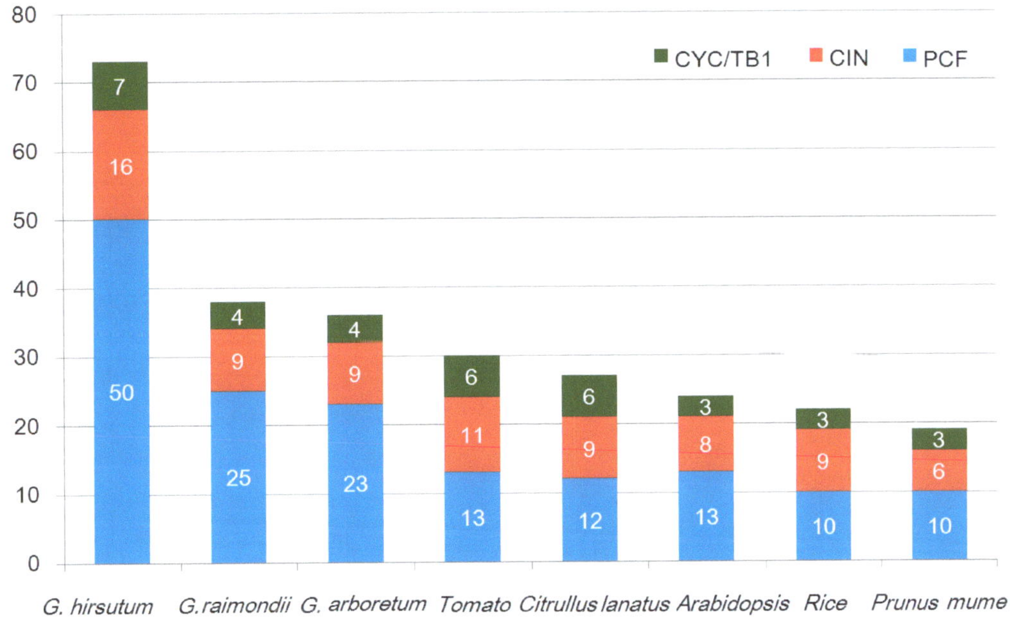

Figure 2. TCP family members of G. hirsutum, G. raimondii, G. arboretum, tomato, Citrullus lanatus, Arabidopsis, rice, and Prunus mume. Different colors represent the different subclasses, and the number of genes in each subclass is shown. Green: CYC/TB1 genes; Red: CIN genes; Blue: PCF genes.

2.2 Genomic Distribution, Gene Structural Organization, and Domain Analysis of GhTCP Genes

The complete genome sequences provided an overview of the chromosomal distribution of these TCP genes. Among the 73 G. hirsutum TCPs, 67 members were located on the 22 chromosomes, and the other six were located at six unmapped scaffolds. GhTCP genes were unevenly distributed on 22 of the 26 G. hirsutum chromosomes, with the number of TCP genes per chromosome in the range of 0–8 (Figure S1). Chromosomes, A12 and D11, contained eight and seven genes, respectively, while chromosomes A02, A06, D03, D06, and D13 had no TCP genes.

To better understand the gene structures of GhTCP family genes, we analyzed their exon–intron organization. Overall, 88% of the GhTCPs contained only one exon (Figure S2). Seven GhTCP genes contained one intron and two exons: GhTCP3, GhTCP23, GhTCP26, GhTCP56, GhTCP64, and GhTCP66. Only GhTCP33 in the CYC/TB1 group possessed four introns and five exons. Losses or gains of exons were identified during the evolution of the PCF group genes. GhTCP19 comprised seven introns and eight exons, whereas GhTCP13 consisted of four introns and five exons.

Comparing their structural patterns showed the loss of an exon in the middle of the GhTCP13 sequence. Two PCF class genes contained one intron and two exons, and the remaining PCF class genes contained only one exon. Analysis of the pattern of exon–intron junctions can provide important understanding into the evolution of gene families. Our results suggested that TCP genes maintained a relatively constant exon–intron composition during evolution of the G. hirsutum genome.

The conserved motif of TCP proteins in G. hirsutum was investigated using Clustal X. The sequences were found to encode a putative TCP-domain protein that contained a bHLH-type motif at the N-terminus (Figure S3). The components of the loop, and helixes I and II, were quite different between class I and II proteins. Within the TCP domain, several putative residues involved in DNA binding were located in the basic region and several putative hydrophobic residues located in helixes I and II. In the basic region, the CIN and CYC/TB1 type proteins contained an insertion of four amino acids. The R domain, an arginine-rich motif of 18–20 residues, was absent from all class I proteins and was mainly present in CYC/TB1 group proteins.

2.3 Expression Analysis of GhTCP Genes in Different Tissues and under Various Stress Conditions

To provide reliable information on the growth and developmental functions of TCP genes in G. hirsutum, their transcript accumulation patterns in mature leaves, stem, root, torus, petal, stamen, pistil, cylycle, ovules, and fibers of G. hirsutum was investigated (Figure 3). We obtained transcriptome data from the NCBI Sequence Read Archive (accession number RJNA248163). Some TCP genes with close phylogenetic relationships showed similar or divergent expression patterns. For instance, the paralogous pair, GhTCP2 and GhTCP5, was expressed highly in both the torus and petal, at moderate levels in the ovules at 20 days post anthesis (DPA), and at low levels in mature leaves. Most CYC/TB1-type genes were only weakly or not expressed in all tissues, suggesting that they were primarily expressed in other organs not tested or under special conditions. In contrast, GhTCP31, GhTCP40, and GhTCP47 were constitutively expressed at very high levels in all tissues tested, indicating that these genes played regulatory roles during multiple development stages. Some TCP genes exhibited tissue-specific expression. GhTCP31 and GhTCP40 were highly expressed only in reproductive organs: Torus, petal, stamen, pistil, and calycles.

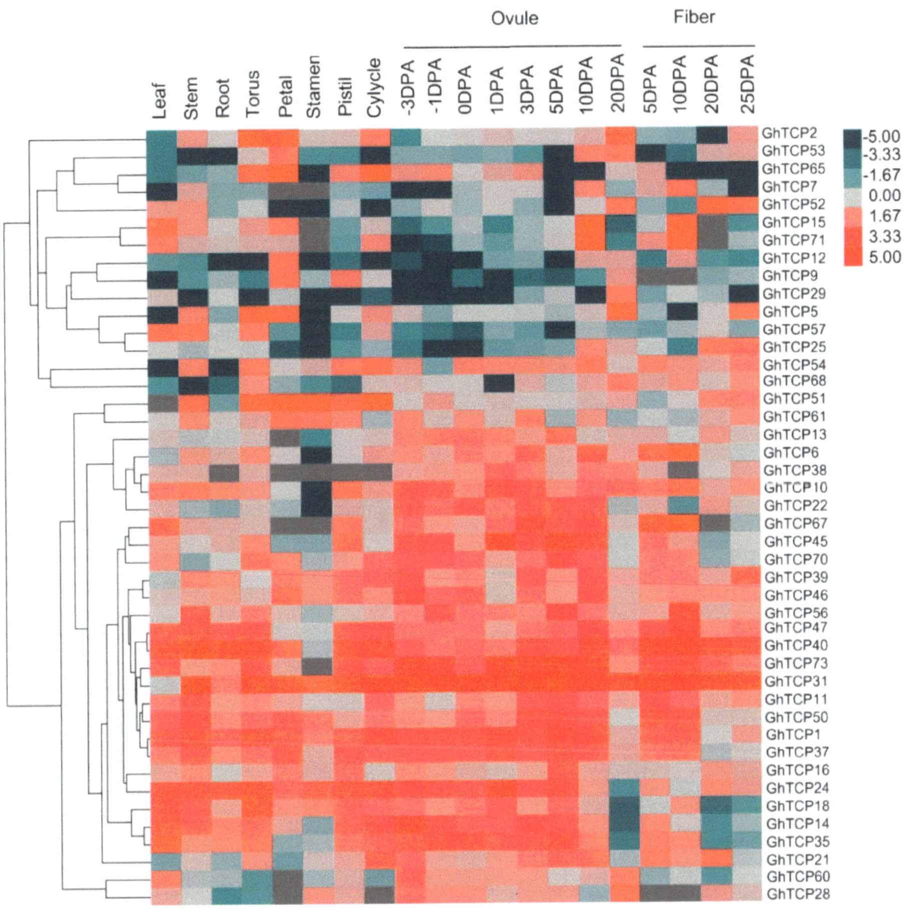

Figure 3. Heat map representation of GhTCP gene expression in different tissues. The tissues used for expression profiling are indicated at the top. The genes are shown on the left of the expression bars and the phylogenetic relationship is shown. The -3 to 20 days post anthesis (DPA) indicate -3, -1, 0, 1, 3,5, 10, and 20 days after pollination.

To predict possible functions of TCP genes in environmental adaptation, we investigated the transcriptional profile of TCP genes under various stress conditions, including heat, salt, and drought stresses. In total, 41 genes exhibited variations in expression (Figure 4). Of the three treatments, heat stress caused relatively more fluctuations in the transcript abundance of TCPs than did salt or drought stress. Under heat stress conditions, 18 TCP genes were downregulated and eight were upregulated. In response to salt treatment, the expression of five GhTCPs (TCP7, 14, 25, 33, and 35) increased instantly, and then decreased slowly during continued salt stress. Six GhTCP genes were selected at random for quantitative RT-PCR (qRT-PCR) analysis to determine the relative expression under salt and drought stresses (Figure 5). The qRT-PCR results indicated that these GhTCP genes showed similar expression patterns to the transcriptome sequencing results.

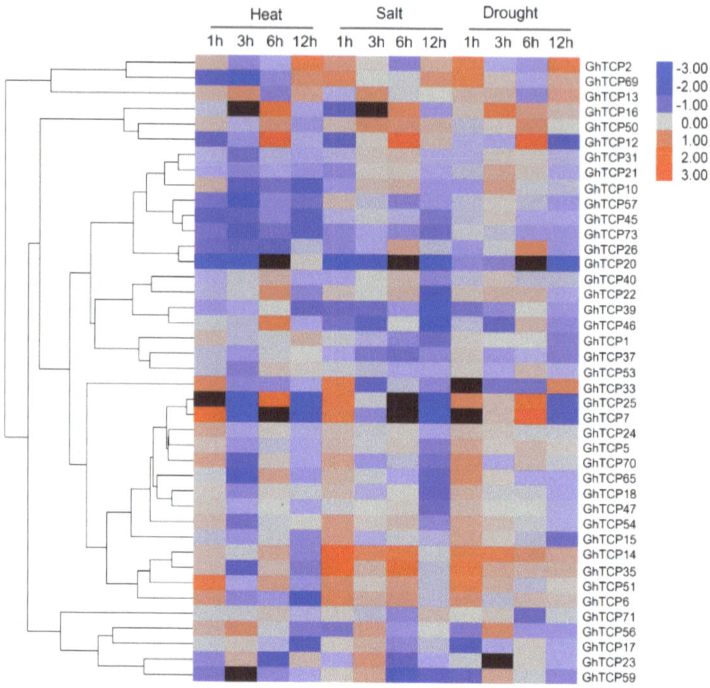

Figure 4. Expression of GhTCP genes under heat, salt, and drought stresses. The genes are shown on the left of the expression bars and the phylogenetic relationship is shown. The abiotic stresses used for expression profiling are indicated at the top. The 1, 3, 6, and 12 h indicate hours after treatment.

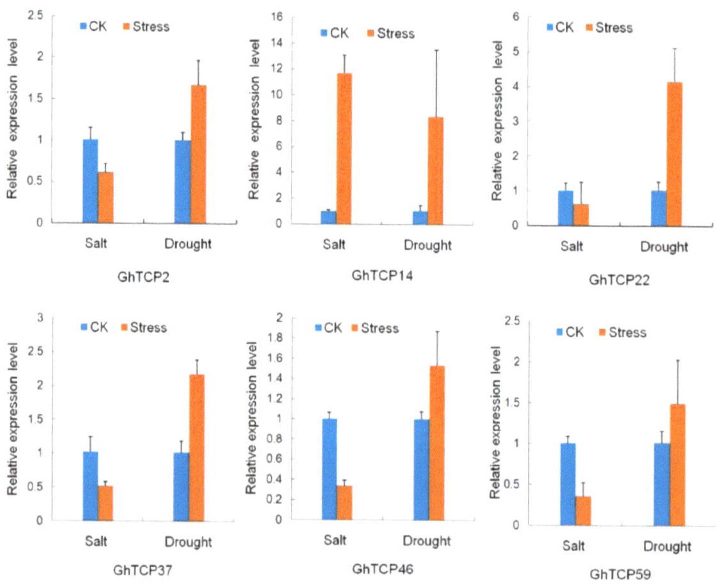

Figure 5. Relative expression levels of six GhTCP genes under salt and drought stress. QRT-PCR analyses were performed using RNA generated from cotton leaves after NaCl and Polyethylene Glycol (PEG) treatment. Error bars represent standard error of the mean.

2.4 Target Sites of miR319 in GhTCP Genes

The miRNAs can cause endonucleolytic cleavage of mRNA by extension, which often perfect complementarity to mRNAs. In plants, miR319 was one of the first characterized and conserved miRNA families, which has been demonstrated to target TCP genes. In G. hirsutum, miR319 had only 1-nt mismatch compared with sequences in Arabidopsis (Figure 6). The predicted hairpin structures of the miR319 precursor had 191 nt. The miR319 sequence was located at the 3t-end of the pre-miRNAs and began with a 5t-uridine. Using a set of strict standards, GhTCP21, GhTCP31, and GhTCP54 were predicted as targets of miR319. These three miR319 target sites were all located in the coding regions, and all miR319-targeted genes belonged to the CIN clade. Similarly, there were five and three TCP genes containing miR319-binding sites in Arabidopsis and P. mume, respectively, and they also belonged to the CIN group [26]. This suggests that miR319 held homologous target interactions during the evolution and diversification of plants.

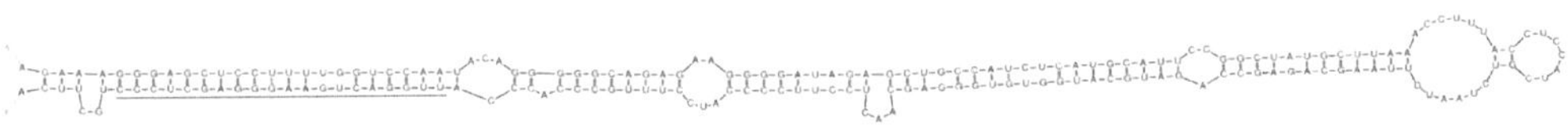

Figure 6. Mature and predicted fold-back structures of miR319 precursors in G. hirsutum. Sequences of mature miR319 are underlined.

Degradome sequencing had been widely used to identify plant miRNA cleavage sites. In this study, the GhTCP mRNA degradation sites were determined by BLASTing the sequenced degraded fragments against the G. hirsutum TCP genes. The degradome sequencing data are available at Gene Expression Omnibus (GEO accession number GSE69820). Using PairFinder software, miR319 was found to cleave GhTCP21 and GhTCP54 mRNA transcripts (Figure 7). The miR319–mRNA pair was at the cleavage site of the two TCP genes. There were both 67 raw reads at the position, with abundance at the position equal to the maximum on the transcript, and with only one maximum on the transcript. The 5t-ends of the mRNA fragments mapped to the nucleotide that paired to the tenth nucleotide of the miR319 sequence. To research the biological function of miR319, a negative-correlation expression test was undertaken for miR319 and its target GhTCP21 and GhTCP54 mRNAs using qRT-PCR. In response to salt and drought treatments, miR319 showed different degrees of upregulation. The transcriptome sequencing analysis showed that expression of GhTCP21 and GhTCP54 was downregulated.

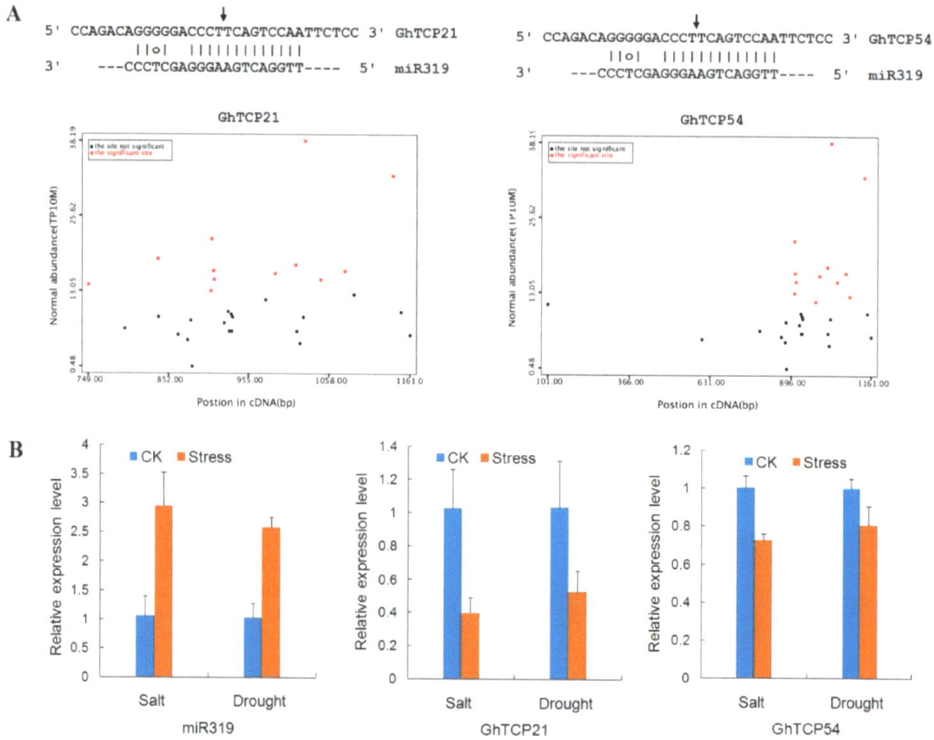

Figure 7. miR319 and its target genes, GhTCP21 and GhTCP54, in G. hirsutum. (A) Target plot (t-plot) for GhTCP21 and GhTCP54, which were targeted by miR319. Arrows indicate the signatures corresponding to the miRNA cleavage site. Partial mRNA sequences of target genes aligned with the miRNAs show perfect matches (straight lines) and G-U wobbles (circles). (B) Relative expression levels of miR319, GhTCP21, and GhTCP54 under salt and drought stresses. QRT-PCR analyses were performed using RNA generated from cotton leaves after NaCl and Polyethylene Glycol (PEG) treatment. Error bars represent standard error of the mean.

3. DISCUSSION

A number of TCP proteins had been recently identified in various plants due to completion of their whole-genome sequence, including Arabidopsis, rice, tomato (Solanum lycopersicum), and watermelon (Citrullus lanatus), Orchis italica, and Populus euphratica [27–31]. The allotetraploid, G. hirsutum, is not only the world's most important fiber crop, but is also a model polyploid crop. Despite being among the largest and most diverse gene families, the TCP gene family has not been systematically identified in the G. hirsutum genome. In this study, we identified 73 TCP genes in the sequenced genome of G. hirsutum. We analyzed their phylogenetic relationship, genomic distribution, conserved protein motif, and exon-intron organization. Over 80% of GhTCP genes were intronless, which was quite similar to the structure of G. raimondii and G. arboretum TCP genes [32,33]. Generally speaking, most GhTCPs within the same subclade showed similar gene structure in terms of numbers and lengths of introns and exons. Furthermore, similar to the exon–intron organization, members of the same subclade also showed similar motif composition, indicating their functional similarities. Additionally, some motifs were only present at specific subclades, such as the R domain, suggesting that they can have subclade-specific functions.

The GhTCP genes possessed an expanded family, with approximately three-fold size compared with Arabidopsis, tomato, and rice, and approximately two-fold compared with G. arboreum and G. raimondii. This suggests that although plant TCP genes may derive from a common ancestor, many had undergone distinct patterns of differentiation with the divergence of different lineages. Based mainly on amino acid sequence differences, especially in the basic region of the TCP domain, the TCP transcription factors are divided into three groups. There were 50 GhTCP genes in the PCF group, 16 in the CIN group, and seven in the CYC/TB1 group. The numbers of genes in each group were approximately twice those in G. arboretum and G. raimondii. According to a recent study, all tetraploid cotton species (AtDt) evolved from A-genome diploid, G. arboretum, and D-genome diploid, G. raimondii, at around 1–2 Mya [34]. In addition, previous studies indicated that gene duplication contributed to increasing the number of gene family members on various scales, including whole-genome duplication [35]. The expansion of regulatory genes is rarely achieved simply through single gene duplication alone, implying that genome duplication contributed to the amplification of the TCP gene family in G. hirsutum.

TCP transcription factors was involved in the regulation of cell growth and proliferation, which performed diverse functions in multiple aspects of plant growth and development [3]. We determined the spatial and temporal expression profiles of G. hirsutum TCP genes in 10 tissues, which included different developmental stages of ovule and fiber, using transcriptome analysis. The expression in different tissues varied widely among GhTCP genes and different organs for individual TCP genes. This implies functional divergence of GhTCP genes during different plant developmental processes. GhTCP15 and GhTCP71 were relatively highly expressed in ovules and fibers at 10 DPA. Previous study demonstrated that GbTCP was preferentially expressed in elongating G. barbadense fiber during5 to 15 DPA [36]. Overexpression of GbTCP enhanced root hair initiation and elongation in Arabidopsis and regulated branching. Both GbTCP in G. barbadense and GhTCP71 are orthologs of AT1G69690 in Arabidopsis (named AtTCP15), compared with which they had only one amino acid difference within the TCP domain [37]. AtTCP15 was expressed in trichomes and rapidly dividing tissues and vascular tissue, and the protein promoted mitotic cell division, but inhibited endo-reduplication by modulating the expression of several key cell-cycle genes [15,38]. In our study,

GhTCP14 was also expressed predominantly in fiber cells, especially at the initiation and elongation stages of development as previously reported. Induced expression of GhTCP14 can increases the density and length of root hairs and trichomes and affects gravitropism of Arabidopsis [39]. These results suggested that cotton fiber and Arabidopsis root hair elongation may have a similar regulatory mechanism for TCP genes.

Many Arabidopsis TCP genes with similar functions tended to cluster in the same clade, implying that TCP genes within the same clade may have similar functions in G. hirsutum. In Arabidopsis, some angiosperm members of the CIN-like clade involved in leaf and flower morphogenesis are targeted by miR319–for example, AtTCP2, 3, 4, 10, and 24 [40]. Loss of function of these genes results in enlarged leaves, due to an excess of cells that are smaller in size, while their gain of function leads to smaller leaves [41,42]. The miR319, previously known as "miR-JAW", was first described in Arabidopsis because its involvement in the control of leaf morphogenesis [43]. Several studies had reported the involvement of miR319 in plants in response to stress conditions via downregulation of its target genes [44]. Transgenic creeping bentgrass overexpressing a rice miR319, Osa-miR319a, exhibited enhanced salt and drought tolerance [45]. In this study, we observed upregulation of miR319 and downregulation of the targets in both salt and drought treatments. To understand the responses of GhTCP genes to stresses, the expression profiles were investigated in response to abiotic stresses, such as heat, salinity, and drought. In total, 40 GhTCP genes exhibited variations in expression. It is noteworthy that some genes showed instantaneous upregulation, and decreased slowly during continued stress. For example, GhTCP6, 14, 35, and 51 exhibited their highest expression at 3 h of dehydration and salinity treatment. However, no significantly upregulated expression was found at late time points. It is plausible to postulate that these genes might be the part of a stress-signaling system. The functions of these stress-responsive GhTCP genes in abiotic stress resistance will be further characterized in future work.

In this study, a total of 73 non-redundant TCP encoding genes were identified in G. hirsutum. Our results provided evidence for the relationship between structure and function in the G. hirsutum TCP gene family, and laid the foundation for further identification of the functions of the GhTCP gene family and their relationship with miR319.

4. MATERIALS AND METHODS

4.1 Plant Materials and Treatments

The G. hirsutum L. accession TM-1 was used in this study. The seeds were provided by the National Mid-term Genebank of the Institute of Cotton Research in China.

Cotton seeds were sterilized, and germinated in vermiculite under greenhouse conditions: 30/22 ◦C day/night temperature, 55–70% relative humidity, and a 14/10 h light/dark cycle under 450 μmol m−2·s−1 light intensity. At the two-leaf stage, healthy seedlings were placed in pots containing aerated nutrient solution. Plants were cultured under normal conditions for 10 d to ensure full establishment before starting the drought and salt stress treatments. The pH was maintained close to 6.9 by adding H2SO4 or KOH as required. The roots of cotton seedlings were irrigated with 20% PEG to test the response to drought. The seedlings were treated with 150 Mm NaCl solution to test the response to salt. After exposing the seedlings to drought and salt stress for 24 h, leaves were harvested directly into liquid nitrogen and stored at -80 ◦C for subsequent use.

4.2 Sequence Retrieval and TCP Gene Identification

To identify TCPs in G. hirsutum, multiple database searches were performed. The completed genome sequence and protein sequences of this species were downloaded from the CottonGen database (http://www.cottongen.org) and the Cotton Genome Project (http://cgp.genomics.org. cn/page/species/index.jsp). A local protein database was constructed using the protein sequences. The TCP proteins from Arabidopsis and rice were used as query sequences, and were collected from published literature and downloaded from The Arabidopsis Information Resource (TAIR release 10, http://www.arabidopsis.org) and the Rice Genome Annotation Project (ftp://ftp.plantbiology.msu. edu), respectively. The BLASTP (http://cgp.genomics.org.cn/) was used to do the BLAST search. The e-value was set at 1e-10. The candidate TCP genes were further aligned to remove redundant sequences. To verify the reliability of the initial results, all non-redundant candidate TCP sequences were analyzed to confirm the presence of the conserved TCP domain using the InterProScan database (https://www.ebi.ac.uk/) and the NCBI's CDD (http://www.ncbi.nlm.nih.gov/Structure/cdd/cdd. shtml). Based on the results, the sequences that did not include the TCP domain were eliminated.

4.3 Analysis of Protein Features and Chromosomal Locations

The Mw and pI of each TCP protein were obtained using the online ExPASy program (http:// web.expasy.org/compute_pi/). Protein pI was calculated using pK values of amino acids. Protein Mw was calculated by the addition of average isotopic masses of amino acids in the protein and the average isotopic mass of one water molecule. The subcellular localization of each GhTCP protein was analyzed using the CELLO v2.5 server (http://cello.life.nctu.edu.tw/). Through BLASTN (http://cgp.genomics.org.cn/) searches against the G. hirsutum whole genome, some information was obtained about the physical locations of each GhTCP genes on chromosomes.

4.4 Phylogenetic Analysis and Gene Structure

To analyzethe phylogenetic relationships between TCP genes in G. hirsutum and other species, the protein sequences of the identified GhTCP genes, Arabidopsis TCP genes, and rice TCP genes, were used to generate aphylogenetic tree. The ClustalX program was used to align the TCP domains. Phylogenetic trees were constructed by MEGA6.0 using the NJ and Minimal Evolution (ME) methods. For both methods, the bootstrap test of phylogeny was performed with 1000 replications. The exon/intron structures for each GhTCP gene was determined by aligning the CDS sequences to their corresponding genomic DNA sequences. The structures were shown using the Gene Structure Display Server 2.0 (http://gsds.cbi.pku.edu.cn/).

4.5 Expression Analyses of the TCP Genes and Search for miR319 Targets

Expression data for GhTCP genes were obtained from transcriptome data. RNA-seq data were obtained from the NCBI Sequence Read Archive (SRA: PRJNA248163). The expression pattern of GhTCP genes was analyzed in leaves, roots, and stems of 2-week-old plants; petals, torus, pistils, stamens, and lower sepals dissected from whole mature flowers; ovules from -3, -1, 0, 1, 3, 5, 10, and 20 days after pollination; fibers from 5, 10, 20, and 25 days; and true leaves of seedlings treated with salt, PEG, and heat. Gene expression levels were calculated according to Fragments Per Kilobase Million (FPKM) values and the default empirical abundance threshold of FPKM > 1 was used to identify the expressed gene.

Degradome sequencing data were used to find miR319 that caused TCP transcript degradation (GEO: GSE69820). We matched the degraded fragments to the GhTCP gene sequences, identified the cDNA sequences expressed, and then calculated normalized expression numbers of each degraded site along every cDNA, blast with miR319 sequences. A t-plot figure was constructed to show the tag distributions. PairFinder software was used to identify the sliced targets for miRNAs.

4.6 RNA Extraction and qRT-PCR Analysis

Total RNA was extracted with TRIzol Reagent (Invitrogen, 15596-026, Dalian, China) according to the manufacturer's instructions. For the first-strand cDNA synthesis experiment of miR319, a One Step PrimeScript® miRNA cDNASynthesis Kit (Takara, Dalian, China) was used. For each sample, 4 μg of total RNA was converted to cDNA in a 20-μL reaction system, which contained 10 μL of 2 miRNA reaction buffer mix, 2 μL of 0.1% BSA, and 2 μL of miRNA PrimeScript®RT Enzyme Mix. qRT-PCR was performed using SYBR® Premix Ex TaqTM II (Takara) and undertaken with a 7500 Fast Real-Time PCR system (Applied Biosystems Inc., Foster City, CA, USA). The specific miR319 and TCP genes primers used are given in Table S2. The reactions were incubated in a 96-well plate at 95 ◦C at 30 s, followed by 40 cycles of 95 ◦C at 15 s and 60 ◦C at 30 s. The 25-μL reaction solutions contained 12.5 μL of SYBR®Premix Ex TaqTMII (2×), 1 μL of PCR forward primer (10 μM), 1 μL of PCR reverse primer (10 μM) and 2 μL of five fold diluted cDNA template. All reactions were performed with three replicates. Relative expression levels were calculated by the comparative threshold cycle (2-ΔΔT) method.

REFERENCES

1. Lee, T.I.; Young, R.A. Transcription of eukaryotic protein-coding genes. Annu. Rev. Genet. 2000, 34, 77–137. [CrossRef] [PubMed]
2. Cubas, P.; Lauter, N.; Doebley, J.; Coen, E. The TCP domain: A motif found in proteins regulating plant growth and development. Plant J. 1999, 18, 215–222. [CrossRef] [PubMed]
3. Martin-Trillo, M.; Cubas, P. TCP genes: A family snapshot ten years later. Trends Plant Sci. 2010, 15, 31–39. [CrossRef] [PubMed]
4. Aggarwal, P.; Das Gupta, M.; Joseph, A.P.; Chatterjee, N.; Srinivasan, N.; Nath, U. Identification of specific DNA binding residues in the TCP family of transcription factors in Arabidopsis. Plant Cell 2010, 22, 1174–1189. [CrossRef] [PubMed]
5. Kosugi, S.; Ohashi, Y. DNA binding and dimerization specificity and potential targets for the TCP protein family. Plant J. 2002, 30, 337–348. [CrossRef] [PubMed]
6. Navaud, O.; Dabos, P.; Carnus, E.; Tremousaygue, D.; Herve, C. TCP transcription factors predate the emergence of land plants. J. Mol. Evol. 2007, 65, 23–33. [CrossRef] [PubMed]
7. Viola, I.L.; Reinheimer, R.; Ripoll, R.; Manassero, N.G.; Gonzalez, D.H. Determinants of the DNA binding specificity of class I and class II TCP transcription factors. J. Biol. Chem. 2012, 287, 347–356. [CrossRef] [PubMed]
8. Sarvepalli, K.; Nath, U. CIN-TCP transcription factors: Transiting cell proliferation in plants. IUBMB Life 2018, 70, 718–731. [CrossRef] [PubMed]
9. Kosugi, S.; Ohashi, Y. PCF1 and PCF2 specifically bind to cis elements in the rice proliferating cell nuclear antigen gene. Plant Cell 1997, 9, 1607–1619. [CrossRef] [PubMed]
10. Li, C.; Potuschak, T.; Colon-Carmona, A.; Gutierrez, R.A.; Doerner, P. Arabidopsis TCP20 links regulation of growth and cell division control pathways. Proc. Nat. Acad. Sci. USA 2005, 102, 12978–12983. [CrossRef] [PubMed]
11. Danisman, S. TCP Transcription Factors at the Interface between Environmental Challenges and the Plant's Growth Responses. Front. Plant Sci. 2016, 7, 1930. [CrossRef] [PubMed]

12. Tatematsu, K.; Nakabayashi, K.; Kamiya, Y.; Nambara, E. Transcription factor AtTCP14 regulates embryonic growth potential during seed germination in Arabidopsis thaliana. Plant J. 2008, 53, 42–52. [CrossRef] [PubMed]

13. Gao, Y.; Zhang, D.; Li, J. TCP1 Modulates DWF4 Expression via Directly Interacting with the GGNCCC Motifs in the Promoter Region of DWF4 in Arabidopsis thaliana. J. Genet. Genom. 2015, 42, 383–392. [CrossRef] [PubMed]

14. Kim, H.B.; Kwon, M.; Ryu, H.; Fujioka, S.; Takatsuto, S.; Yoshida, S.; An, C.S.; Lee, I.; Hwang, I.; Choe, S. The regulation of DWARF4 expression is likely a critical mechanism in maintaining the homeostasis of bioactive brassinosteroids in Arabidopsis. Plant Physiol. 2006, 140, 548–557. [CrossRef] [PubMed]

15. Lucero, L.E.; Uberti-Manassero, N.G.; Arce, A.L.; Colombatti, F.; Alemano, S.G.; Gonzalez, D.H. TCP15 modulates cytokinin and auxin responses during gynoecium development in Arabidopsis. Plant J. 2015, 84, 267–282. [CrossRef] [PubMed]

16. Zhou, Y.; Zhang, D.Z.; An, J.X.; Yin, H.J.; Fang, S.; Chu, J.F.; Zhao, Y.D.; Li, J. TCP Transcription Factors Regulate Shade Avoidance via Directly Mediating the Expression of Both PHYTOCHROME INTERACTING FACTORs and Auxin Biosynthetic Genes. Plant Physiol. 2018, 176, 1850–1861. [CrossRef] [PubMed]

17. Mukhopadhyay, P.; Tyagi, A.K. OsTCP19 influences developmental and abiotic stress signaling by modulating ABI4-mediated pathways. Sci. Rep. 2015, 5, 9998. [CrossRef] [PubMed]

18. Carrington, J.C.; Ambros, V. Role of microRNAs in plant and animal development. Science 2003, 301, 336–338. [CrossRef] [PubMed]

19. Schommer, C.; Palatnik, J.F.; Aggarwal, P.; Chetelat, A.; Cubas, P.; Farmer, E.E.; Nath, U.; Weigel, D. Control of jasmonate biosynthesis and senescence by miR319 targets. PLoS Biol. 2008, 6, e230. [CrossRef] [PubMed]

20. Zhou, M.; Li, D.; Li, Z.; Hu, Q.; Yang, C.; Zhu, L.; Luo, H. Constitutive expression of a miR319 gene alters plant development and enhances salt and drought tolerance in transgenic creeping bentgrass. Plant Physiol. 2013, 161, 1375–1391. [CrossRef] [PubMed]

21. Chen, Z.J.; Scheffler, B.E.; Dennis, E.; Triplett, B.A.; Zhang, T.; Guo, W.; Chen, X.; Stelly, D.M.; Rabinowicz, P.D.; Town, C.D.; et al. Toward sequencing cotton (Gossypium) genomes. Plant Physiol. 2007, 145, 1303–1310. [CrossRef] [PubMed]

22. Zhang, T.; Hu, Y.; Jiang, W.; Fang, L.; Guan, X.; Chen, J.; Zhang, J.; Saski, C.A.; Scheffler, B.E.; Stelly, D.M.; et al. Sequencing of allotetraploid cotton (Gossypium hirsutum L. acc. TM-1) provides a resource for fiber improvement. Nat. Biotechnol. 2015, 33, 531–537. [CrossRef] [PubMed]

23. Li, F.; Fan, G.; Lu, C.; Xiao, G.; Zou, C.; Kohel, R.J.; Ma, Z.; Shang, H.; Ma, X.; Wu, J.; et al. Genome sequence of cultivated Upland cotton (Gossypium hirsutum TM-1) provides insights into genome evolution. Nat. Biotechnol. 2015, 33, 524–530. [CrossRef] [PubMed]

24. Marchler-Bauer, A.; Zheng, C.; Chitsaz, F.; Derbyshire, M.K.; Geer, L.Y.; Geer, R.C.; Gonzales, N.R.; Gwadz, M.; Hurwitz, D.I.; Lanczycki, C.J.; et al. CDD: Conserved domains and protein three-dimensional structure. Nucleic Acids Res. 2013, 41, D348–352. [CrossRef] [PubMed]

25. Qin, L.J.; Guo, X.Z.; Feng, X.Z.; Weng, L.; Yan, J.; Hu, X.H.; Luo, D. Cloning of LjCYC1 gene and nuclear localization of LjCYC1 protein in Lotus japonicus. Zhi Wu Sheng Li Yu Fen Zi Sheng Wu Xue Xue Bao 2004, 30, 523–532. [PubMed]

26. Zhou, Y.; Xu, Z.; Zhao, K.; Yang, W.; Cheng, T.; Wang, J.; Zhang, Q. Genome Wide Identification, Characterization and Expression Analysis of the TCP Gene Family in Prunus mume. Front. Plant Sci. 2016, 7, 1301. [CrossRef] [PubMed]

27. Parapunova, V.; Busscher, M.; Busscher-Lange, J.; Lammers, M.; Karlova, R.; Bovy, A.G.; Angenent, G.C.; de Maagd, R.A. Identification, cloning and characterization of the tomato TCP transcription factor family. BMC Plant Biol. 2014, 14, 157. [CrossRef] [PubMed]

28. De Paolo, S.; Gaudio, L.; Aceto, S. Analysis of the TCP genes expressed in the inflorescence of the orchid Orchis italica. Sci. Rep. 2015, 5, 16265. [CrossRef] [PubMed]

29. Ma, X.; Ma, J.; Fan, D.; Li, C.; Jiang, Y.; Luo, K. Genome-wide Identification of TCP Family Transcription Factors from Populus euphratica and Their Involvement in Leaf Shape Regulation. Sci. Rep. 2016, 6, 32795. [CrossRef] [PubMed]

30. Shi, P.; Guy, K.M.; Wu, W.; Fang, B.; Yang, J.; Zhang, M.; Hu, Z. Genome-wide identification and expression analysis of the ClTCP transcription factors in Citrullus lanatus. BMC Plant Biol. 2016, 16, 85. [CrossRef] [PubMed]

31. Yao, X.; Ma, H.; Wang, J.; Zhang, D. Genome-Wide Comparative Analysis and Expression Pattern of TCP Gene Families in Arabidopsis thaliana and Oryza sativa. J. Integrat. Plant Biol. 2007, 49, 32795. [CrossRef]

32. Ma, J.; Wang, Q.; Sun, R.; Xie, F.; Jones, D.C.; Zhang, B. Genome-wide identification and expression analysis of TCP transcription factors in Gossypium raimondii. Sci. Rep. 2014, 4, 6645. [CrossRef] [PubMed]

33. Ma, J.; Liu, F.; Wang, Q.; Wang, K.; Jones, D.C.; Zhang, B. Comprehensive analysis of TCP transcription factors and their expression during cotton (Gossypium arboreum) fiber early development. Sci. Rep. 2016, 6, 21535. [CrossRef] [PubMed]

34. Paterson, A.H.; Wendel, J.F.; Gundlach, H.; Guo, H.; Jenkins, J.; Jin, D.; Llewellyn, D.; Showmaker, K.C.; Shu, S.; Udall, J.; et al. Repeated polyploidization of Gossypium genomes and the evolution of spinnable cotton fibres. Nature 2012, 492, 423–427. [CrossRef] [PubMed]

35. Kurosaki, M.; Bolis, M.; Fratelli, M.; Barzago, M.M.; Pattini, L.; Perretta, G.; Terao, M.; Garattini, E. Structure and evolution of vertebrate aldehyde oxidases: From gene duplication to gene suppression. Cell Mol. Life Sci. 2013, 70, 1807–1830. [CrossRef] [PubMed]

36. Hao, J.; Tu, L.; Hu, H.; Tan, J.; Deng, F.; Tang, W.; Nie, Y.; Zhang, X. GbTCP, a cotton TCP transcription factor, confers fibre elongation and root hair development by a complex regulating system. J. Exp. Bot. 2012, 63, 6267–6281. [CrossRef] [PubMed]

37. Kieffer, M.; Master, V.; Waites, R.; Davies, B. TCP14 and TCP15 affect internode length and leaf shape in Arabidopsis. Plant J. 2011, 68, 147–158. [CrossRef] [PubMed]

38. Li, Z.Y.; Li, B.; Dong, A.W. The Arabidopsis transcription factor AtTCP15 regulates endoreduplication by modulating expression of key cell-cycle genes. Mol. Plant 2012, 5, 270–280. [CrossRef] [PubMed]

39. Wang, M.Y.; Zhao, P.M.; Cheng, H.Q.; Han, L.B.; Wu, X.M.; Gao, P.; Wang, H.Y.; Yang, C.L.; Zhong, N.Q.; Zuo, J.R.; et al. The cotton transcription factor TCP14 functions in auxin-mediated epidermal cell differentiation and elongation. Plant Physiol. 2016, 162, 1669–1680. [CrossRef] [PubMed]

40. Mendez-Vigo, B.; de Andres, M.T.; Ramiro, M.; Martinez-Zapater, J.M.; Alonso-Blanco, C. Temporal analysis of natural variation for the rate of leaf production and its relationship with flowering initiation in Arabidopsis thaliana. J. Exp. Bot. 2010, 61, 1611–1623. [CrossRef] [PubMed]

41. Koyama, T.; Furutani, M.; Tasaka, M.; Ohme-Takagi, M. TCP transcription factors control the morphology of shoot lateral organs via negative regulation of the expression of boundary-specific genes in Arabidopsis. Plant Cell 2007, 19, 473–484. [CrossRef] [PubMed]

42. Efroni, I.; Blum, E.; Goldshmidt, A.; Eshed, Y. A protracted and dynamic maturation schedule underlies Arabidopsis leaf development. Plant Cell 2008, 20, 2293–2306. [CrossRef] [PubMed]

43. Palatnik, J.F.; Allen, E.; Wu, X.; Schommer, C.; Schwab, R.; Carrington, J.C.; Weigel, D. Control of leaf morphogenesis by microRNAs. Nature 2003, 425, 257–263. [CrossRef] [PubMed]

44. Thiebaut, F.; Rojas, C.A.; Almeida, K.L.; Grativol, C.; Domiciano, G.C.; Lamb, C.R.; de Engler, J.A.; Hemerly, A.S.; Ferreira, P.C. Regulation of miR319 during cold stress in sugarcane. Plant Cell Environ. 2012, 35, 502–512. [CrossRef] [PubMed]

45. Zhou, M.; Luo, H. Role of microRNA319 in creeping bentgrass salinity and drought stress response. Plant Signal. Behav. 2014, 9, e28700. [CrossRef] [PubMed]

14 Chapter

EXTRACTION OF LATE EMBRYOGENESIS ABUNDANT (LEA) FAMILY GENES IN CLEISTOGENES SONGORICA, A XEROPHYTIC PERENNIAL DESERT FLORA

1. INTRODUCTION: AN OVERVIEW

Abiotic stressors resulting from elevated temperature or salinity can impair optimal plant performance and lead to substantial crop production reductions [1]. To sustain optimal homeostasis for normal growth, plants have developed several mechanisms to counteract adverse conditions by activating a diverse array of stress-responsive genes [2]. Proteins expressed by late embryogenesis abundant (LEA) family genes are recognized for their protective functions in plants under abiotic stressors [3,4]. The LEA family genes were initially examined in cotton seeds during the later stages of seed development [4]. The LEA family genes were subsequently found in diverse tissues of numerous plant species, and the proteins encoded by these genes were demonstrated to play a crucial role during cold, drought, and/or high salinity stressors [5,6]. LEA proteins are not exclusive to plants; they are also present in invertebrates, fungi, and bacteria [7,8]. Typical LEA proteins have significant hydrophilicity owing to elevated levels of charged amino acid residues, alongside the presence of threonine, serine, and alanine in their sequences [9]. LEA proteins are proposed to possess molecular shielding properties that can mitigate protein aggregation and inhibit enzyme degradation, therefore facilitating normal cellular homeostasis under stress conditions. LEA proteins are flexible entities capable of undergoing conformational alterations and engaging with various macromolecules, including proteins, membranes, and nucleic acids, under diverse adverse stress circumstances [10].

LEA proteins are divided into at least eight distinct subfamilies, based on their conserved protein domains in the Pfam database: LEA (1–6), Dehydrin and Seed Maturation Protein (SMP) [13]. Motif structures within subfamily genes are mostly conserved, except the genes in the LEA_2/LEA5C subfamily [9]. In addition, proteins in the LEA_2/LEA5C subfamily are known to have other non-canonical LEA protein properties like high hydrophobicity and at least one atypical LEA domain known as Water stress and Hypersensitive response (WHy) domain. The presence of atypical LEA domain(s) in LEA proteins indicate that these proteins may function differently from typical LEA proteins [14–16]. Proteins in the Dehydrin subfamily are featured with at least one K-segment, a 15 amino acid residue rich in lysine (i.e., EKKGIMDKIKEKLPG) and can function like chaperones to protect peripheral membrane and proteins during dehydration [2,17–20]. Numerous earlier studies have demonstrated that the LEA family genes are potential abiotic stress responsive genes, important for enhancing plant stress tolerance. For instance, transgenic Arabidopsis [21,22], maize [23], alfalfa [24] bacteria [25], yeast and tobacco [26] expressing different LEA genes exhibited improved abiotic stress tolerance compared with their respective wild-type plant, bacteria or yeast.

C. songorica is a xerophytic C4 desert plant extensively found in the arid regions of northwestern China, where the annual precipitation averages approximately 100 mm [27]. Prior genome-wide investigations of LEA family genes have been conducted for many plant species, excluding perennial monocotyledons. This work examined the LEA family genes in C. songorica and studied the responses of four selected LEA genes to heat stress and abscisic acid (ABA) treatment in leaves and shoots. This study's results offer novel insights into the development of LEA family proteins, their structures, and the possible applications of these genes for enhancing agricultural tolerance to abiotic challenges.

2. RESULTS

2.1 Identification of CsLEA Genes and Phylogenetic Analysis

A total of 44 putative C. songorica LEA proteins were identified in this study (Table S1). These proteins were named from CsLEA1 to CsLEA44 and grouped into eight different subfamilies (Figure 1): CsLEA_1, CsLEA_2/LEA5C (Battaglia classification), CsLEA_3, CsLEA_4, CsLEA_5, CsLEA_6, SMP and Dehydrin, based on their Pfam conserved protein domains and their homology with the published LEA proteins of A. thaliana [28]. Two atypical LEA stress related domains, Water Stress and Hypersensitive response (WHy) and LEA14-like desiccation related protein (COG5608), were detected in the proteins from CsLEA_2 subfamily (Figure S1).

2.2 Structures, Physiochemical Properties and Subcellular Localizations of CsLEA Proteins

Most proteins within the same family exhibited similar structures and properties (Table S1). Over one third of the CsLEA proteins were classified as unstable proteins with an instability index value higher than 40. Furthermore, this property varied significantly among the proteins within the Dehydrin subfamily, ranging from 3.83 to 56.11. All the proteins in the LEA_5 subfamily showed instability index values greater than 47 and were considered as the most unstable and/or potentially disordered proteins [29–31]. The GRAVY (grand average of hydropathicity index) values of more than 90% CsLEA proteins were below 0, stressing that CsLEA proteins are likely to have low hydrophobicity features. CsLEA_2 subfamily proteins were the most hydrophobic proteins, while Cs_LEA5 proteins were the highest hydrophilic proteins. These results are consistent with previous studies [28,32] and reinforce the structural disordered properties of LEA proteins by which they are capable of interacting with other molecules and mitigating the collision of enzymes during plant stress conditions [11]

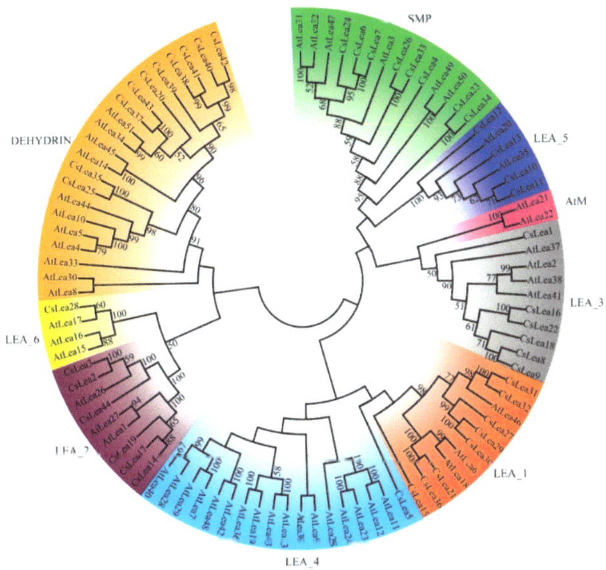

Figure 1. Phylogenetic analysis of C. songorica LEA proteins. Full-length amino acid sequences of the 44 CsLEA proteins were analyzed using the unrooted method in the ClustalW software.

2.3 Gene and Motif Structure Analyses

LEA genes within the same subfamily showed similar exon and intron architectures (Figure 2, right panel). Further investigation of structures of paired genes at the short-end branches in the phylogenetic tree revealed that six of them (e.g., CsLEA 43–37, 35–25, 2–3, 8–9, 23–34 and 6–7) might have experienced exon-intron gain/loss events during their evolutionary history. Similar situations have been reported for Brassica LEA genes [21,33]. In total, 18 motifs were identified in 43 CsLEA proteins (Figure 2, left panel). No motif was found in CsLEA28 protein. Except for LEA_2 and LEA_3 subfamily proteins, motif structures and compositions were nearly identical among the proteins in the same subfamily, but differed significantly between the proteins belonging to different subfamilies, implying functional specificities of different CsLEA subfamily proteins [14,34].

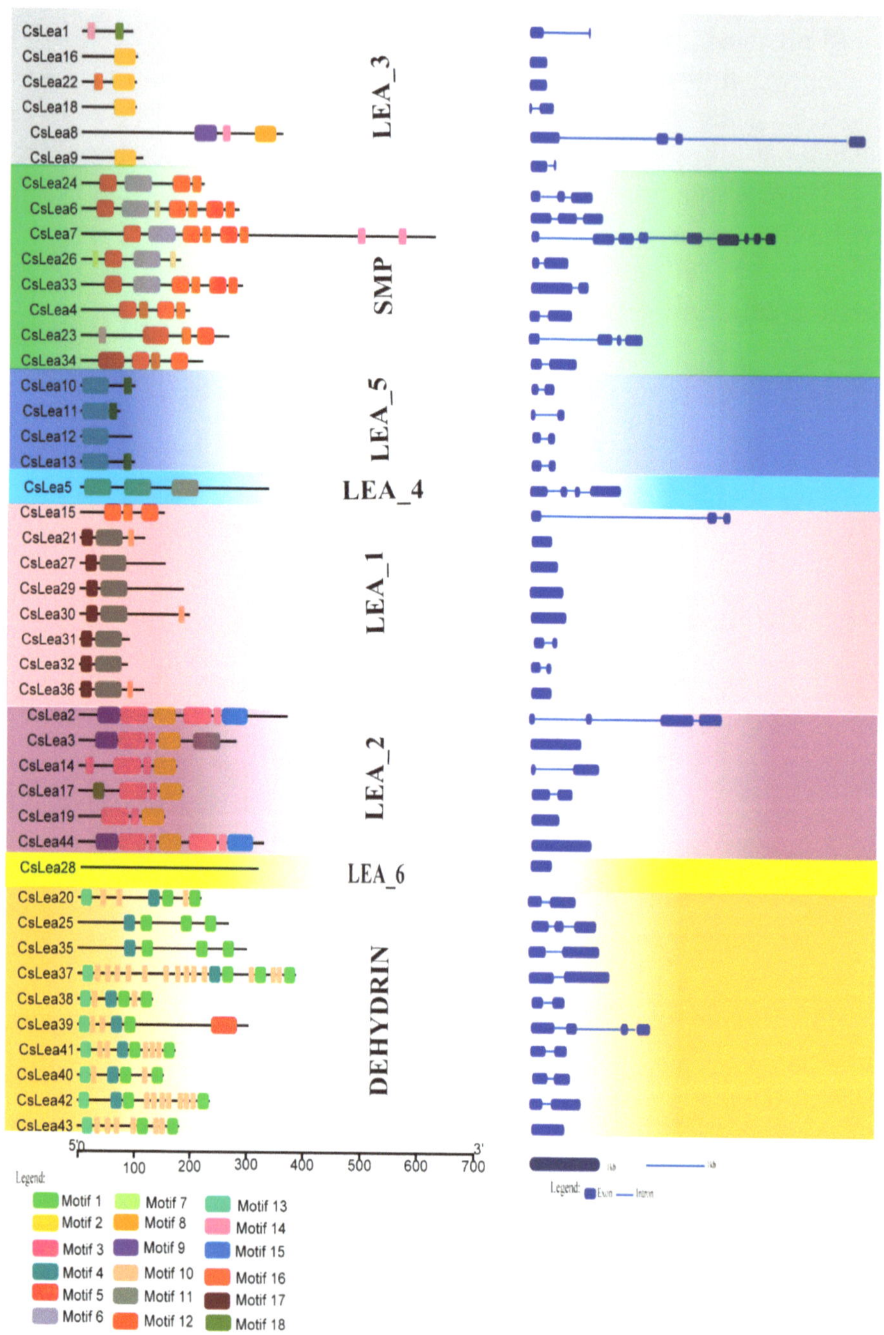

Figure 2. Motif structure and exon-intron organizations of the 44 CsLEA genes. The 18 motifs discovered in this study are shown on these CsLEA genes (left). The blue boxes represent exons and the blue lines represent introns (right).

2.4 Sequence Alignment of C. Songorica and Arabidopsis Dehydrin Proteins

Multiple sequence alignment using C. songorica Dehydrin protein sequences and their Arabidopsis counterparts revealed the conservation stress response related segments including Y, K and S segments (YKS) in C. songorica Dehydrin proteins (Figure 3).

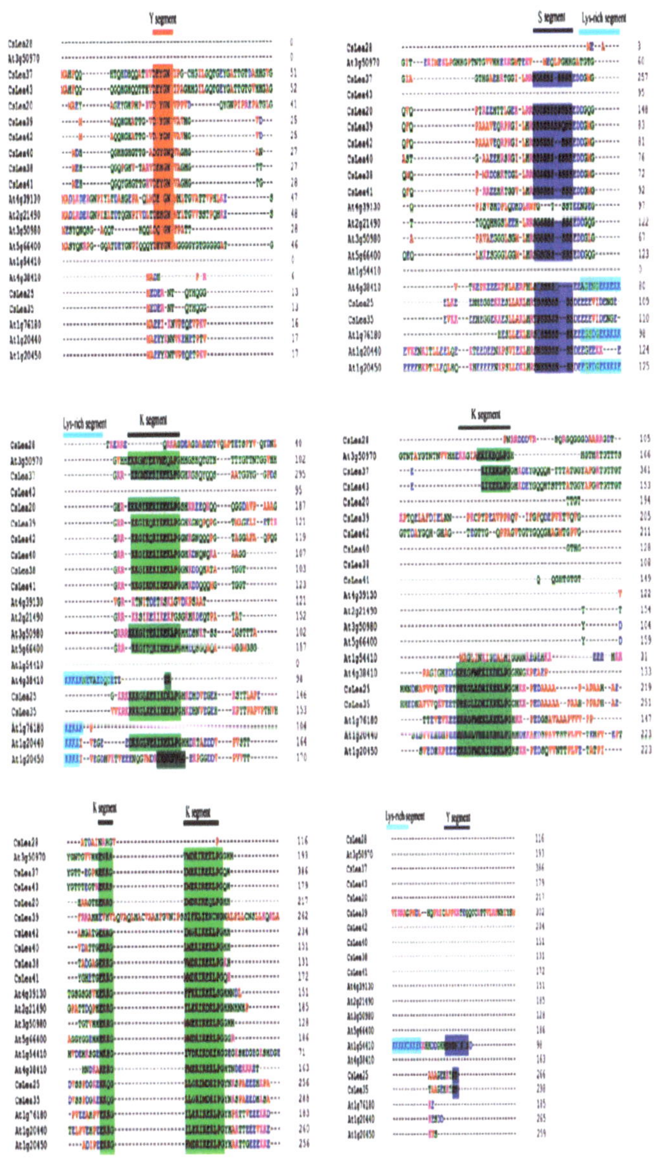

Figure 3. Multiple sequence alignment using C. songorica and Arabidopsis Dehydrin protein sequences. The identified Y segment, K segment and S segment are indicated by different colors. Y segment = red, K segment = green and S segment = purple.

2.5 Cis-Regulatory Element in C. Songorica LEA Gene Promoters

Cis-regulatory elements control expression patterns of stress responsive genes in various tissues and organs. These elements are located upstream of gene coding sequences and provide binding sites for transcription factors (TFs) [35]. More than three G-box cis-elements were recorded for each Dehydrin, LEA_2 and SMP genes and nearly more than two MBS elements were detected for each Dehydrin and LEA_2 subfamily gene (Table 1).

Table 1. Stress-related cis-regulatory elements in 44 C. songorica LEA gene promoters.

CsLEA Subfamilies	Gene Names	Functional cis-Element Names and Sequences			
		MBS (CGGTC)	G-Box (GTGCAT/ CACGAC)	ABRE (GACACGTACGT)	CGTCA Motif
		Functions			
		Drought Responsive (MYB Binding Site)	Light Responsive	Abscisic Acid Responsive	MeJA Responsive
LEA_1	CsLEA29	3	1	1	0
	CsLEA30	2	3	1	4
	CsLEA31	4	1	1	0
	CsLEA32	2	2	1	0
	CsLEA36	4	5	2	2
LEA_2	CsLEA2	2	1	0	3
	CsLEA3	0	3	2	2
	CsLEA14	2	6	1	0
	CsLEA17	3	9	3	0
	CsLEA19	5	0	0	1
	CsLEA44	0	0	0	2
LEA_3	CsLEA1	4	2	0	2
	CsLEA16	1	2	0	3
	CsLEA22	1	3	1	3
	CsLEA18	8	3	3	0
	CsLEA8	0	7	1	2
	CsLEA9	3	0	0	0
LEA_4	CsLEA5	1	5	4	1
LEA_5	CsLEA10	0	2	0	1
	CsLEA11	0	3	0	3

	CsLEA12	2	7	1	1
	CsLEA13	2	1	3	5
LEA_6	CsLEA28	0	9	5	3
SMP	CsLEA24	0	4	1	0
	CsLEA6	2	0	2	0
	CsLEA7	2	6	0	0
	CsLEA26	1	5	1	1
	CsLEA33	0	5	1	4
	CsLEA4	2	1	0	0
	CsLEA23	2	1	0	0
	CsLEA34	1	4	0	1
	CsLEA42	0	3	1	2
	CsLEA43	0	0	1	5
Dehydrin	CsLEA20	4	5	0	0
	CsLEA25	4	2	0	1
	CsLEA35	3	8	2	1
	CsLEA37	0	6	3	4
	CsLEA38	2	10	2	3
	CsLEA39	3	1	0	1
	CsLEA41	2	0	1	2
	CsLEA40	0	2	0	1
	CsLEA42	0	3	1	2
	CsLEA43	0	0	1	5

2.6 Chromosomal Mapping of CsLEA Genes

C. songorica genome has in total twenty chromosomes. Positions of the 44 CsLEA genes on 15 different C. songorica chromosomes were estimated (Figure 4). Genes from the same subfamily were mostly found on different chromosomes, suggesting a strategy to exert their functions across the whole C. songorica genome. However, genes in the LEA_5 and Dehydrin subfamily were mostly found in clusters on the 14th, 15th and 18th chromosome.

2.7 Gene Expression Analysis qRT-PCR Validation

The expression levels of the CsLEA 14, CsLEA 19, CsLEA 37 and CsLEA 38 genes were induced after 24 h of ABA or heat treatment but were not tissue specific (Figure 5). To validate results from expression profile analysis, qRT-PCR was carried out for CsLEA 14, CsLEA 19 (from the LEA_2 subfamily) and for CsLEA37 and CsLEA 38 (from the Dehydrin subfamily) as these genes showed a relatively high number of stress related cis acting elements and motifs.

We carried out qRT-PCR analyses using CsLEA14 and CsLEA19 (LEA_2 subfamily), and CsLEA37 and CsLEA 38 (Dehydrin family). Results showed that after 24 h heat treatment the expression levels of these four genes up-regulated by 156.5 (CsLEA 19), 95.8 (CsLEA14), 52.6 (CsLEA38) and 14.6 fold (CsLEA37) in C. songorica leaves compared with the untreated plant leaf samples. The expression levels of these four genes were slightly up-regulated after the ABA treatment, especially CsLEA 38 (14.6 fold, Figure 6).

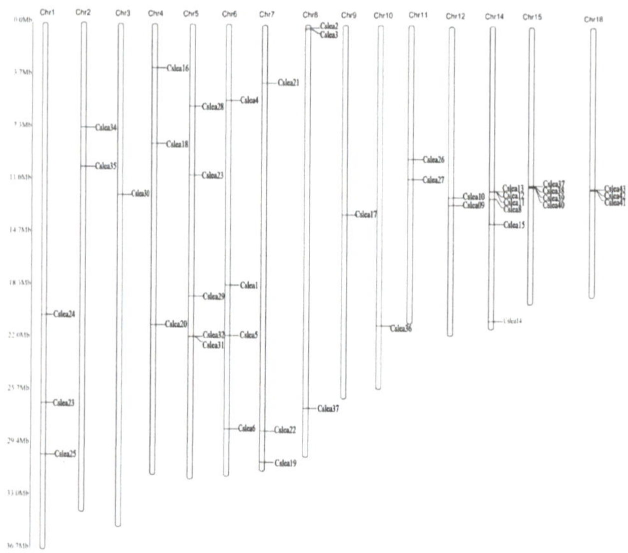

Figure 4. Locations of the 44 CsLEA genes on 15 chromosomes of C. songorica.

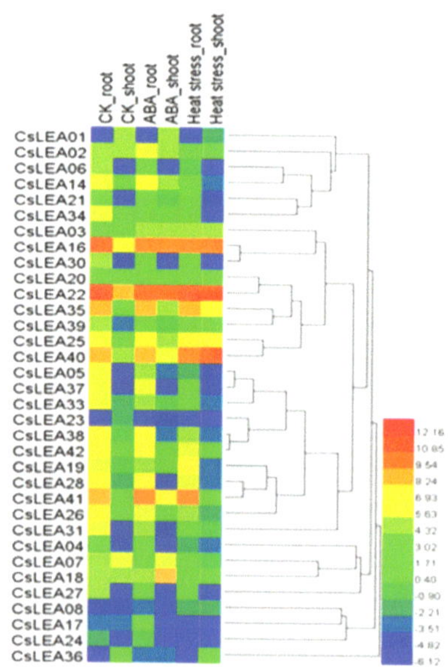

Figure 5. Hierarchical clustering of CsLEA gene expression profiles in root and shoot tissues after 24 h heat or ABA treatment. The log transformed values for

the relative expressions of CsLEA genes were used for the hierarchical clustering analysis. The blue scale means low transcript expression and the red scale means high transcript expression.

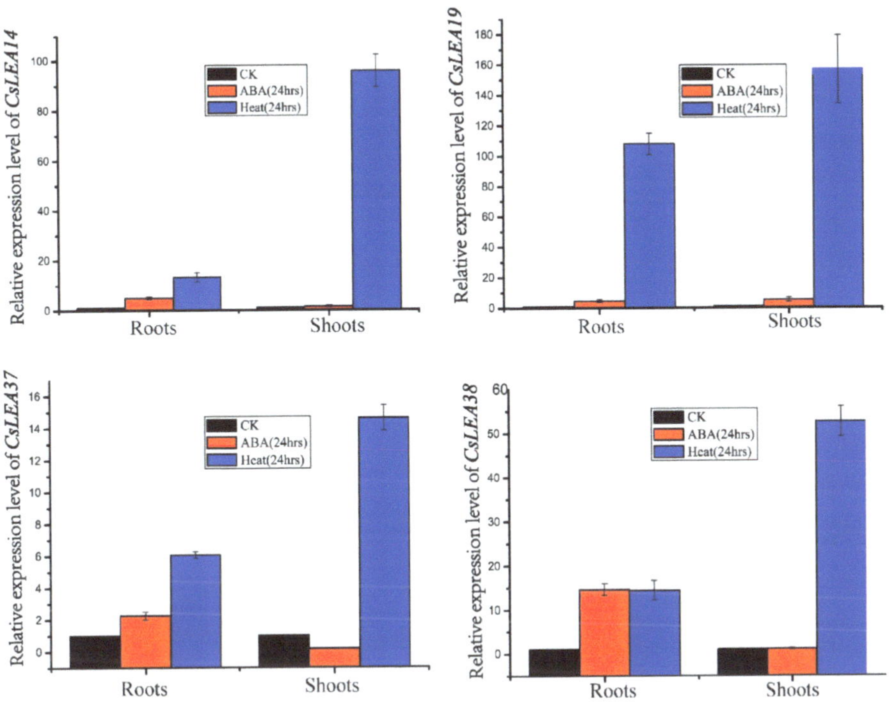

Figure 6. Hierarchical clustering of CsLEA gene expression profiles in root and shoot tissues after 24 h heat or ABA treatment. The log transformed values for the relative expressions of CsLEA genes were used for the hierarchical clustering analysis. The blue scale means low transcript expression and the red scale means high transcript expression.

3. DISCUSSION

3.1 Phylogeny Analysis and Protein Sequence Analysis

Genome surveys of LEA subfamily proteins (e.g., LEA_1 to LEA_6, SMP and DEHYDRIN) were done for A. thaliana [28], rice [36] and maize [37]. C. songorica is a perennial monocotyledonous desert plant with a high tolerance for drought stress. Investigation of stress responsive proteins encoded by LEA family genes in a desert plant should benefit crop improvement for drought and other abiotic stresses. Phylogenetic analysis grouped the CsLEA proteins into eight different subfamilies. Several C. songorica and Arabidopsis Dehydrin subfamily proteins were clustered together with high bootstrap values, which implied potential significant functional similarities between C. songorica and Arabidopsis DEHYDRIN proteins. Many of the Arabidopsis Dehydrin proteins are known as stress regulatory proteins such as RAB18 (AtLea51) and COR47 (AtLea4), two ABA and cold inducible proteins [38,39],

ERD14 (AtLea4) and ERD10 (AtLea5), two disordered chaperon proteins [40] and Dehydrin Xero2 (AtLea33), a disordered cold responsive protein with membrane binding activity [41].

Sequence alignment using CsDehydrin proteins and their Arabidopsis counterparts revealed that all C. songorica Dehydrin proteins contained YKS segments but lacked the lysine rich segment. The K segment is critical for the formation of structural disordered alpha-helical compounds that can enhance bindings between proteins and their targeted molecules [10,42]. The presence of the K segment in C.songorica DEHYDRIN proteins emphasizes their role in limiting aggregation of molecules and thence promoting proper cellular homeostasis during dehydration stresses [11]. Additionally, the detection of the S segment in the C. songorica DEHYDRIN protein also suggests their implication in enhancing plant tolerance against abiotic stresses through protein phosphorylation, as previous studies indicated that the S segment participates in calcium binding through protein phosphorylation [43]. The findings above support the fact that disordered LEA proteins are flexible proteins, capable of adjusting their conformation to maintain proper cellular homeostasis during detrimental stress conditions [43,44].

3.2 Protein Domain Analysis

Two additional non-LEA conserved protein domains that are associated with stress response were detected in the CsLEA_2 subfamily proteins (Supplementary Figure S1). At least one WHy (Water stress and Hypersensitive response) [15,16] and one COG5608 domain (LEA14-like desiccation related protein) was spotted within each single protein sequence from the CsLEA_2 subfamily. The COG5608 domain was previously detected in the Arabidopsis LEA14 protein, a well characterized abiotic stress marker protein, which suggests that the CsLEA_2 subfamily proteins may function similarly to Arabidopsis LEA14 protein (Atlea1). On the other hand, a thorough functional characterization of the WHy domain has only been elucidated in a few bacterial genes, dwhy1 [15] and drwh [45]. Studies in vivo indicated that dwhy1 confers cold and freeze damage resistance. Furthermore, in E. coli, the function of drwh is related to oxidative stress tolerance and salinity stresses. Silencing this gene triggered reduced activity of antioxidant enzymes such as lactate dehydrogenase (LDH) malate dehydrogenase (MDH) [15,45]. All the CsLea_2 subfamily genes contained the WHy domain which could explain the functional importance of this domain during low water availability in a typical desert grass, C. songorica.

3.3 C. songorica Gene Promoter and Gene Expression Analysis

ABA and stress responsive cis-regulatory elements, such as ABRE, MBS/MYB and G-Box were found to be abundant in the promoters of C. songorica Dehydrin and the LEA_2 subfamilies genes. These regulatory elements are known to provide binding sites for transcription factors like ABEF (a member of the bZIPTFs family), BHLH and ERF for the transcription of downstream stress responsive genes [46]. Expression analysis of CsLEA37 and CsLEA38 (DEHYDRIN genes), and CsLEA14 and CsLEA19 (LEA_2 subfamily genes) with qRT-PCR indicated that the expression levels of these four genes in root and shoot tissues were significantly up-regulated after the drought or ABA treatment.

The relevant role of DEHYDRIN or the LEA_2 subfamily proteins during plant stress tolerance has been reported in earlier studies using various transgenic plants. For example, transgenic tobacco plant overexpressing the CaLEA6 gene showed an enhanced dehydration and salt tolerance [47]. Additionally, sweet potato plants overexpressing thc IbLEA14 gene exhibited an improved salinity and dehydration tolerance [48]. A Foxtail millet plant overexpressing the SiLEA4 gene displayed salt and drought resilience [49]. For DEHYDRIN proteins, transgenic Arabidopsis plant expressing a Dehydrin gene from an olive showed an enhanced osmotic stress tolerance [50]. Similarly, a wheat Dhn-5 gene increased salinity and dehydration stress tolerance in transgenic Arabidopsis plants [51].

C. songorica DEHYDRIN and LEA_2 gene transcripts accumulation during water deficit and ABA treatment, reinforcestheir functional importance under detrimental stress conditions.

4. MATERIALS AND METHODS

4.1 Mining LEA Genes in the C. Songorica Genome

C. songorica LEA genes were mined based on their protein sequence homology with the previously published A. thaliana [28], Oryza sativa (rice) [36] and Zea mays (maize) [37] LEA protein sequences. The published full length A. thaliana, rice and maize LEA protein sequences or coding sequences were retrieved from (https://phytozome.jgi.doe.gov/pz/portal.html) [28,32]. The obtained LEA protein sequences were used as queries to blast search the whole C. songorica genome sequence retrieved from the BMK cloud: http://www.biocloud.net/ using a local blast tool [52,53].

The resulting non-redundant sequences were further examined with the Hidden Markov Model available in the Pfam database (http://pfam.sanger.ac.uk/search) [13] and then submitted to the SMART database (http://smart.embl-heidelberg.de/) [54] and the NCBI Conserved Domain Search database (http://www.ncbi.nlm.nih.gov/Structure/cdd/wrpsb.cgi) [55], respectively, to confirm CsLEA Pfam domain families. The obtained LEA nucleotide and protein sequences were then submitted to the Genbank to obtain respective accession numbers (Table S1).

4.2 Multiple Sequence Alignment and Phylogenetic Analysis of CsLEA Family Proteins.

The alignment of the C. songorica and Arabidopsis LEA protein sequences was performed using the ClustalW software [56] in the MEGA 6 program with a default parameter setting. After sequence alignment and pair-wise deletion of gaps as previously described [57], a phylogenetic tree was constructed using the Neighbor Joining (NJ) algorithm with bootstrap analysis of 1000 trials [57,58]. Multiple sequence alignment and sequence homology analysis of C. songorica and Arabidopsis Dehydrin proteins were performed using the ClustalW algorithm embedded in the DNAMAN version 6 program as instructed (Lynnon Corporation, Quebec, Canada).

4.3 In Silico Analyses of CsLEA Proteins.

Determination of GRAVY (grand average of hydropathicity index) values and pI (theoretical isoelectric point) were carried out by the ProtParam Tool (web.expasy.org/protparam/) [59]. Protein Prowler Subcellular Localization Predictor version 1.2 (http://bioinf.scmb.uq.edu.au/ pprowler_webapp_1--2/) 53 and TargetP1.1 (http://www.cbs.dtu.dk/services/TargetP/) servers [60] were used to predict the subcellular locations of C. songorica LEA proteins. All of the prediction servers were run under the default settings. To determine the conserved motifs in different C. songorica and Arabidopsis LEA proteins, protein sequences were analyzed using MEME (The Multiple Expectation Maximization for Motif Elicitation) platform (http://alternate.meme-suite.org/) [61]. MEME parameters were then customized to detect a maximum of 40 motifs with a width covering 6 to 50 amino acid residues.

4.4 Analysis of Cis-Regulatory Elements and Motifs

Sequences of 2000 bp from promoters of the 44 identified C. songorica LEA genes were analyzed for potential cis-regulatory elements and motifs by querying them through the PlantCARE database (http://bioinformatics.psb.ugent.be/webtools/plantcare/html/) [62]. Stress- and ABA-related cis-regulatory elements, including MYB binding site (drought responsive) [63], G-box (light inducible) [64], ABRE (Abscisic acid responsive) [65] and CGTCA-motif (Methyl jasmonate responsive) [66] were recorded.

4.5 Plant Material Preparation and Transcriptomic Data Analysis

C. songorica seeds were sown in vermiculite medium supplied with 1/4 diluted Hoagland's nutrient solution, pH 5.8. Growth chamber conditions were set at 75–80% relative humidity, 30/28◦C (day/night), and 16/8 h (day/night) light at 200 mmol photons m-2s-1. One-month old seedlings were treated with 40◦C or with 100 μM ABA. Root and shoot tissue samples were collected at 0 and 24 h post the treatment and kept at -80 ◦C till RNA extraction. Three root and shoot samples were collected from each treatment. Total RNA was isolated from the samples using the Shengong RNA isolation kit as instructed (Shengong Ltd., Shanghai, China). RNA pools were constructed following Illumina sequencing guidelines and then sequenced conferring to RNA-seq procedure. In total 24 million 250-bp raw reads were produced from the 12 samples. To eliminate adapter sequences from raw reads, the FASTX version 0.0.13 toolkit (http://hannonlab.cshl.edu/fastxtoolkit/) was used. Additionally, the FastQC server tool (http://www.bioinformatics.babraham.ac.uk/projects/fastqc/) was utilized to assess the quality of sequences. The resulting clean reads were aligned with the C. songorica genome by means of Tophat v.2.0.10 server (http://tophat.cbcb.umd.edu/) [67] and the produced alignment files were used as Cufflinks inputs to create transcriptome assemblies [68]. The C. songorica gene expression levels were estimated based on fragments per kilobase of exon model per million mapped reads (FPKM) for root and shoot tissues. Following this, a random sampling model built on read count for each individual gene was applied to determine differentially expressed [69]. Gene expression levels were normalized with Pearson coefficients to generate hierarchical clustering with average linkage using HemI toolkit [70].

4.6 Gene Expressions Analysis.

The isolated RNA was then used to synthesize first strand cDNAs using an oligo dT primer and the cDNA synthesis kit (Shengong Ltd, Shanghai, China). The resulting cDNA samples were individually diluted to 100 ng/μL prior to qPCR using gene specific primers (Table S2). For each PCR reaction, three biological replicates with three technical replicates were each used. qPCR reactions were 40 cycles of 95 □C for 5 s, 60 ◦C for 15 s, and 72 ◦C for 34 s using a SYBR Green Master (Shengong Ltd, Shanghai, China). The relative gene expression levels were determined using the comparative ΔΔCt method [71]. The expression level of the C. songorica GADPH gene was used as an internal control. Two-way analysis of variance and Duncan's multiple range test (DMRT) were used for multiple mean comparisons. SPSS (IBM Corp. 2013, IBM SPSS Statistics for Windows, Version 21.0, Armonk, NY, USA) was used to determine the significant differences between means ($p < 0.005$).

5. Concise Summary

This study systematically examined LEA proteins at a genome-wide level in the monocot perennial desert plant, Cleistogenes songorica. A total of 44 identified genes were categorized into eight distinct subfamilies and were seen to be unevenly distributed across the chromosomes of C. songorica. The analysis of the physicochemical features, motifs, gene structures, homology, and evolutionary relationships revealed significant similarities within the same groupings, but notable differences were observed among various subfamilies. Our research specifically investigated the CsLEA_2 and CsDEHYDRIN protein subfamilies and clarified their significant associations with the regulatory mechanisms of plant abiotic stress tolerance. This study provides a thorough overview of the development of the C. songorica LEA genes and offers significant insights into the functional activities of this gene family, which may serve as a crucial basis for enhancing crop tolerance to abiotic stress..

REFERENCES

1. Shahbaz, M.; Ashraf, M. Critical reviews in plant sciences improving salinity tolerance in cereals improving salinity tolerance in cereals. Crit. Rev. Plant Sci. 2013, 32, 237–249. [CrossRef]
2. Verma, G.; Dhar, Y.V.; Srivastava, D.; Kidwai, M.; Chauhan, P.S.; Bag, S.K.; Asif, M.H.; Chakrabarty, D. Genome-wide analysis of rice dehydrin gene family: Its evolutionary conservedness and expression pattern in response to peg induced dehydration stress. PLoS ONE 2017, 12. [CrossRef] [PubMed]
3. Huang, Z.; Zhong, X.-J.; He, J.; Jin, S.-H.; Guo, H.-D.; Yu, X.-F.; Zhou, Y.-J.; Li, X.; Ma, M.-D.; Chen, Q.-B.; et al. Genome-wide identification, characterization, and stress-responsive expression profiling of genes encoding lea (late embryogenesis abundant) proteins in moso bamboo (phyllostachys edulis). PLoS ONE 2016, 11, e0165953. [CrossRef] [PubMed]
4. Dure, L., 3rd; Greenway, S.C.; Galau, G.A. Developmental biochemistry of cottonseed embryogenesis and germination: Changing messenger ribonucleic acid populations as shown by in vitro and in vivo protein synthesis. Biochemistry 1981, 20, 4162–4168. [CrossRef] [PubMed]
5. Wang, F.; Zhu, H.; Cheng, W.; Liu, Y.; Cheng, X.; Sun, J.; Gill, S.S.; Tuteja, N. Polyamines and fabiotic stress tolerance in plants. Plant Signal Behav. 2010, 5, 26–33.
6. Thomashow, M.F. Plant cold acclimation: Freezing tolerance genes and regulatory mechanisms. Annu. Rev. Plant Physiol. Plant Mol. Biol. 1999, 50, 571–599. [CrossRef] [PubMed]
7. Kikawada, T.; Nakahara, Y.; Kanamori, Y.; Iwata, K.-I.; Watanabe, M.; McGee, B.; Tunnacliffe, A.; Okuda, T. Dehydration-induced expression of lea proteins in an anhydrobiotic chironomid. Biochem. Biophys. Res. Commun. 2006, 348, 56–61. [CrossRef] [PubMed]
8. Hand, S.C.; Menze, M.A.; Toner, M.; Boswell, L.; Moore, D. Lea proteins during water stress: Not just for plants anymore. Annu. Rev. Physiol. 2011, 73, 115–134. [CrossRef] [PubMed]

9. Battaglia, M.; Olvera-Carrillo, Y.; Garciarrubio, A.; Campos, F.; Covarrubias, A.A. The enigmatic lea proteins and other hydrophilins. Plant Physiol. 2008, 148, 6–24. [CrossRef] [PubMed]

10. Boucher, V.; Buitink, J.; Lin, X.; Boudet, J.; Hoekstra, F.A.; Hundertmark, M.; Renard, D.; Leprince, O. Mtpm25 is an atypical hydrophobic late embryogenesis-abundant protein that dissociates cold and desiccation-aggregated proteins. Plant Cell Environ. 2010, 33, 418–430. [CrossRef] [PubMed]

11. Chakrabortee, S.; Tripathi, R.; Watson, M.; Schierle, G.S.; Kurniawan, D.P.; Kaminski, C.F.; Wise, M.J.; Tunnacliffe, A. Intrinsically disordered proteins as molecular shields. Mol. Biosyst. 2012, 8, 210–219. [CrossRef] [PubMed]

12. Olveracarrillo, Y.; Campos, F.; Reyes, J.L.; Garciarrubio, A.; Covarrubias, A.A. Functional analysis of the group 4 late embryogenesis abundant proteins reveals their relevance in the adaptive response during water deficit in arabidopsis. Plant Physiol. 2010, 154, 373–390. [CrossRef] [PubMed]

13. Finn, R.D.; Bateman, A.; Clements, J.; Coggill, P.; Eberhardt, R.Y.; Eddy, S.R.; Heger, A.; Hetherington, K.; Holm, L.; Mistry, J.; et al. Pfam: The protein families database. Nucleic Acids Res. 2014, 42, D222–D230. [CrossRef] [PubMed]

14. Bies-Etheve, N.; Gaubier-Comella, P.; Debures, A.; Lasserre, E.; Jobet, E.; Raynal, M.; Cooke, R.; Delseny, M. Inventory, evolution and expression profiling diversity of the lea (late embryogenesis abundant) protein gene family in arabidopsis thaliana. Plant Mol. Biol. 2008, 67, 107–124. [CrossRef] [PubMed]

15. Ciccarelli, F.D.; Bork, P. The Why Domain Mediates the Response to Desiccation in Plants and Bacteria; Oxford University Press: Oxford, UK, 2005; pp. 1304–1307.

16. Jaspard, E.; Hunault, G. Comparison of amino acids physico-chemical properties and usage of late embryogenesis abundant proteins, hydrophilins and why domain. PLoS ONE 2014, 9, e109570. [CrossRef] [PubMed]

17. Mouillon, J.-M.; Gustafsson, P.; Harryson, P. Structural investigation of disordered stress proteins. Comparison of full-length dehydrins with isolated peptides of their conserved segments. Plant Physiol. 2006, 141, 638–650. [CrossRef] [PubMed]

18. Tolleter, D.; Jaquinod, M.; Mangavel, C.; Passirani, C.; Saulnier, P.; Manon, S.; Teyssier, E.; Payet, N.; Avelange-Macherel, M.H.; Macherel, D. Structure and function of a mitochondrial late embryogenesis abundant protein are revealed by desiccation. Plant Cell 2007, 19, 1580–1589. [CrossRef] [PubMed]

19. Koag, M.-C.; Wilkens, S.; Fenton, R.D.; Resnik, J.; Vo, E.; Close, T.J. The k-segment of maize dhn1 mediates binding to anionic phospholipid vesicles

and concomitant structural changes. Plant Physiol. 2009, 150, 1503–1514. [CrossRef] [PubMed]

20. Rahman, L.N.; Chen, L.; Nazim, S.; Bamm, V.V.; Yaish, M.W.; Moffatt, B.A.; Dutcher, J.R.; Harauz, G. Interactions of intrinsically disordered thellungiella salsuginea dehydrins tsdhn-1 and tsdhn-2 with membranes—synergistic effects of lipid composition and temperature on secondary structure. Biochem. Cell Biol. 2010, 88, 791–807. [CrossRef] [PubMed]
21. Liang, J.; Zhou, M.; Zhou, X.; Jin, Y.; Xu, M.; Lin, J. Jclea, a novel lea-like protein from jatropha curcas, confers a high level of tolerance to dehydration and salinity in arabidopsis thaliana. PLoS ONE 2014, 8, e83056. [CrossRef] [PubMed]
22. Zhang, J.; Kong, L.; Liu, Z.; Jahufer, Z.; Duan, Z.; Huo, Y.; Di, H.; Wang, Y. Stress-induced expression in arabidopsis with a dehydrin lea protein from cleistogenes songorica, a xerophytic desert grass. Plant Omics 2015, 8, 485–492.
23. Waie, B.; Rajam, M.V. Effect of increased polyamine biosynthesis on stress responses in transgenic tobacco by introduction of human s -adenosylmethionine gene. Plant Sci. 2003, 164, 727–734. [CrossRef]
24. Zhang, J.; Duan, Z.; Zhang, D.; Zhang, J.; Di, H.; Wu, F.; Wang, Y. Co-transforming bar and cslea enhanced tolerance to drought and salt stress in transgenic alfalfa (medicago sativa l.). Biochem. Biophys. Res. Commun. 2016, 472, 75–82. [CrossRef] [PubMed]
25. Gao, J.; Lan, T. Functional characterization of the late embryogenesis abundant (lea) protein gene family from pinus tabuliformis (pinaceae) in Escherichia coli. Sci. Rep. 2016, 6. [CrossRef] [PubMed]
26. Liu, Y.; Wang, L.; Xing, X.; Sun, L.; Pan, J.; Kong, X.; Zhang, M.; Li, D. Zmlea3, a multifunctional group 3 lea protein from maize (zea mays l.), is involved in biotic and abiotic stresses. Plant Cell Physiol. 2013, 54, 944–959. [CrossRef] [PubMed]
27. Zhang, J.; John, U.P.; Wang, Y.; Li, X.; Gunawardana, D.; Polotnianka, R.M.; Spangenberg, G.C.; Nan, Z. Targeted mining of drought stress-responsive genes from est resources in Cleistogenes songorica. J. Plant Physiol. 2011, 168, 1844–1851. [CrossRef] [PubMed]
28. Hundertmark, M.; Hincha, D.K. Lea (late embryogenesis abundant) proteins and their encoding genes in Arabidopsis thaliana. BMC Genomics 2008, 9. [CrossRef] [PubMed]
29. Baker, J.; Dennsteele, C.V.; Iii, L.D. Sequence and characterization of 6 lea proteins and their genes from cotton. Plant Mol. Biol. 1988, 11, 277–291. [CrossRef] [PubMed]

30. Galau, G.A.; Wang, H.Y.; Hughes, D.W. Cotton Lea5 and Lea14 encode atypical late embryogenesis-abundant proteins. Plant Physiol. 1993, 101, 695–696. [CrossRef] [PubMed]

31. He, S.; Tan, L.; Hu, Z.; Chen, G.; Wang, G.; Hu, T. Molecular characterization and functional analysis by heterologous expression in e. Coli under diverse abiotic stresses for oslea5, the atypical hydrophobic lea protein from oryza sativa l. Mol. Genet. Genom. 2012, 287, 39–54. [CrossRef] [PubMed]

32. Lan, T.; Gao, J.; Zeng, Q.Y. Genome-wide analysis of the lea (late embryogenesis abundant) protein gene family in Populus trichocarpa. Tree Genet. Genom. 2013, 9, 253–264. [CrossRef]

33. Cheng, F.; Wu, J.; Wang, X. Genome triplication drove the diversification of brassica plants. Hortic. Res. 2014, 1. [CrossRef] [PubMed]

34. Liang, Y.; Xiong, Z.; Zheng, J.; Xu, D.; Zhu, Z.; Xiang, J.; Gan, J.; Nadia, R.; Yin, Y.; Li, M. Genome-wide identification, structural analysis and new insights into late embryogenesis abundant (lea) gene family formation pattern in Brassica napus. Sci. Rep. 2016, 6. [CrossRef] [PubMed]

35. Yamaguchi-Shinozaki, K.; Shinozaki, K. Organization of cis-acting regulatory elements in osmotic- and cold-stress-responsive promoters. Trends Plant Sci. 2005, 10, 88–94. [CrossRef] [PubMed]

36. Wang, X.-S.; Zhu, H.-B.; Jin, G.-L.; Liu, H.-L.; Wu, W.-R.; Zhu, J. Genome-scale identification and analysis of lea genes in rice (oryza sativa l.). Plant Sci. 2007, 172, 414–420. [CrossRef]

37. Li, X.; Cao, J. Late embryogenesis abundant (lea) gene family in maize: Identification, evolution, and expression profiles. Plant Mol. Biol. Rep. 2015, 34, 15–28. [CrossRef]

38. Lång, V.; Palva, E.T. The expression of a rab-related gene, rab18, is induced by abscisic acid during the cold acclimation process of arabidopsis thaliana (l.) heynh. Plant Mol. Boil. 1992, 20, 951–962. [CrossRef]

39. Zhu, Y.; Wang, B.; Tang, K.; Hsu, C.C.; Xie, S.; Du, H.; Yang, Y.; Tao, W.A.; Zhu, J.K. An Arabidopsis nucleoporin NUP85 modulates plant responses to ABA and salt stress. PLoS Genet. 2017, 13, e1007124. [CrossRef] [PubMed]

40. Kovacs, D.; Kalmar, E.; Torok, Z.; Tompa, P. Chaperone activity of erd10 and erd14, two disordered stress-related plant proteins. Plant Physiol. 2008, 147, 381–390. [CrossRef] [PubMed]

41. Eriksson, S.K.; Kutzer, M.; Procek, J.; Grobner, G.; Harryson, P. Tunable membrane binding of the intrinsically disordered dehydrin lti30, a cold-induced plant stress protein. Plant Cell 2011, 23, 2391–2404. [CrossRef] [PubMed]

42. Candat, A.; Macherel, D. The ubiquitous distribution of late embryogenesis abundant proteins across cell compartments in Arabidopsis offers tailored protection against abiotic stress. Plant Cell 2014, 26, 3148–3166. [CrossRef] [PubMed]

43. Hanin, M.; Brini, F.; Ebel, C.; Toda, Y.; Takeda, S.; Masmoudi, K. Plant dehydrins and stress tolerance: Versatile proteins for complex mechanisms. Plant Signal Behav. 2011, 6, 1503–1509. [CrossRef] [PubMed]

44. Close, T.J. Dehydrins: A commonalty in the response of plants to dehydration and low temperature. Physiologia Plantarum 1997, 100, 291–296. [CrossRef]

45. Jiang, S.; Wang, J.; Liu, X.; Liu, Y.; Guo, C.; Zhang, L.; Han, J.; Wu, X.; Xue, D.; Gomaa, A.E.; et al. Drwh, a novel why domain-containing hydrophobic lea5c protein from deinococcus radiodurans, protects enzymatic activity under oxidative stress. Sci. Rep. 2017, 7. [CrossRef] [PubMed]

46. Franco-Zorrilla, J.M.; López-Vidriero, I.; Carrasco, J.L.; Godoy, M.; Vera, P.; Solano, R. DNA-binding specificities of plant transcription factors and their potential to define target genes. Proc. Nati. Acad. Sci. USA 2014, 111, 2367–2372. [CrossRef] [PubMed]

47. Kim, H.S.; Lee, J.H.; Kim, J.J.; Kim, C.H.; Jun, S.S.; Hong, Y.N. Molecular and functional characterization of CaLea6, the gene for a hydrophobic LEA protein from capsicum annuum. Gene 2005, 344, 115–123. [CrossRef] [PubMed]

48. Park, S.-C.; Kim, Y.-H.; Jeong, J.C.; Kim, C.Y.; Lee, H.-S.; Bang, J.-W.; Kwak, S.-S. Sweetpotato late embryogenesis abundant 14 (iblea14) gene influences lignification and increases osmotic- and salt stress-tolerance of transgenic calli. Planta 2011, 233, 621–634. [CrossRef] [PubMed]

49. Wang, M.; Li, P.; Li, C.; Pan, Y.; Jiang, X.; Zhu, D.; Zhao, Q.; Yu, J. SiLEA14, a novel atypical lea protein, confers abiotic stress resistance in foxtail millet. BMC Plant Biol. 2014, 14. [CrossRef] [PubMed]

50. Chiappetta, A.; Muto, A.; Bruno, L.; Woloszynska, M.; Lijsebettens, M.V.; Bitonti, M.B. A dehydrin gene isolated from feral olive enhances drought tolerance in arabidopsis transgenic plants. Front. Plant Sci. 2015, 6. [CrossRef] [PubMed]

51. Brini, F.; Hanin, M.; Lumbreras, V.; Amara, I.; Khoudi, H.; Hassairi, A.; Pages, M.; Masmoudi, K. Overexpression of wheat dehydrin dhn-5 enhances tolerance to salt and osmotic stress in Arabidopsis thaliana. Plant Cell Rep. 2007, 26, 2017–2026. [CrossRef] [PubMed]

52. Finn, R.D.; Mistry, J.; Tate, J.; Coggill, P.; Heger, A.; Pollington, J.E.; Gavin, O.L.; Gunasekaran, P.; Ceric, G.; Forslund, K.; et al. The pfam protein families database. Nucleic Acids Res. 2010, 38, D211–D222. [CrossRef] [PubMed]

53. Altschul, S.F.; Gish, W.; Miller, W.; Myers, E.W.; Lipman, D.J. Basic local alignment search tool. J. Mol. Biol. 1990, 215, 403–410. [CrossRef]

54. Letunic, I.; Copley, R.R.; Schmidt, S.; Ciccarelli, F.D.; Doerks, T.; Schultz, J.; Ponting, C.P.; Bork, P. Smart 4.0: Towards genomic data integration. Nucleic Acids Res. 2004, 32, D142–D144. [CrossRef] [PubMed]

55. Marchler-Bauer, A.; Derbyshire, M.K.; Gonzales, N.R.; Lu, S.; Chitsaz, F.; Geer, L.Y.; Geer, R.C.; He, J.; Gwadz, M.; Hurwitz, D.I.; et al. Cdd: Ncbi's conserved domain database. Nucleic Acids Res. 2015, 43, D222–D226. [CrossRef] [PubMed]

56. Feng, D.-F.; Doolittle, R.F. Progressive sequence alignment as a prerequisitetto correct phylogenetic trees. J. Mol. Evol. 1987, 25, 351–360. [CrossRef] [PubMed]

57. Tamura, K.; Stecher, G.; Peterson, D.; Filipski, A.; Kumar, S. Mega6: Molecular evolutionary genetics analysis version 6.0. Mol. Biol. Evol. 2013, 30, 2725–2729. [CrossRef] [PubMed]

58. Higgins, D.G.; Sharp, P.M. Clustal: A package for performing multiple sequence alignment on a microcomputer. Gene 1988, 73, 237–244. [CrossRef]

59. Gasteiger, E.; Gattiker, A.; Hoogland, C.; Ivanyi, I.; Appel, R.D.; Bairoch, A. Expasy: The proteomics server for in-depth protein knowledge and analysis. Nucleic Acids Res. 2003, 31, 3784–3788. [CrossRef] [PubMed]

60. Emanuelsson, O.; Brunak, S.; von Heijne, G.; Nielsen, H. Locating proteins in the cell using targetp, signalp and related tools. Nat. Protoc. 2007, 2, 953–971. [CrossRef] [PubMed]

61. Bailey, T.L.; Boden, M.; Buske, F.A.; Frith, M.; Grant, C.E.; Clementi, L.; Ren, J.; Li, W.W.; Noble, W.S. Meme suite: Tools for motif discovery and searching. Nucleic Acids Res. 2009, 37, W202–W208. [CrossRef] [PubMed]

62. Rombauts, S.; Dehais, P.; Van Montagu, M.; Rouze, P. Plantcare, a plant cis-acting regulatory element database. Nucleic Acids Res. 1999, 27, 295–296. [CrossRef] [PubMed]

63. Yue, R.; Lu, C.; Sun, T.; Peng, T.; Han, X.; Qi, J.; Yan, S.; Tie, S. Identification and expression profiling analysis of calmodulin-binding transcription activator genes in maize (Zea mays L.) under abiotic and biotic stresses. Front. Plant Sci. 2015, 6. [CrossRef] [PubMed]

64. Petrov, V.; Vermeirssen, V.; De, C.I.; Van, B.F.; Minkov, I.; Vandepoele, K.; Gechev, T.S. Identification of cis-regulatory elements specific for different types of reactive oxygen species in arabidopsis thaliana. Gene 2012, 499, 52–60. [CrossRef] [PubMed]

65. Passricha, N.; Saifi, S.; Ansari, M.W.; Tuteja, N. Prediction and validation of cis-regulatory elements in 5' upstream regulatory regions of lectin receptor-

like kinase gene family in rice. Protoplasma 2017, 254, 669–684. [CrossRef] [PubMed]

66. Acharya, B.R.; Jeon, B.W.; Zhang, W.; Assmann, S.M. Open stomata 1 (OST1) is limiting in abscisic acid responses of arabidopsis guard cells. New Phytol. 2013, 200, 1049–1063. [CrossRef] [PubMed]
67. Trapnell, C.; Pachter, L.; Salzberg, S.L. Tophat: Discovering splice junctions with RNA-seq. Bioinformatics 2009, 25, 1105–1111. [CrossRef] [PubMed]
68. Trapnell, C.; Roberts, A.; Goff, L.; Pertea, G.; Kim, D.; Kelley, D.R.; Pimentel, H.; Salzberg, S.L.; Rinn, J.L.; Pachter, L. Differential gene and transcript expression analysis of rna-seq experiments with tophat and cufflinks. Nat. Protoc. 2012, 7. [CrossRef] [PubMed]
69. Wang, L.; Feng, Z.; Wang, X.; Wang, X.; Zhang, X. Degseq: An r package for identifying differentially expressed genes from rna-seq data. Bioinformatics 2010, 26, 136–138. [CrossRef] [PubMed]
70. Deng, W.; Wang, Y.; Liu, Z.; Cheng, H.; Xue, Y. Hemi: A toolkit for illustrating heatmaps. PLoS ONE 2014, 9, e111988. [CrossRef] [PubMed]
71. Fujisawa, M.; Takita, E.; Harada, H.; Sakurai, N.; Suzuki, H.; Ohyama, K.; Shibata, D.; Misawa, N. Pathway engineering of brassica napus seeds using multiple key enzyme genes involved in ketocarotenoid formation. J. Exp. Bot. 2009, 60, 1319–1332. [CrossRef] [PubMed]

15 Chapter

TRANSCRIPTOMIC EXAMINATION OF BETULA HALOPHILA UNDER SALT STRESS CONDITIONS

1. INTRODUCTION: AN OVERVIEW

Soil salinization is a global issue that can modify the soil's osmotic potential to such an extent that it obstructs water absorption by plants, significantly affecting agricultural output and the ecological environment. Reports indicate that over 6% of the Earth's surface is impacted by salinity, and escalating salinization could result in the loss of 30% of arable land within the next 25 years. Reports indicate that about 6% of the global land is impacted by salinity, with projections suggesting a potential loss of 30% of arable land during the next 25 years and up to 50% by 2050. Consequently, soil salinization poses a significant hazard to plant growth and development. Currently, it is imperative to identify salinity-resistant plant types and to screen for salt-tolerant gene alleles or to genetically alter them to facilitate growth and reproduction under escalating salinity stress [4]. Furthermore, comprehending the mechanisms of salt tolerance in plants might yield critical insights for effective engineering approaches. The mechanism of salt resistance in plants is intricate and encompasses a series of processes at the molecular, cellular, metabolic, physiological, and whole-plant levels. Upon experiencing salt stress, many signal transduction pathways are engaged to mitigate the effects of salinity. Despite considerable advancements in research on ion uptake and transport, osmotic control, hormone metabolism, antioxidant metabolism, and stress signaling in recent years [4–11], the molecular mechanisms underlying salt tolerance have yet to be clarified. Moreover, next-generation high-throughput sequencing using RNA-seq analysis has been extensively employed to elucidate expression patterns under abiotic stress, offering a thorough approach for finding and examining differentially expressed genes [12–15]. Betula halophila is a haloduric plant native to China, classified within the family Betulaceae. In 1956, Professor Renchang Qin originally discovered it in a highly saline marsh in Xinjiang province [16]. B. halophila is a critically endangered plant characterized by its great salt resistance and significant ecological and economic importance in facilitating the afforestation of saline soils in arid and semi-arid regions. B. halophila serves as a significant reservoir of salt-tolerant genes. To the best of our knowledge, there is no available literature on genes linked to salt tolerance in B. halophila. Comprehending the molecular underpinnings of salt tolerance is potentially crucial for the cultivation of salt-tolerant cultivars. To identify the genes responsive to elevated salt concentration and to elucidate the molecular processes of salt tolerance in B. halophila, we created transcriptome libraries from the leaves of both control B. halophila plants and those treated to salt treatment. The objectives were to identify salt-responsive genes in B. halophila and investigate their functions in relation to salt stress. Our findings elucidate the molecular pathways underlying salt tolerance in B. halophila. An enhanced comprehension of these tolerance processes can facilitate the breeding of crops with superior yield performance under saline stress..

2. RESULTS

2.1 Transcriptome Sequencing and Assembly

In order to explore the salt tolerant genes of B. halophila, six cDNA libraries were constructed from leaves of eight-months-old control plants (untreated) and plants treated with 200 mM of NaCl for 24 h, and sequenced using the Illumina deep sequencing platform. A total of 68,348,352, 75,684,144, 70,510,750, 86,101,536, 77,402,824, and 84,837,142 raw reads were generated by Illumina sequencing, respectively, after adapter sequence and low quality sequences had been removed, a total of 66,834,236, 73,359,338, 69,082,978, 84,155,542, 75,577,004, and 82,931,756 clean reads were obtained, and the Q20 percentage (proportion of nucleotides with a quality value larger than 20) for each data was 96.5%, 96.48%, 96.25%, 96.54%, 96.49%, and 96.45%, respectively. The GC (%) ratio for each library was 46.88%, 47.4%, 47.29%, 47.28%, 47.33% and 47.29% (Table 1). Transcriptome assembly was accomplished based on the left.fq and right.fq using Trinity with the min_kmer_cov set to 2 by default and all other parameters set to default [17]. As a result, a total of 117,091 unigenes with lengths ranging from 201 bp to 15,762 bp were obtained. The size distribution of the unigenes is shown in Figure 1. The size distribution showed that the unigenes ranged from 200 bp to 1 kbp was the majority (Figure 1). The average length, median length, and N50 of the assembled unigenes were 956 bp, 631 bp and 1408 bp, separately. The total length of 117,091 unigenes was 110 Mb, which suggests that most of the sequencing data had been successfully assembled into relatively long unigenes.

Table 1. Summary of the sequencing data of the Betula halophila transcriptome.

Sample	Raw Reads	Clean Reads	Clean Bases	Error (%)	Q20 (%)	Q30 (%)	GC Content(%)
CK_1	68,348,352	66,834,236	10.03G	0.03	96.5	94.17	46.88
CK_2	75,684,144	73,359,338	11G	0.03	96.48	94.14	47.4
CK_3	70,510,750	69,082,978	10.36G	0.03	96.25	93.83	47.29
SC_1	86,101,536	84,155,542	12.62G	0.03	96.54	94.23	47.28
SC_2	77,402,824	75,577,004	11.34G	0.03	96.49	94.16	47.33
SC_3	84,837,142	82,931,756	12.44G	0.03	96.45	94.12	47.29

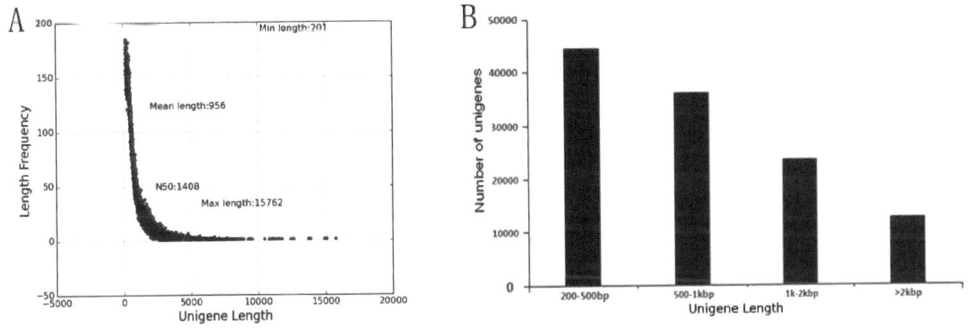

Figure 1. Length distribution of the assembled unigenes. (A) The number of contigs with same length. (B) The number of contigs with length 200–500 bp, 500 bp–1 kb, 1–2 kb and larger than 2 kb are shown.

2.2 Functional Annotation and Classification of the Unigenes

Functional annotation of unigenes were performed to search for homologues against the NCBI non-redundant protein sequence database (Nr), NCBI nucleotide sequences (Nt), Pfam (Protein family), Kyoto Encyclopedia of Genes and Genomes (KEGG), swiss-prot sequence databases (SwissProt), Gene ontology (GO), and Eukaryotic Orthologous Groups (KOG) using the Basic Local Alignment Search Tool (BLAST) [18]. An e-value cut-off of 10-5 was applied to the homologue recognition. The results were shown in Table 2. 64551 (55.12%) total unigenes were annotated in at least one database and 8973 unigenes (7.66%) were annotated in all databases. 51,105 (43.64%) total unigenes were annotated in the Nr protein database.

Table 2. Summary of function annotation of the Betula halophila transcriptome.

	Number of Unigenes	Percentage (%)
Annotated in NR	51,105	43.64
Annotated in NT	45,933	39.22
Annotated in KO	18,876	16.12

Annotated in SwissProt	40,624	34.69
Annotated in PFAM	40,661	34.72
Annotated in GO	41,116	35.11
Annotated in KOG	15,572	13.29
Annotated in all Databases	8973	7.66
Annotated in at least one Database	64,551	55.12
Total Unigenes	117,091	100

The GO analysis indicated that a total of 41,116 unigenes were summarized into the three main GO categories (biological process, cellular component, and molecular function) and 56 sub-categories (Figure 2). In the biological process category, genes involved in cellular process, metabolic process, and single-organism process were dominant. As for the cellular component category, genes involved in cell, cell part, and organelle were highly represented. The molecular function category mainly included genes involved in binding and catalytic activity.

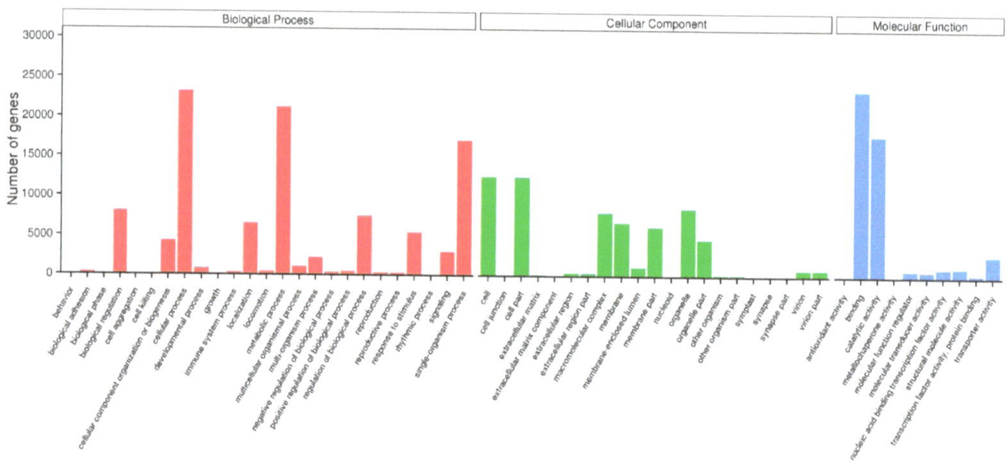

Figure 2. Gene ontology (GO) classification of unigenes. The GO terms are summarized into three main categories: biological process, cellular component, and molecular function.

The KOG analysis showed that all of the 15,572 unigenes were divided into 26 different functional classes, which were represented by A to Z (Figure 3). Among the 26 categories, the largest group was 'Post-translational modification, protein turnover, chaperon' (2104, 13.51%) followed by 'General function prediction' (1906, 12.24%), 'Translation, ribosomal structure, and biogenesis' (1552, 9.97%), 'RNA processing and modification' (1272, 8.17%) and 'Signal Transduction' (1248, 8.01%). The smallest group was 'Cell motility' (7, 0.04%) and 'Unnamed protein' (2, 0.01%).

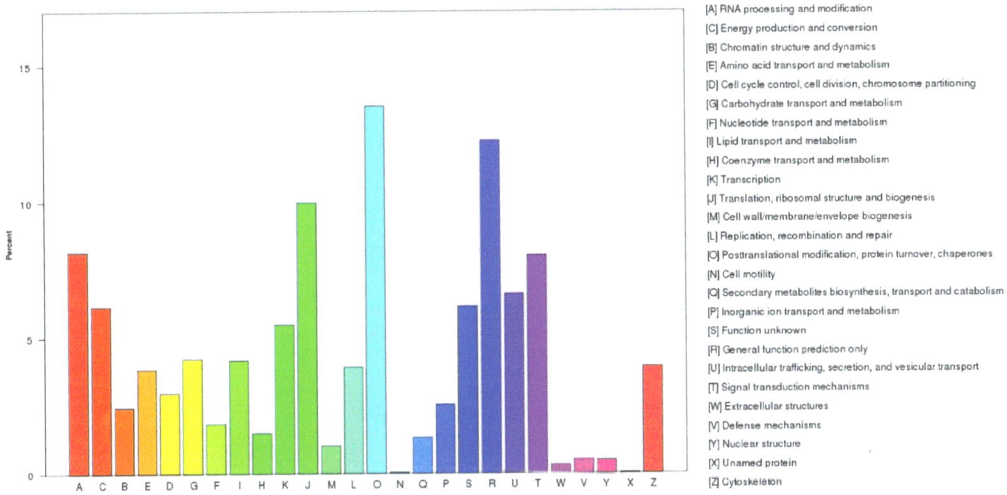

Figure 3. Eukaryotic Orthologous Groups (KOG) classification of the unigenes.

The KEGG pathway analysis revealed that 18876 (16.12%) of the unigenes could be mapped to the KEGG database and referred to 129 pathways (Figure 4). The pathway involved the highest number of unigenes was 'Translation' (1915, 10.14%), followed by 'Folding, sorting, and degradation' (1465, 7.76%), 'Carbohydrate metabolism' (1388, 7.35%) and 'Overview' (1012, 5.36%). These results are very important for studying the mechanism in B. halophila response to salt.

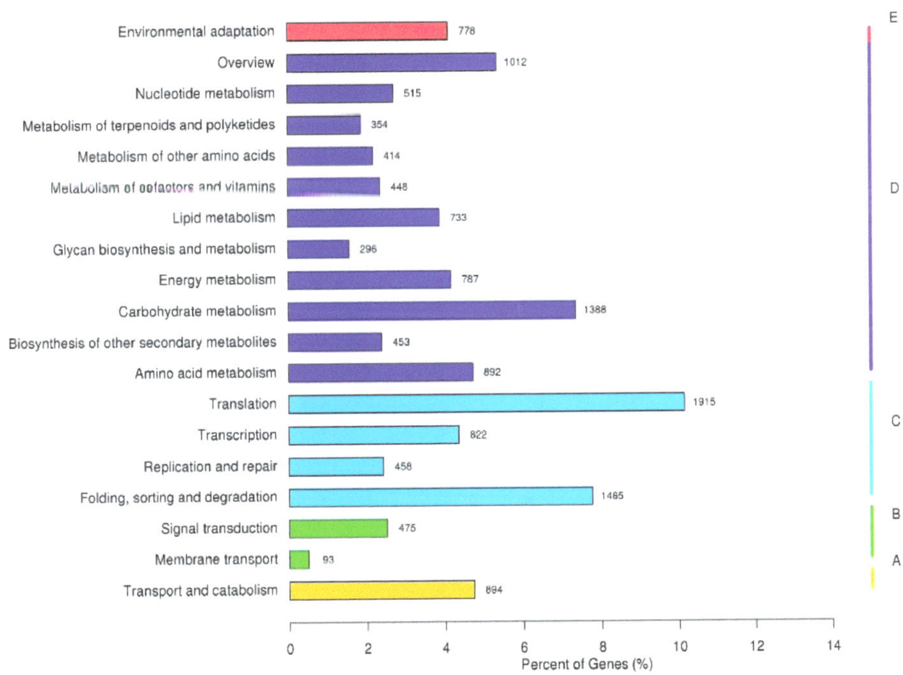

Figure 4. Kyoto Encyclopedia of Genes and Genomes (KEGG) classification of KO annotated unigenes.

2.3 Differential Expression Genes in B. halophila Response to Salt

To obtain the differential expression genes' response to salt in B. halophila, we compared the differentially expressed tags of two libraries. As a results, a total of 519 differentially expressed genes (DEGs) with q value < 0.05 and |log2 (fold change)| >1 were identified in the two libraries (Table S1). As shown in Figure 5a, there were more down-regulated genes (351) than up-regulated genes (168). Among these DEGs, 332 DEGs were present in both libraries, (Figure 5b). 66 DEGs were only detected in the salt stress library (Figure 5b) and 121 DEGs were only detected in the control library. In this study, the transcription factor AT-Hook Motif Nuclear Localized gene (AHL) was the most up-regulate gene in leaves after the salt stress. Conversely, a dehydrin (DHNs) was the most down-regulated gene. These results suggest that the two genes may have a high correlation with salt resistance of B. halophila. The GO and KEGG classification of the 519 DEGs were analyzed (Figure S1). GO enrichment and KEGG enrichment were performed for further analysis of the functions of 519 DEGs.

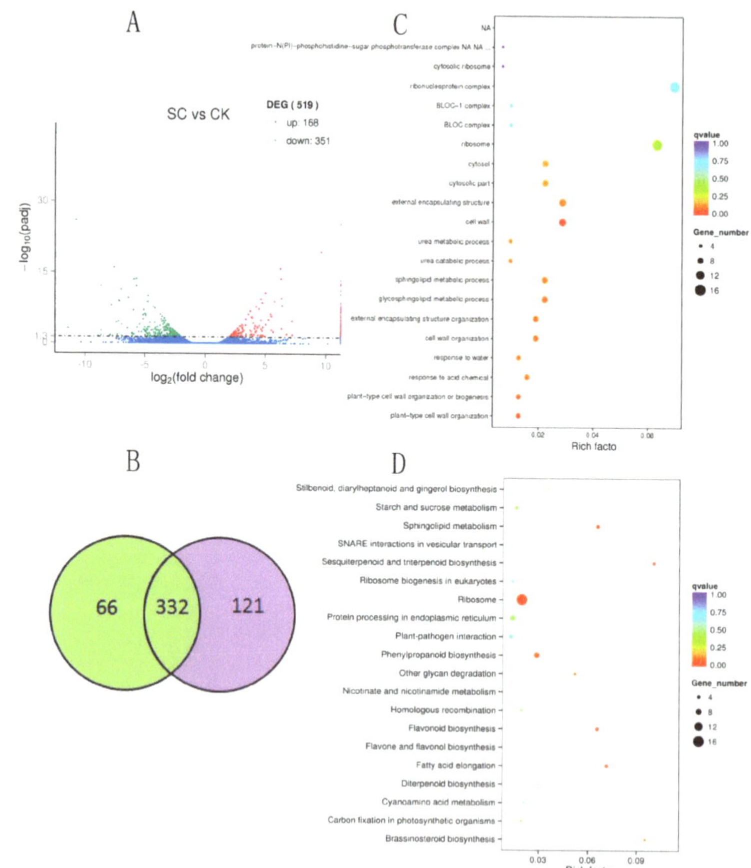

Figure 5. A. Up-regulated and down-regulated differentially expressed genes in SC vs. CK; B. Venn diagrams showing unique and shared differentially expressed

genes (DEGs) in SC (green) vs. CK (purple); C. Scatterplot of GO category enrichment of DEGs in SC vs. CK; D. Scatterplot of enriched KEGG pathways for DEGs in SC vs. CK. Rich factor is the ratio of the differentially expressed gene number to the total gene number in a certain pathway. The size and color of dot represent the gene number and the range of the q value, respectively.

2.4 GO category Enrichment of DEGs Under Salt Stress

To characterize the function of the DEGs under salt stress, the GO category enrichment analysis was performed using Fisher's exact test with p value ≤0.05 as the cutoff. GO category enrichment analysis for 519 DEGs under salt stress showed that these DEGs were mainly involved in a plant-type cell wall organization biological process, plant-type cell wall organization or biogenesis biological process, cell wall cellular component and structural constituent of cell wall molecular function (Figure 5c, Table S2). For the up-regulated DEGs, metalloendopeptidase activity molecular function was most highly enriched (Table S3). For down-regulated DEGs (Figure 5c), in the BP category, 'plant-type cell wall organization biological process', 'plant-type cell wall organization or biogenesis biological process', 'cell wall organization biological process', and 'external encapsulating structure organization biological process' were most highly enriched. In the CC category, 'cell wall cellular component', 'cytosolic part cellular component', 'cytosol cellular component', and 'external encapsulating structure cellular component' were the main enriched terms. In MF, the most enriched term was structural constituent of cell wall molecular function (Table S4).

2.5 KEGG Enrichment of DEGs under Salt Stress

The KEGG pathway enrichment analysis were performed to identify the candidate pathways involved in salt stress using KOBAS 2.0 [19]. The results showed that genes with KO number within 519 DEGs under salt stress were enriched in 48 KEGG pathways (Table S5). The top-four enriched pathways for DEGs in SC vs CK were 'Fatty acid elongation', 'Ribosome', 'Sphingolipid metabolism' and 'Flavonoid biosynthesis'. For up-regulated DEGs (Table S6), the most highly enriched pathways were 'Sphingolipid metabolism', 'Other glycan degradation', 'Brassinosteroid biosynthesis' and 'Citrate cycle (TCA cycle)' (Figure 5d). For down-regulated DEGs (Table S7), 'Ribosome', 'Fatty acid elongation', 'Flavonoid biosynthesis' and 'Phenylpropanoid biosynthesis' were the top-four enriched pathways (Figure 5d).

2.6 qRT-PCR Analysis

In order to validate the RNA-seq data and confirm the differential expression genes, we performed qRT-PCR on sixteen candidate DEGs associated with salt stress. The results revealed that these DEGs include AHL, dehydrin-1, highly ABA-induced PP2C gene, proline transporter 2-like, sodium/hydrogen exchanger 4, serine/threonine-protein kinase, heat shock protein, Cation/H(+) antiporter, wall-associated receptor kinase-like and dehydrin-3 were up-regulated in the leaves with salt treatment (Figure 6), whereas major allergen variant Cor, dehydrin-2, phosphoglycerate kinase, potassium transporter, peroxidase 4 and WRKY transcription factor were down- regulated in the leaves with salt treatment. The results indicated that these sixteen candidate DEGs had the same expression patterns compared with the sequencing data, suggesting the reliability of the RNA-seq data.

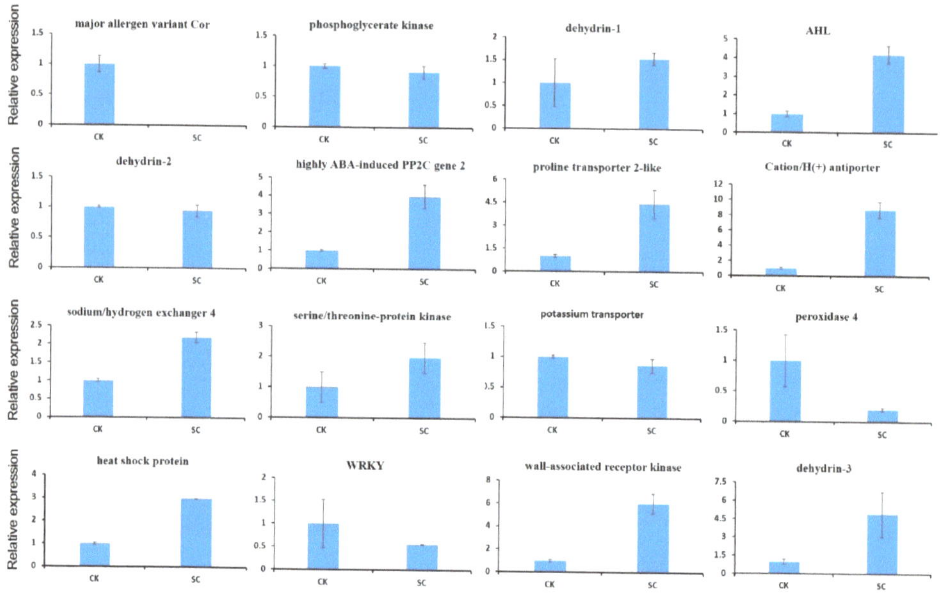

Figure 6. qRT-PCR validation of sixteen selected DEGs in leaves. Fold changes of the DEGs are shown. The expression levels in CK were arbitrarily set to 1. Error bars represent the standard deviations of three technical PCR replicates.

3. DISCUSSION

B. halophila is a plant with high salt tolerance, so it is important to discover the salt tolerant genes of B. halophila for breeding salinity resistant varieties of trees. However, to our best knowledge, there is no report about genes associated with salt tolerance in B. halophila. In this study, we analyzed the transcriptomic data from the leaves of wild type B. halophila plants and plants with salt treatment. As a result, approximately 460 M raw reads were generated and were further assembled into 117,091 unigenes, among these unigenes, 64551 unigenes (55%) were annotated with gene descriptions, while the other 45% were unknown. This is the first report of transcriptome data from B. halophila. This transcriptome data provides an important genus resource for insight into the molecular mechanism of salt tolerance and facilitates discovery of novel genes responsive to salt stress in B. halophila.

In plants, salt stress responsive mechanisms are very complicated, which involve a complex interaction of physiological processes, metabolic pathways, and regulation at the molecular and cellular levels. Although plant response to salt stress has been extensively studied at different levels, the mechanisms underlying salinity tolerance are far from being completely understood. In addition, salt stress responsive mechanisms in different plants are also different. At present, the main mechanisms for which plants respond to salt stress include ion homeostasis and compartmentalization, ion transport and uptake, biosynthesis of osmoprotectants and compatible solutes, activation of antioxidant enzyme, and synthesis of antioxidant compounds [4–7,11]. In this study, 168 up-regulated genes and 351 down-regulated genes were identified in B. halophila under salt stress, respectively. These DEGs include dehydrin proteins, sodium/hydrogen exchanger, potassium transporter, sarcoplasmic/endoplasmic reticulum calcium ATPase, Ca2+ antiporter/cation exchanger, Nodulin MtN21/EamA-like transporter, heat shock protein, phosphoenolpyruvatecarboxykinase, NADH dehydrogenase, highly ABA-induced PP2C gene, homeobox-leucine zipper protein, phosphoglycerate kinase, WRKY transcription factor, AP2/ERF and B3 domain-containing transcription factor, flavonoid 3,,5,-hydroxylase, which is consistent with the other plants that are reported to be responsive to salt stress [20–24].

The analysis of GO enrichments suggested that the 519 DEGs response to salt stress was mainly involved in plant-type cell wall organization biological process, plant-type cell wall organization or biogenesis biological process, cell wall cellular component and structural constituent of cell wall molecular function. KEGG pathway enrichment results showed that the top-four enriched pathways for DEGs was 'Fatty acid elongation', 'Ribosome', 'Sphingolipid metabolism', and 'Flavonoid biosynthesis'. The expression patterns of sixteen of these DEGs were analyzed by qRT-PCR to verify the RNA-seq results. It revealed that the qRT-PCR results were consistent with RNA-seq data.

Based on the functional annotations of the 519 DEGs and the physiological evidence of B. halophila in response to salt stress [25], the possible mechanism of salt tolerance in the leaves of B. halophila was summarized in Figure 7. The possible salt tolerance mechanism is coordinately linked with ion homeostasis, osmotic protection, antioxidant regulation, ABA signal pathway, transcription factors and chaperons. When the plant is treatment with 200mM NaCl, multiple signal pathways are activated to cope with salt stress such as the SOS pathway, antioxidant pathway and ABA signal pathway and so on. Meanwhile, the osmoprotectants such as proline and polyols were accumulated to protect the cell. In addition, the transcription factors (WRKY, ERF, ZIP and AHL) and (LEAs, HSPs and AQPs) were activated to regulate the genes involved in the above pathways [2–6]. Overall, the salt tolerance mechanism in B. halophila is a complex network that involved the interactions at multiple levels. This information will be useful in elucidating the salt tolerance mechanisms in B. halophila.

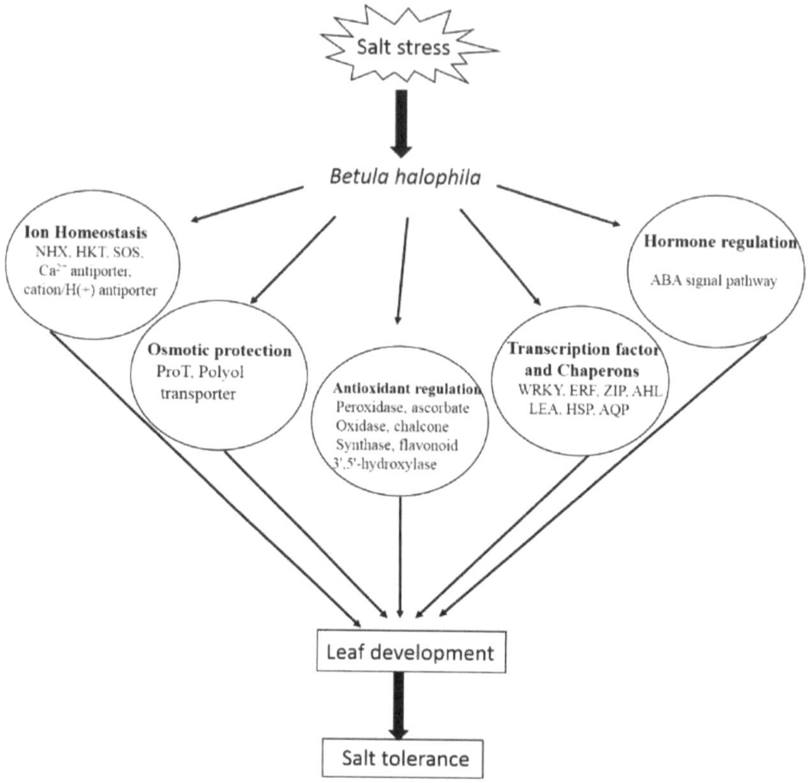

Figure 7. The possible mechanism of salt tolerance in the leaves of B. halophila.

In the present study, we observed that one dehydrin (DHNs) which is the most down-regulated gene among these DEGs and two other dehydrins showed a distinct salt responsive expression, suggesting that these dehydrin proteins may play different roles in response to salt stress in B. halophila. Dehydrins, also known as group 2 LEA (Late Embryogenesis Abundant) proteins, play a fundamental role in plant response to abiotic stresses [26–28]. Their expression is often induced under salinity, dehydration, cold and frost stress. Dehydrins are divided into five structural subgroups: Kn, SKn, KnS, YxKn and YxSKn [28]. The three dehydrins protein features of B. halophila were all SK3 subclass. It has been shown that SK3 dehydrins play an important protective role in plant stress tolerance, including drought, cold, and salinity [27]. For example, the expression of the durum wheat DHN-5 in thaliana led to an increase in salt and osmotic stress tolerance [28]. Rab16A in salt-tolerant Indica rice variety Pokkali can enhance tolerance to drought and salt stress in tobacco plants [29]. Similarly, overexpression of the wheat dehydrin PMA80 (as well as the LEAI protein PMA1959) enhances rice tolerance to drought and salt stress [30]. Although experimental evidence suggests that dehydrins have diverse roles (membrane protection, cryoprotection of enzymes, and protection from reactive oxygen species) in response to stresses [27–31], further efforts are still needed to precisely confirm the roles of these dehydrins and explore the regulatory mechanism underlying these functions in plant adaptive response to abiotic stresses.

In addition, our results indicated that the transcription factor AT-Hook Motif Nuclear Localized gene (AHL) was the most up-regulated gene in leaves after salt stress, implying that it might play an important role in response to salt stress in B. halophila. Previous studies showed that the AHL genes regulate diverse aspects of growth and development in plants. Such as the homeostasis of phytohormones [32], and defense responses [33–40]. However, there is no report about the function of AHL genes associated with salt stress. Further studies are still needed to understand the function of AHL genes in salt stress.

Therefore, our results provide a list of candidate genes for further investigation to determine whether they have a role in allowing B. halophila to tolerate high salt levels, and may be helpful in the understanding of the molecular mechanisms of salt stress response in B. halophila.

4. MATERIALS AND METHODS

4.1 Plant Materials

The seeds of B. halophila were obtained from Xinjiang Academy of Forestry. After germination, the seedlings of B. halophila were grown in the greenhouse in Chinese Academy of Forestry. Leaves were collected from eight-months-old plants, the fourth or fifth leaf from top to bottom was used for sampling and RNA extraction. Three independent biological replicates were performed for each experiment. All samples were frozen and stored in liquid nitrogen until use.

4.2 Salt stress Treatment

Based on Betula halophila physiological response to salt stress as described by Zhang et al. [25], plantlets were treated with 200 mM of NaCl for 24h and then leaves were collected from stressed plants, plantlets treated with water were used as controls. Three independent biological replicates were performed for each experiment. All samples were frozen and stored in liquid nitrogen until use.

4.3 Library Construction and Sequencing for RNA-seq

In order to construct cDNA libraries, total RNAs were extracted from the control and the NaCl treated plant using Trizol RNA extraction kit (Life Technology, Beijing, China) according to the manufacturer's instruction. Six samples were sequenced by Novogene (Tianjin, China) using Illumina HiSeq2500 system.

4.4 Transcriptome Assembly and Bioinformatics Analysis

Transcriptome assembly was accomplished based on the left.fq and right.fq using Trinity [17] with min_kmer_cov set to 2 by default and all other parameters set default. In brief, the left files (read1 files) from all libraries/samples were pooled into one big left.fq file, and right files (read2 files) into one big right.fq file. Gene function was annotated based on Nr (NCBI non-redundant protein sequences), Nt (NCBI non-redundant nucleotide sequences), Pfam (Protein family), KOG/COG (Clusters of Orthologous Groups of proteins), Swiss-Prot (A manually annotated and reviewed protein sequence database), KO (KEGG Ortholog database), GO (Gene Ontology) databases. Gene expression levels were estimated by RSEM [41] for each sample. Clean data were mapped back onto the assembled transcriptome. Read counts for each gene were obtained from the mapping results. For differential expression analysis, prior to differential gene expression analysis, for each sequenced library, the read counts were adjusted by edgeR program package [42] through one scaling normalized factor. Differential expression analysis of six samples was performed using the DEGseq R package [43]. All transcripts were searched against the latest versions (as of August 2018) of Nr (nonredundant) database (http://www.ncbi.nlm.nih.gov/) and the Swiss-Prot database (http://www.gpmaw.com/html/swiss-prot.html) using the BLAST program with an e < 10-5. The transcripts with the top hits were selected as unigenes. Open reading frames (ORFs) were predicted using the GetORF program contained in the EMBOSS software package. The Blast2GO program was used for GO annotation (http: //www.geneontology.org), and the unigenes were aligned to the eggNOG (evolutionary genealogy of genes: non-supervised orthologous groups) database (http://www.ncbi.nlm.nih.gov/COG/) to identify functional categories. The KEGG database (http://www.genome.jp/kegg/) was used for pathway annotation. All searches were conducted using an e-value cut-off of 10-5. GO terms were downloaded from the GO Analysis Toolkit and Database for Agriculture Community (AGRI go, http: //bioinfo.cau.edu.cn/agriGO/download.php). All the genes identified with significant differential expression ($p < 0.05$) and FC >2 in this study were used as inputs to carry out GO enrichment analysis. Gene Ontology (GO) enrichment analysis of the differentially expressed genes (DEGs) was implemented by the GOseq R packages based Wallenius non-central hyper-geometric distribution [44] that can adjust for gene length bias in DEGs. KEGG pathway enrichment analysis used KOBAS [19] software to test the statistical enrichment of the differential expression genes in the KEGG pathways. In the scatterplot, the rich factor is the ratio of the differentially expressed gene number to the total gene number in a certain pathway.

4.5 Quantitative RT-PCR

The candidate DEGs in response to salt stress were selected to validate the reliability of the RNA-seq data using quantitative RT-PCR following the previously reported procedures [45,46]. Gene-specific primers were listed in Table S8. BhActin was used as a reference gene. Three independent biological replicates were performed. The results from gene-specific amplification were analyzed using the comparative Cq method, which uses an arithmetic formula, 2-ΔΔCq, to achieve results for relative quantification [47]. Cq represents the threshold cycle.

5. CONCISE SUMMARY

We sequenced and conducted a comparative analysis of the transcriptomes from the leaves of wild-type B. halophila plants and those subjected to salt treatment. This study allowed us to delineate gene expression profiles and identify functional genes associated with salt tolerance. A total of 519 genes exhibited altered expression in response to salt stress. These differentially expressed genes (DEGs) seem to participate in several processes, including the SOS pathway, ion transport and uptake, antioxidant enzymes, and the ABA signaling pathway, among others. Research indicates that a gene encoding the AT-Hook Motif Nuclear Localized transcription factor and three genes producing dehydrins may significantly contribute to the response to salt stress in B. halophila. The findings identify promising candidate genes for the cultivation of salt-tolerant plants and will enhance the comprehension of the molecular pathways underlying salt stress in B. halophila..

REFERENCES

1. Flowers, T.J. Improving crop salt tolerance. J. Exp. Bot. 2004, 55, 307–319. [CrossRef] [PubMed]
2. Munns, R.; Tester, M. Mechanisms of salinity tolerance. Annu. Rev. Plant Biol. 2008, 59, 651–681. [CrossRef] [PubMed]
3. Wang, M.C.; Peng, Z.Y.; Li, C.L.; Li, F.; Liu, C.; Xia, G.M. Proteomic analysis on a high salt tolerance introgression strain of Triticumaestivum/ Thinopyrumponticum. Proteomics 2008, 8, 1470–1489. [CrossRef] [PubMed]
4. Roy, S.; Chakraborty, U. Salt tolerance mechanisms in Salt Tolerant Grasses (STGs) and their prospects in cereal crop improvement. Bot. Stud. 2014, 55, 31. [CrossRef] [PubMed]
5. Gupta, B.; Huang, B. Mechanism of Salinity Tolerance in Plants: Physiological, Biochemical, and Molecular Characterization. Int. J. Genom. 2014, 2014, 701596. [CrossRef] [PubMed]
6. Ji, H.; Pardo, J.M.; Batelli, G.; Van Oosten, M.J.; Bressan, R.A.; Li, X. The Salt Overly Sensitive (SOS) pathway: Established and emerging roles. Mol. Plant 2013, 6, 275–286. [CrossRef] [PubMed]

7. Barragán, V.; Leidi, E.O.; Andrés, Z.; Rubio, L.; De Luca, A.; Fernández, J.A.; Cubero, B.; Pardo, J.M. Ion exchangers NHX1 and NHX2 mediate active potassium uptake into vacuoles to regulate cell turgor and stomatal function in Arabidopsis. Plant Cell 2012, 24, 1127–1142. [CrossRef] [PubMed]

8. Dugasa, M.T.; Cao, F.; Ibrahim, W.; Wu, F. Genotypic difference in physiological and biochemical characteristics in response to single and combined stresses of drought and salinity between the two wheat genotypes (Triticumaestivum) differing in salt tolerance. Physiol. Plant 2018. [CrossRef] [PubMed]

9. James, R.A.; Blake, C.; Byrt, C.S.; Munns, R. Major genes for Na+ exclusion, Nax1 and Nax2 (wheatHKT1;4 and HKT1;5), decrease Na+ accumulation in bread wheat leaves under saline and waterlogged conditions. J. Exp. Bot. 2011, 62, 2939–2947. [CrossRef] [PubMed]

10. Xue, H.W.; Chen, X.; Mei, Y. Function and regulation of phospholipid signalling in plants. Biochem. J. 2009, 421, 145–156. [CrossRef] [PubMed]

11. Apel, K.; Hirt, H. Reactive oxygen species: Metabolism, oxidative stress, and signal transduction. Annu. Rev. Plant Biol. 2004, 55, 373–399. [CrossRef] [PubMed]

12. Wang, X.C.; Zhao, Q.Y.; Ma, C.L.; Zhang, Z.H.; Cao, H.L.; Kong, Y.M.; Yue, C.; Hao, X.Y.; Chen, L.; Ma, J.Q.; et al. Global transcriptome profiles of Camellia sinensis during cold acclimation. BMC Genom. 2013, 14, 415. [CrossRef] [PubMed]

13. Zhang, Q.; Cai, M.; Yu, X.; Wang, L.; Guo, C.; Ming, R.; Zhang, J. Transcriptome dynamics of Camellia sinensis in response to continuous salinity and drought stress. Tree Genet. Genomes 2017, 13, 78. [CrossRef]

14. Zeng, A.; Chen, P.; Korth, K.L.; Ping, J.; Thomas, J.; Wu, C.; Srivastava, S.; Pereira, A.; Hancock, F.; Brye, K.; et al. RNA sequencing analysis of salt tolerance in soybean (Glycine max). Genomics 2018. [CrossRef] [PubMed]

15. Yu, J.; Chen, S.; Zhao, Q.; Wang, T.; Yang, C.; Diaz, C.; Sun, G.; Dai, S. Physiological and Proteomic Analysis of Salinity Tolerance in Puccinelliatenuiflora. J. Proteome Res. 2011, 10, 3852–3870. [CrossRef] [PubMed]

16. Wei, G.S.; Jian, D.Y.; Fu, C.Z.; Yu, F.Z.; Xue, W.D. Research on introduction and salt tolerance of Betulahalophila. J. Gansu Agric. Univ. 2011, 5, 101–105.

17. Grabherr, M.G.; Haas, B.J.; Yassour, M.; Levin, J.Z.; Thompson, D.A.; Amit, I.; Adiconis, X.; Fan, L.; Raychowdhury, R.; Zeng, Q.; et al. Full-length transcriptome assembly from RNA-Seq data without a reference genome. Nat. Biotechnol. 2011, 29, 644–652. [CrossRef] [PubMed]

18. Altschul, S.F.; Madden, T.L.; Schäffer, A.A.; Zhang, J.; Zhang, Z.; Miller, W.; Lipman, D.J. Gapped BLAST and PSI-BLAST: A new generation of protein

database search programs. Nucleic Acids Res. 1997, 25, 3389–3402. [CrossRef] [PubMed]

19. Xie, C.; Mao, X.; Huang, J.; Ding, Y.; Wu, J.; Dong, S.; Kong, L.; Gao, G.; Li, C.Y.; Wei, L. KOBAS 2.0: A web server for annotation and identification of enriched pathways and diseases. Nucleic Acids Res. 2011, 39, W316–W322. [CrossRef] [PubMed]

20. Zhang, J.; Feng, J.; Lu, J.; Yang, Y.; Zhang, X.; Wan, D.; Liu, J. Transcriptome differences between two sister desert poplar species under salt stress. BMC Genom. 2014, 15, 337. [CrossRef] [PubMed]

21. Zhou, Y.; Yang, P.; Cui, F.; Zhang, F.; Luo, X.; Xie, J. Transcriptome Analysis of Salt Stress Responsiveness in the Seedlings of Dongxiang Wild Rice (OryzarufipogonGriff.). PLoS ONE 2016, 11, e0146242. [CrossRef]

22. Wang, W.S.; Zhao, X.Q.; Li, M.; Huang, L.Y.; Xu, J.L.; Zhang, F.; Cui, Y.R.; Fu, B.Y.; Li, Z.K. Complex molecular mechanisms underlying seedling salt tolerance in rice revealed by comparative transcriptome and metabolomic profiling. J. Exp. Bot. 2016, 67, 405–419. [CrossRef] [PubMed]

23. Villarino, G.H.; Hu, Q.; Scanlon, M.J.; Mueller, L.; Bombarely, A.; Mattson, N.S. Dissecting Tissue-Specific Transcriptomic Responses from Leaf and Roots under Salt Stress in Petunia hybrida Mitchell. Genes 2017, 8, 195. [CrossRef] [PubMed]

24. Villarino, G.H.; Bombarely, A.; Giovannoni, J.J.; Scanlon, M.J.; Mattson, N.S. Transcriptomic analysis of Petunia hybrida in response to salt stress using high throughput RNA sequencing. PLoS ONE 2014, 9, e94651. [CrossRef] [PubMed]

25. Zhang, H.B.; Zeng, Y.L.; Lan, H.Y.; Zhang, F.C. Physiological response of Betulahalophila (Betaluceae) to salt stress. Acta Bot. Yunnanica 2009, 31, 260–264. [CrossRef]

26. Liu, Y.; Song, Q.; Li, D.; Yang, X.; Li, D. Multifunctional Roles of Plant Dehydrins in Response to Environmental Stresses. Front. Plant Sci. 2017, 8, 1018. [CrossRef] [PubMed]

27. Kumar, M.; Lee, S.C.; Kim, J.Y.; Kim, S.J.; Aye, S.S.; Kim, S.R. Over-expression of dehydrin gene, OsDhn1, improves drought and salt stress tolerance through scavenging of reactive oxygen species in rice (Oryza sativa L.). J. Plant Biol. 2014, 57, 383–393. [CrossRef]

28. Brini, F.; Hanin, M.; Lumbreras, V.; Amara, I.; Khoudi, H.; Hassairi, A.; Pagès, M.; Masmoudi, K. Overexpression of wheat dehydrin DHN5 enhances tolerance to salt and osmotic stress in Arabidopsis thaliana. Plant Cell Rep. 2007, 26, 2017–2026. [CrossRef] [PubMed]

29. Roy Choudhury, A.; Sengupta, D.N. Transgenic tobacco plants overexpressing the heterologous lea gene Rab16A from rice during high salt and water deficit display enhanced tolerance to salinity stress. Plant Cell Rep. 2007, 26, 1839–1859. [CrossRef] [PubMed]

30. Cheng, Z.; Targolli, J.; Huang, X.; Wu, R. Wheat LEA genes, PMA80 and PMA1959, enhance dehydration tolerance of transgenic rice (Oryza sativa L.). Mol. Breed. 2002, 10, 71–82. [CrossRef]

31. Perdiguero, P.; Collada, C.; Soto, A. Novel dehydrins lacking complete K-segments in Pinaceae. The exception rather than the rule. Front. Plant Sci. 2014, 5, 682. [CrossRef] [PubMed]

32. Close, T.J. Dehydrins: A commonalty in the response of plants to dehydration and low temperature. Physiol. Plant 1997, 100, 291–296. [CrossRef]

33. Liu, C.C.; Li, C.M.; Liu, B.G.; Ge, S.J.; Dong, X.M.; Li, W.; Zhu, H.; Wang, B.; Yang, C. Genome-wide identification and characterization of a dehydrin gene family in poplar (Populustrichocarpa). Plant Mol. Biol. Rep. 2012, 30, 848–859. [CrossRef]

34. Du, H.; Liu, H.; Xiong, L. Endogenous auxin and jasmonic acid levels are differentially modulated by abiotic stresses in rice. Front. Plant Sci. 2013, 4, 397. [CrossRef] [PubMed]

35. Kim, H.B.; Oh, C.J.; Park, Y.C.; Lee, Y.; Choe, S.; An, C.S.; Choi, S.B. Comprehensive analysis of AHL homologous genes encoding AT-hook motif nuclear localized protein in rice. BMB Rep. 2011, 44, 680–685. [CrossRef] [PubMed]

36. Matsushita, A.; Furumoto, T.; Ishida, S.; Takahashi, Y. AGF1, an AT-hook protein, is necessary for the negative feedback of AtGA3ox1 encoding GA3-oxidase. Plant Physiol. 2007, 143, 1152–1162. [CrossRef] [PubMed]

37. Endt, D.V.; Silva, M.S.; Kijne, J.W.; Pasquali, G.; Memelink, J. Identification of a bipartite jasmonate-responsive promoter element in the Catharanthusroseus ORCA3 transcription factor gene that interacts specifically with AT-hook DNA-binding proteins. Plant Physiol. 2007, 144, 1680–1689. [CrossRef] [PubMed]

38. Street, I.H.; Shah, P.K.; Smith, A.M.; Avery, N.; Neff, M.M. The AT-hook-containing proteins SOB3/AHL29 and ESC/AHL27 are negative modulators of hypocotyl growth in Arabidopsis. Plant J. 2008, 54, 1–14. [CrossRef] [PubMed]

39. Lim, P.O.; Kim, Y.; Breeze, E.; Koo, J.C.; Woo, H.R.; Ryu, J.S.; Park, D.H.; Beynon, J.; Tabrett, A.; Buchanan-Wollaston, V.; et al. Overexpression of a chromatin architecture-controlling AT-hook protein extends leaf longevity and increases the post-harvest storage life of plants. Plant J. 2007, 52, 1140–1153. [CrossRef] [PubMed]

40. Lu, H.; Zou, Y.; Feng, N. Overexpression of AHL20 negatively regulates defenses in Arabidopsis. J. Int. Plant Biol. 2010, 52, 801–808. [CrossRef] [PubMed]

41. Li, B.; Dewey, C. RSEM: Accurate transcript quantification from RNA-Seq data with or without a reference genome. BMC Bioinform. 2011, 12, 323. [CrossRef] [PubMed]

42. Robinson, M.D.; McCarthy, D.J.; Smyth, G.K. edgeR: A Bioconductor package for differential expression analysis of digital gene expression data. Bioinformatics 2010, 26, 139–140. [CrossRef] [PubMed]

43. Wang, L.; Feng, Z.; Wang, X.; Wang, X.; Zhang, X. DEGseq: An R package for identifying differentially expressed genes from RNA-seq data. Bioinformatics 2010, 26, 136–138. [CrossRef] [PubMed]

44. Young, M.D.; Wakefield, M.J.; Smyth, G.K.; Oshlack, A. Gene ontology analysis for RNA-seq: Accounting for selection bias. Genome Biol. 2010, 11, R14. [CrossRef] [PubMed]

45. Shao, F.; Lu, Q.; Wilson, I.W.; Qiu, D. Genome-wide identification and characterization of the SPL gene family in Ziziphusjujuba. Gene 2017, 627, 315–321. [CrossRef] [PubMed]

46. Lu, Q.; Shao, F.; Macmillan, C.; Wilson, I.W.; Van der Merwe, K.; Hussey, S.G.; Myburg, A.A.; Dong, X.; Qiu, D. Genomewide analysis of the lateral organ boundaries domain gene family in Eucalyptus grandis reveals members that differentially impact secondary growth. Plant Biotechnol. J. 2018, 16, 124–136. [CrossRef] [PubMed]

47. Livak, K.J.; Schmittgen, T.D. Analysis of relative gene expression data using real-time quantitative PCR and the 2 (-Delta Delta, C.(T.)) method. Methods 2001, 25, 402–408. [CrossRef] [PubMed]

www.ingramcontent.com/pod-product-compliance
Lightning Source LLC
LaVergne TN
LVHW071245210926
848486LV00008B/355

* 9 7 8 1 7 8 7 1 5 6 2 3 4 *